AF522254

MOLECULAR GENETICS

ENCYCLOPAEDIA OF GENE EVOLUTION

Vol. II

MOLECULAR GENETICS

By

Dr. M. Prakash

DISCOVERY PUBLISHING HOUSE

NEW DELHI-110002

First Published-2006

ISBN 81-8356-169-1 (Set)

Published by

DISCOVERY PUBLISHING HOUSE
4831/24, Ansari Road, Prahlad Street,
Darya Ganj, New Delhi-110002 (India)
Phone: 23279245 • Fax: 91-11-23253475
E-mail:dphtemp@indiatimes.com

Printed at:
Amit Enterprises
Delhi-53

PREFACE

The present title **Encyclopaedia of Gene Evolution** encompasses diverse areas of study from the origin of life and is subsequent record to the molecular and population concepts that explain these events. All biological phenomena derive from evolutionary relationships and past interactions. As the great evolutionary geneticist Theodosius Dobzhansky remarked, 'Nothing in biology makes sense except in the light of evolution" yet unifying all biology under an evolutionary theme is still difficult, although the explosive increase in molecular, organismic, and populations information makes the realization of this goal more possible now than ever before.

The aim in writing this text was to bring together the highly dispersed literature on gene structure, function and expression, to integrate this with our emerging knowledge of chromosome and genome structure, and to draw comparison both within and between paralogous and orthologous gene sequences in order to establish the nature of the mutational mechanisms responsible for their evolutionary divergence. Such an approach, encompassing a large number of human genes and their extended families, is essential for any attempt to discern underlying evolutionary principles.

The purpose of this book is to bring together some prevailing knowledge and ideas about evolution, to provide an informed framework of together for under graduates, post-graduate and those of related fields.

Since evolution is the broadest of biological fields, covering the greatest range of disciplines, even the brief survey of evolution offered here has errors and ambiguities. To the extent that this book has been spared many such failings the author owes thanks to many reviewers we commented on one or more sections of this text.

There can be no claim to originality except in the manner of treatment and much of the information has been obtained from the books and scientific journals available in the different libraries.

The author expresses his thanks to his friends and colleagues whose continue inspirations have initiated him to bring out this book.

The author expresses his gratitude to Mr. Wasan and staff of M/s Discovery Publishing House for their whole hearted co-operation in the publication of this book.

Author

Contents

1

INTRODUCTION

Evolution, at the molecular level, is observable as nucleotide (or base) changes in the DNA and amino acid changes in proteins. Both can be studied by examining the differences between species—if, for example, one species has nucleotide *A* at a certain site and another species has nucleotide *G*—and in polymorphisms within a species—if, for example, some individuals of a species have nucleotide *A* at a certain site while other individuals have *G*. Both polymorphism and evolutionary change between species can be explained by two processes we have already encountered: natural selection and neutral drift.

The main factors controlling the relative importance of neutral drift and natural selection are population size and the selection coefficients of the different genotypes. If the population is small and the selection coefficients low, genetic drift dominates, whereas natural selection dominates if the population and the selection coefficients are large. In this chapter we will mainly, though not completely, ignore the effects of population size. We will adopt this tactic not because population size in unimportant, but rather to simplify the discussion, and because reliable measurements are rarely available. Whenever we consider neutral drift in this chapter, we will assume that population sizes are small enough for the process to operate.

We will also employ a slightly different notation for selection coefficients. The genotype with the highest fitness was given a fitness of 1 and the other genotypes were given fitnesses like (1 – s). Here we will be interested in whether one form of a molecule has a higher, lower, or equal fitness with another form. It will therefore be more convenient to talk about selection coefficients that are +, 0, or –: a

+ve selection coefficient means natural selection favours the variant; -ve means it is selected against; 0 means it is neutral.

The evolution of modern species has incorporated millions of molecular changes. Let us think about them all, in the abstract. Natural selection and neutral drift could logically have produced any proportions of the changes (though the proportions of the two must add to one). The subject is normally discussed in terms of a controversy between two extreme positions, however. In 1968 and 1969, Kimura, and King and Jukes, respectively, suggested that most evolutionary change at the molecular level is driven by random drift rather than natural selection. Kimura called this view *the natural theory of molecular evolution*. The neutral theory does not suggest that random drift explains all evolutionary change; natural selection is still needed to explain adaptation. It is possible that the adaptations we observe in organisms required only a minority of all evolutionary changes that have taken place in the DNA. The neutral theory states that evolutionary at the level of DNA and protein, but not at the level of adaptation, is dominated by random processes; most evolution at the molecular level would then be nonadaptive. We can contrast the neutral theory with its converse: the idea that almost all molecular evolution has been driven by natural selection.

The difference between the two ideas can be understood in terms of the frequency distribution for the selection coefficients of mutations, or genetic variants. (It does not matter here whether we talk about new mutations or the set of genetic variants existing in a population at a genetic locus. "Genetic variant" could be substituted for "mutation" throughout this paragraph.) Given a mutation of a certain selection coefficient, the theory of neutral drift or selection applies in a mathematically automatic way: if the selection coefficient is positive, the mutation increases in frequency; if it is negative, it is eliminated; if it is zero, the gene frequencies drift. The dispute is about how likely it is that mutations will arise with a certain selection coefficient.

Consider the nucleotide sequence of a gene in a living organism. The gene codes for reasonably well-adapted protein. After all, the protein is unlikely to be a complete failure if the organism containing it is alive. Now consider all of the mutations that can be made in the gene. You could work down the gene, altering one nucleotide at a time, and ask, for each change, whether the new version was better, worse, or equally as good as the original gene. In a population of organisms in nature, mutations will occur and cause those kinds of change, in certain frequencies.

Almost all biologists agree that most of the changes will be for the worse; most will have negative selection coefficients. Adaptation is an unlikely physical state, and if a random change is made in an adapted protein it is likely to be for the worse.

On the other hand, biologists disagree about the relative frequency of the other two classes of mutations: neutral and selectively advantageous mutations. If natural selection has produced most evolutionary change, many advantageous mutations must have occurred, whereas mutations must have been rare. If neutral drift has produced most evolutionary change, the relative frequencies are the reverse.

Imagine a gene of about 1000 nucleotides (corresponding to a protein of about 300 amino acids). There are 4^{1000} (or about 10^{600}) possible sequences of the gene. The protein encoded by the gene will have some function—for example, carrying oxygen in the blood (actually done by hemoglobin, which is made up of four polypeptides of slightly less than 150 amino acids each). The neutralist view of evolution suggests that, of the 10^{600} possible molecules, the great majority would fail to carry oxygen at all, and many would do so poorly. A minority of perhaps several hundred different sequences, all very similar to one another, would then code for proteins that carried oxygen equally well. What we observe as evolution would consists of shuffling around within this limited set of equivalent sequences. The selectionist alternative is that the few hundred variants are not equivalent, but that one works better in one environment, another in another environment, and so on; evolution consists of the substitution of one variant for another when the environment changes. The different in viewpoint exists between

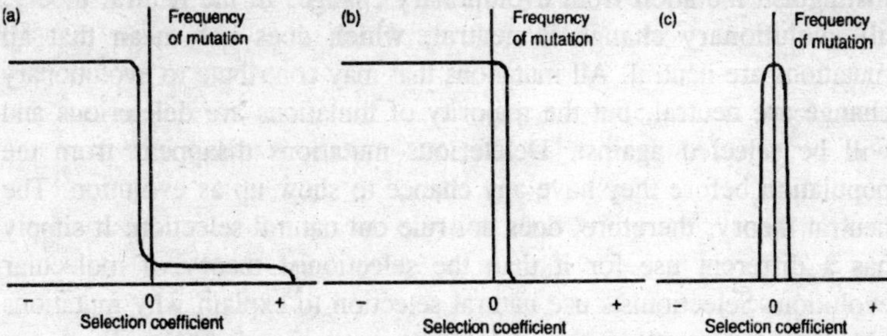

Fig. 1.1. The neutral and selectionist postulate different frequency distributions for the rates of mutation with various selection coefficients. (a) According to selectionists, exactly neutral mutations are rare and enough favourable mutations arise to account for all molecular evolution. (b) Neutralists believe more neutral, and hardly any selectively favoured, mutations arise. (c) The theory of pan-neutralism, according to which all mutations are selectively neutral.

people who think that a protein could have a fairly large number of equivalent forms, and those who think that any change in the amino acid sequence, no matter how small, is likely to make a difference in the protein's function (and its bearer's fitness) under some conditions.

Graphs can be drawn because each genetic variant has a selection coefficient, which determines its evolutionary fate. However, the selection coefficient of a genetic variant will often depend on the environment in which it finds itself. Two genetic variants of an enzyme, for example, may have identical fitnesses in one environment, but different fitnesses in another environment. In nature, during the life of an organism, a genetic variant has a fitness given by the average of all the environments it experiences, and the formulation is therefore theoretically valid.

When measuring fitnesses, or considering the biological reason why some variant exists, it may be necessary to think more subtly, however. It has been argued, for instance, that many mutants are "conditionally neutral" or "conditionally advantageous". That is, the mutants are neutral in some environments, but make a difference in other (perhaps rare) environments. In this chapter, we will continue to talk about "the" fitness, or selection coefficient, of a genetic variant, but the use of that term should not be taken to mean that a genetic variant always unconditionally has the same fitness.

Another idea with which the neutral theory is often confused. Pan-neutrality means that all forms of a protein (or, at least, the majority of them) have equal fitness—that all mutations are neutral. How does this idea differ from the neutral theory? First, we need to distinguish mutation from evolutionary change. In the neutral theory, all evolutionary change is neutral, which does not mean that all mutations are neutral. All mutations that may contribute to evolutionary change are neutral, but the majority of mutations are deleterious and will be selected against. Deleterious mutations disappear from the population before they have any chance to show up as evolution. The neutral theory, therefore, does not rule out natural selection. It simply has a different use for it than the selectionist theory of molecular evolution. Selectionists use natural selection to explain why mutations are lost (when they are deleterious) or fixed (when they are advantageous). Neutralists, on the other hand, use selection to explain only why deleterious mutations are lost; they use drift to explain how new mutations are fixed.

Pan-neutralism is almost certainly false. Good empirical reasons exist to reject this theory. For example, pan-neutralism cannot explain why different genes evolve at different rates, or why different species have different biases in codon usage. Nor is it theoretically plausible. It is absurd to suggest that hardly any mutations are disadvantageous. Organisms, including their molecules, are adapted to their environments; we need reflect on only the efficiency of digestive enzymes—or any other biological molecule—for supporting life to realize that fact. If molecules are affected by adaptation, many (or most) changes in them will be for the worse.

It the crucial difference between the selective and neutral theories involves the relative frequencies of neutral and selectively advantageous mutations, the direct way to test between them would simply be to measure the fitnesses of many genetic variants at a locus, and count the numbers with negative, neutral, or positive selection coefficients under certain environmental conditions. It has proved impossible to settle the controversy in this way. Measuring the fitness of even one common genetic variant represents a major research program; it would be completely impractical to measure the fitnesses of many rare variants.

Testing between the theories has been attempted by less direct means, however. We shall look at four of the main kinds of tests (although these four by no means exhaust the list of evidence that has been brought into this controversy). The four tests include the following:

1. The absolute rate of molecular evolution and degree of polymorphism, both of which have been argued to be too high to be explained by natural selection.
2. The constancy of molecular evolution, which has been argued to be inconsistent with natural selection
3. The observation that functionally less constrained parts of molecules evolve at a higher rate, which has been argued to be the opposite of what the theory of natural selection would predict.
4. The observation that the degree of polymorphism and the rate of evolution of a molecule are not correlated in exactly the way the neutral theory predicts.

Measurement of Genetic Variation

In the 1960s, when the sequences of particular proteins became known for several species, biologists could work out the number of amino acid changes that occurred during the evolutionary divergence of a pair of species. The same calculation has been made with many DNA sequences, as they have become increasingly available since the 1980s. For any two species, the approximate age of their common ancestor can be estimated from the fossil record. The rate of protein

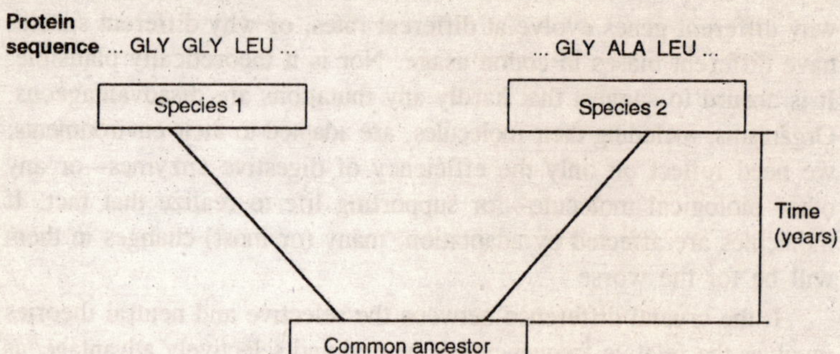

Fig. 1.2. Imagine that some regions of a protein has the illustrated sequences in two species.

evolution can then be calculated as the number of amino acid differences between the protein of the two species divided by 2 × the time to their common ancestor. If the species are humans and mice, their common ancestor probably lived about 80 million years ago. If we look at the sequence of a 100 amino acid protein in the two species and it differs at 16 sites, then the rate of evolution is estimated at $16/(100 \times 160 \times 10^6) \approx 1 \times 10^{-9}$ per amino acid site per year. The same calculation can be made per nucleotide site for the rate of DNA evolution. In practice, a correction has also to be made for "multiple hits." If a sequence differs between two species at a site, more than one substitution could have occurred in the two lineages joining them and their common ancestor. Various techniques have been developed to correct the observed number of differences and estimate the underlying number of evolutionary changes.

Table 1.1. Rates of evolution in eight proteins.

Protein	*Number of amino acid changes × 10^9/yr*
Fibrinopeptides	8.3
Pancreatic ribonuclease	2.1
Lysozyme	2.0
α-globin	1.2
Myoglobin	0.89
Insulin	0.44
Cytochrome c	0.3
Histone H4	0.01

As can be seen, different proteins evolve at different rates. For example, histones evolve slowly, while fibrinopeptides evolve rapidly.

We will discuss why different proteins evolve at different rates, but the argument we will consider first is more general and uses only an approximate average evolution rate for proteins as a whole. An approximate figure suggested by Table 1.1. is that amino acids are substituted at a rate of around 1.5×10^{-9} per amino acid site per year.

Another important figure relates to the amount of genetic variation within a species at a particular time. In 1966, Harris, and Hubby and Lewontin were the first to use the method of gel electrophoresis to estimate the average amount of genetic variation in humans and in the fruitfly *Drosophila*. The amount of variation can be described by two main indexes. One index is the chance that two randomly drawn alleles differ at an average locus, or heterozygosity. We previously dealt with H as a property of one locus; here it is measured for a number of loci and we consider the average for all loci. The other measure is the percentage of polymorphic loci. If, for example, 20 loci are studied by gel electrophoresis, and 16 shows no variation and 4 have more than one band on the gel, then the percent polymorphism would be $4/20 \times 100 = 20\%$. Gel electrophoretic evidence suggests that 10-20% of loci are polymorphic in species in nature.

Kimura found these numbers—for the rate of evolution and the level of genetic variation—remarkably high, if proteins evolved by natural selection. To see why, we need to understand the concept of "genetic load." Genetic load can be examined as both a general concept within evolutionary biology as a whole and in its neutralist application.

Table 1.2. Amount of variant in natural populations.

Species	*Number of Loci*	*P (%)*	*H (%)*
Phlox cuspidata	16	11	12
Liatris cylindracea	27	56	5.7
Limulus polyphemus	25	25	5.7
Balanus eburneus	14	67	6.7
Homarus americanus	28-42	18	3.8
Gryllus bimaculatus	25	58	6.3
Drosophila robusta	40	39	11
Bombus americanorum	12	0	0
Salmo gairdneri	23	15	3.7
Bufo americanus	14	26	11.6
Passer domesticus	15	33	9.8
Homo sapiens	71	28	6.7

GENETIC LOAD

At a particular time, there may be a single optimal genotype that an individual in a particular population could possess. We can avoid philosophical worries about imaginary omnipotent organisms by confining attention to the genotypes that actually exist in the population. Suppose a variety of genotypes segregate out in the population, each with its characteristic fitness. One genotype has a higher fitness than the rest (at this particular time and place) and we call it fitness w_{opt}. We can also measure the average fitness of the whole population; it consists of the fitness of each genotype multiplied by its frequency and is called mean fitness (i.e., the frequency of each genotype multiplied by its fitness and is conventionally symbolized by $\overline{w}$. The general formula for genetic load (L) is as follows:

$$L = \frac{w_{opt} - \overline{w}}{w_{opt}} \quad \text{...(1)}$$

If all individuals in the population have the optimal genotype, $\overline{w} = w_{opt}$ and the load is zero; if all but one have a genotype of zero fitness, $\overline{w} = 0$ and L = 1. Genetic load is a number between 0 and 1 that measures the extent to which the average individual in a population is inferior to the best possible kind of individual, given the range of genotypes in the population. To be exact, the genetic load equals the relative chance that an average individual will die before reproducing because of the deleterious genes that it possesses.

Genetic load can exist for several reasons. The balance between selection against a deleterious gene and its production by mutation is a simple example, called *mutational load*. At the equilibrium, the deleterious gene has (if it is dominant) a frequency of m/s, where m = the mutation rate and s = the selective disadvantage of the mutation. What is the mutational load in this case? We first calculate the mean fitness, by the standard procedure; it is simpler to write the frequency of A as p to begin with and then substitute p = m/s after simplifying.

Genotype	AA	Aa	aa
Fitness	1- s	1- s	1
Frequency	p^2	2pq	q^2

$$\text{mean fitness} = \overline{w} = p^2(1 - s) + 2pq(1 - s) + q^2$$
$$= 1 - p^2s - 2pqs.$$

(Because $p^2 + 2pq + q^2 = 1$.) We can simplify this by two approximations. The frequency of A will be low because it is a deleterious gene, and only maintained by recurrent mutation. The term with p^2 will therefore be tiny and can be ignored. Likewise, in the 2pqs term, p will be low and $q \approx 1$; we can therefore simplify it as $2pqs \approx 2ps$. Then

$$\overline{w} = 1 - 2\,ps.$$

Now we substitute $p = m/s$, and

$$\overline{w} = 1 - 2m.$$

We next put the formula for $\overline{w} = I - 2m$ into the general formula for genetic load (equation 1):

$$L = \frac{1-(1-2m)}{1}$$

$$L = 2m$$

In this case, the mutational load equals twice the mutation rate. The load expresses the fact that individuals are dying because of the deleterious mutations that arise. The population carries a load of deleterious mutations, which reduce the average fitness of its members.

Genetic load has a real meaning and can be measured, but its main value is as a theoretical concept. We can express the consequences of many theoretical processes as a genetic load, and the concept can then be used as a heuristic reasoning device. Because a population contains only a finite number of individuals, a limit exists for the genetic load it can endure. This fact can be used to estimate the upper limit on the processes (like mutation) that cause genetic loads, because if the rate at which deleterious mutations arose were too high, every individual in the population would be dead. Thus, the deleterious mutation rate for real populations cannot be higher than a certain value. These kinds of arguments make various assumptions, however, and we will discuss them in due course.

Cost of Natural Selection

When a gene is substituted by natural selection, the bearers of the inferior gene must die at a higher rate or have lower fertility than bearers of the superior gene. The higher the intensity of natural selection, the higher the amount of selective death or infertility there must be for every generation. (It does not matter whether the selection takes place by differential mortality or fertility, or some combination of both; we shall discuss selection here in terms of mortality, but only as a shorthand—"mortality" stands for both real mortality and differential fertility.) A population cannot tolerate an indefinitely large amount of selective death, however. If selection becomes too strong) it will drive the population extinct. Haldane's point was not that selection could never be very strong, but that, if it ever has been extremely strong, the species would have gone extinct. Species that have survived cannot have suffered too great a selective cost during their evolutionary histories. Haldane referred to the selective death that must occur for a gene to be substituted as the *cost of selection*. In a 1957 paper, he investigated the upper limit that it would place on the rate of evolution.

The principle can be discussed most easily in a haploid case. Suppose a population contains two genes, A and A', with fitnesses 1 and (1 – s) and frequencies p and q (= 1 – p), respectively. In any generation, of the q A'-bearers, s will die without reproducing and (1 – s) will survive like A-bearers. A proportion sq of the population dies without reproducing because of selection at this locus.

We can now define a ratio of the proportion of individuals in the population that survive to the proportion that die. In this case, sq die, and p + q(1 – s) = 1 – sq survive. The ratio in one generation is sq/(1 – sq). The ratio is the same every generation until A' is eliminated. As selection operates for each generation, more and more selective death accumulates. We can now define the total cost of natural selection as

$$C = \sum \frac{sq}{1 - sq}$$

The summation occurs over all of the generations it takes to fix the A gene. Although not very relevant to the neutralist—selectionist controversy, the most interesting property of Haldane's cost is that the total cost of selection is almost independent of the selection coefficient 5 when s is small. If selection is stronger (i.e., s is higher), more individuals die within each generation, but the gene itself is substituted more rapidly and the cost has to be added over fewer generations. The two factors—cost within a generation and the number of generations taken to substitute the gene—approximately cancel; the total amount of selective death needed to substitute a gene is, therefore, more or less independent of the gene's selective advantage.

If a population is to maintain itself, the survivor individuals must produce sufficient extra offspring to make up for those that die before reproduction. If, for example, the ratio sq/(1 – sq) was 0.1/0.9, then each survivor would have to produce $1^1/_9$ offspring on average, which should be easily manageable. However, an upper limit exists to the possible cost of natural selection. If the ratio was 0.999/0.001, each survivor would have to leave 1000 *surviving* offspring, which would be much more difficult. Haldane's approximate estimate for the upper limit of cost that could be borne was derived for a diploid rather than a haploid model. The algebra for the diploid case is, in principle, the same, only more complex, and we will not go into the details here. The upper limit suggested by Haldane was one gene substitution per 300 generations. He arrived at the figure by guessing the cost of a typical gene substitution was about 30 and that a species had about 10% excess reproductive capacity available for selective processes. Each gene substitution will then take 300 generations. These figures are very crude, because we have no evidence that a typical cost of a gene substitution is 20 or 40 rather than 30, but they illustrate Haldane's abstract argument.

The cost of selection can be expressed as a genetic load. While selection is fixing the favoured allele, the population will contain inferior alleles as well as the superior allele that is being fixed. We can, therefore, define a ***substitutional load*** with the standard formula $L = w_{opt} - \overline{w}/w_{opt}$. The mean fitness ($\overline{w}$) varies with gene frequency, moving closer to w_{opt} as the favoured allele moves to fixation. Regardless of whether the concept is formulated as a load or a cost, the key idea is that the substitution of one gene for another by selection requires death (or non-reproduction); without that death, gene frequencies cannot change.

Rates of Molecular Evolution

When the earliest measurements of rates of molecular evolution became available, Kimura noticed that they implied much higher rates of change than Haldane's estimate of the maximum rate of evolution by natural selection (i.e., one substitution per 300 generations). The figures for rates of change are in units of changes per amino acid site per year. Kimura multiplied these numbers to find the total number of changes in the genome per generation. In a mammal, the genome contains perhaps 3×10^9 nucleotides. Kimura used an estimate of one change in 28×10^8 years per amino acid site. Multiplying that rate by the number of amino acid triplets in the genome (10^9) gives an estimate of 1/2.8 changes per year, or about one change in the genome every two or three years—more than 100 times as fast as Haldane's upper limit. If Haldane's argument is correct, natural selection could not drive evolution as fast as it has been occurring in proteins. If natural selection operated that aggressively on a population, everyone in it would be dead before long. Although Haldane's estimate was very crude, the real rate is so much higher than the theoretical limit that the exact figures hardly matter. Kimura duly concluded that natural selection could not account for the majority of molecular evolutionary change.

With neutral drift, no difficulty arises. No cost is incurred in substituting one allele for another by random drift. Evolution by random drift takes place as rapidly as neutral mutations arise. The rate at which neutral mutations occur (u) is unknown, so the neutral theory can hypothetically explain almost any observed rates of evolution by postulating appropriate values for u. All we can conclude is that the rate of molecular evolution is not inconsistent with neutralism. The only upper limit on u is the total mutation rate, because the neutral mutation rate must be less (probably much less) than the total mutation rate. This upper limit, however, poses no problem, because the observed rates of molecular evolution are all much less than total mutation rates. Consider, for instance, the figure for the rate of molecular evolution of one change in 28 $\times$ 10^8 years per amino acid site. A typical total mutation rate for a whole gene might be about 10^{-6} per generation, or 10^{-8} per amino acid site for a gene of 100 amino acids. This mutation rate would permit one change in 10^8 years, which is nearly 30 times as fast as the observed figure of one change in 28 $\times$ 10^8 years.

Thus, the neutral theory is not contradicted by what we know about mutation rates. The neutralist argument, in summary, is that the observed rate of evolution is too high for evolution by natural selection, but not for neutral drift.

Degree of Genetic Variation

Gel electrophoresis revealed a large amount of protein variation in natural populations, even though the method does not detect all protein variation. For the discussion in this section, it is useful to have a rough figure for the amount of variation in a natural population. In a fruitfly, an approximate figure for the percentage of loci that are polymorphic may be around 30% and the average heterozygosity is about 10%. These figures probably underestimate the real figures because gel electrophoresis does not detect all protein variability. A low estimate of the number of loci in a fruitfly (*Drosophila*) is 10,000; a population of fruitflies will therefore have about 3000 polymorphic loci. Their average heterozygosity will be 0.1/0.3 = 0.33 (0.1 is the average heterozygosity of the whole genome, but the heterozygosity is concentrated in 0.3 of the loci; the actually polymorphic loci will therefore have a heterozygosity of about 0.33). The first question is whether the selective and neutral theories can account for this degree of variation.

At the time the controversy began, heterozygous advantage was thought to be the only general explanation of polymorphism. Dobzhansky had previously (e.g., in 1951) suggested that a large number of selectively maintained polymorphisms were found in natural populations. When large amounts of variation were indeed discovered, heterozygous advantage was the first explanation to be considered. For one or two loci, the idea is plausible. However, for larger numbers of loci it soon runs into the problem of *segregational load*. Kimura and Crow first put forward this argument, in 1964. We begin with the standard selection scheme for heterozygous advantage:

genotype	AA	Aa	aa
fitness	1 – s	1	1 – t

We have seen how to calculate the equilibrial gene frequency. After this calculation, it is easy to calculate the proportions of the three genotypes at equilibrium. Thus, the equilibrial gene frequencies at the locus are:

$$p = \frac{t}{s+t} \text{ and } q = \frac{s}{s+t}$$

The proportion of heterozygotes in the population at the locus is:

$$H = \frac{2st}{(s+t)^2}$$

The mean fitness of the population (for this one locus) is:

$$\overline{w} = 1 - \frac{st}{(s+t)}$$

Mean fitness is found, as usual, by multiplying the frequency of each genotype by its fitness, as in $\overline{w} = p^2(1 - s) + 2pq + q^2(1 - t)$. To reduce that equation to the formula above, first multiply it out, then note that $p^2 + 2pq + q^2 = 1$, and substitute for p and q from the formulas in terms of s and t above. Note also that $p^2s + q^2t = ps = qt$. The algebraic simplification is then easy.

The formula for mean fitness can be further simplified if we substitute the formula for heterozygosity into the one for mean fitness. Then,

$$\overline{w} = 1 - H\left(\frac{s+t}{2}\right)$$

$$\overline{w} = 1 - H\overline{s}$$

where $\bar{s} = (s + t)/2$. If $\bar{s} = 0.1$ and H = 0.33, the average member of the population has a fitness of 0.967 as compared with 1 for a heterozygote. Mean fitness is less than one, because s and t are both positive. With heterozygous advantage, the population mean fitness is lower than if all individuals were heterozygotes. In addition, $\bar{w}$ is lower if H is higher because selection against the homozygotes must be stronger (higher s) when a higher proportion of heterozygotes is present. Thus, more homozygotes must die each generation and the population "suffers" more reduction in mean fitness than when the homozygotes and heterozygotes have similar fitnesses.

The load exists because with Mendelian heredity it is impossible to create a population of pure-breeding heterozygotes. Even in the extreme cases of s = t = 1, when all homozygotes die before reproduction, the matings between heterozygotes will still generate the inferior homozygous forms; in practice, s and $t < 1$ and the homozygote crosses contribute as well. The population inevitably produces homozygotes every generation, and the average fitness of its members will therefore fall below that of a heterozygote.

So far we have considered only one locus. A fruitfly population, however, includes about 3000 polymorphic loci, with heterozygosity averaging 0.33 each. If $\bar{s} = 0.1$ for each of them, then at every locus the fitness of an average individual is about 0.967 compared with 1 for a heterozygote. We will assume for now that natural selection operates independently on each locus; homozygotes at each of the 3000 loci contribute independently to the loss of average fitness in the population. The average fitness for all loci is found by multiplying the average fitness of each of the 3000 polymorphic loci:

$$\bar{w} = (1 - 0.33 \times 0.1)^{3000} = 0.967^{3000} \approx 10^{-43}$$

This absurdly low number means that if we compare the fitness of an imaginary individual who was heterozygous at all 3000 loci (who has $w_{opt} = 1$) with that of a random member of the population, the full heterozygote would be 10^{43} times fitter than the randomly picked individual. We have derived an impossibility, because, in the words of Lewontin (1974), "one can hardly imagine a *Drosophila* female, no matter how many loci she was heterozygous for, laying 10^{43} eggs."

Something has gone wrong. Kimura's conclusion was that the error lies in the idea that polymorphism is maintained by natural selection. He deduced that natural selection cannot account for the observed levels of natural variation in proteins. As in the case of evolutionary rates, no such difficulty arises if the variability is due to neutral drift.

As we have seen, neutral drift will preserve a degree of heterozygosity. The heterozygotes are transient. New alleles appear first as unique mutations, and then exist as transient polymorphisms until drift either fixes or eliminates the allele. The heterozygosity maintained is given by

$$H = \frac{4Nu}{4Nu + 1}$$

As we saw when we tried to explain the rate of evolution by neutral drift, almost any result could be explained if we put the right numbers into the formula. The same is true here. For example, H = 9% if N = 2.5×10^5 and u = 10^{-7}. This u may be approximately correct, though the N is low. N = 2.5×10^5 does, however, lie within the bounds of possibility, and a single observation of H = 9% does not contradict the neutral theory.

We have now completed the first part of Kimura's original argument. It states that impossible substitutional and segregational loads would be incurred if selection had produced the observed rates of evolution and observed ranges of heterozygosity. Neutral drift, by contrast, runs into no problems with genetic loads.

Before we move on, let us make an observation for later discussion. The neutral theory would be more comfortable with higher

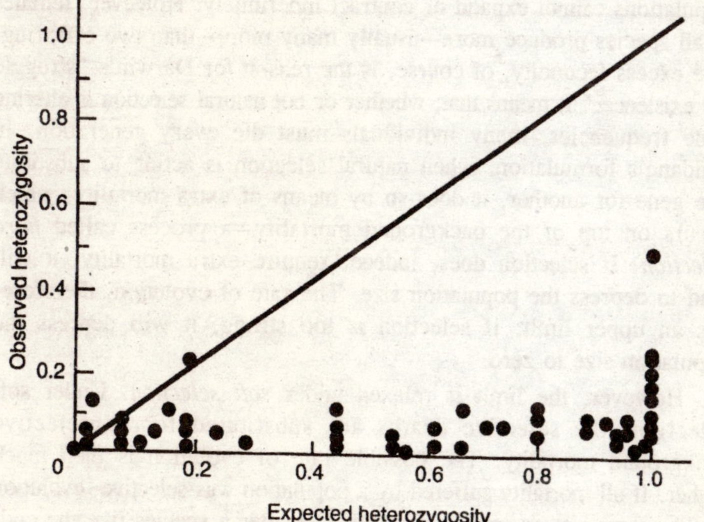

Fig. 1.3. Observed heterozygosities are probably too low for the neutral theory, at least in its simplest form.

heterozygosities than are found in nature. A figure of 9% heterozygosity, we have seen, is predicted by the neutral theory if the product of population size and mutation rate, Nu = 2.5×10^{-2}. Several critics have argued that natural values for Nu will be higher. The neutral theory should therefore predict higher heterozygosities than are observed. The population size of Drosophila, for example, may be 100 times larger than that required by the neutral theory (H ≈ 10% in *Drosophila*). Moreover, as heterozygosities have been measured in further species, it has become clear that they are consistently too low for the neutral theory. Because N and u are both unknown numbers, this observation is not strong evidence against the theory, but it is a problem to worry about discusses one modification to the neutral theory to account for the low levels of heterozygosity, and one or two other awkward facts.

Operation of Natural Selection

In the next few sections, we will examine three reasons to doubt whether the rates of evolution and levels of polymorphism pose any theoretical difficulty for natural selection. The first concerns a distinction between what is called "hard" and "soft" selection. Haldane's cost of natural selection assumes that there are two kinds of death (or, in more abstract terms, population control processes). On average, each pair must produce two offspring, because natural populations cannot expand or contract indefinitely. However, females in all species produce more—usually many more—than two offspring. The excess fecundity, of course, is the reason for Darwin's "struggle for existence." It means that, whether or not natural selection is altering gene frequencies, many individuals must die every generation. In Haldane's formulation, when natural selection is acting to substitute one gene for another, it does so by means of extra mortality, which occurs on top of the background mortality—a process called *hard selection*. If selection does, indeed, require extra mortality, it will tend to depress the population size. The rate of evolution, therefore, has an upper limit: if selection is too strong, it will depress the population size to zero.

However, the limit is relaxed under *soft selection*. Under soft selection, the selective deaths are substituted for nonselective background mortality. The possible rate of evolution is then much higher. If all mortality suffered by a population was selective, evolution could proceed at an immense speed. Consider a species like the cod, in which a female can produce more than 5,000,000 eggs. In theory, a single mutation in a population of 2,500,000 cod could increase from a

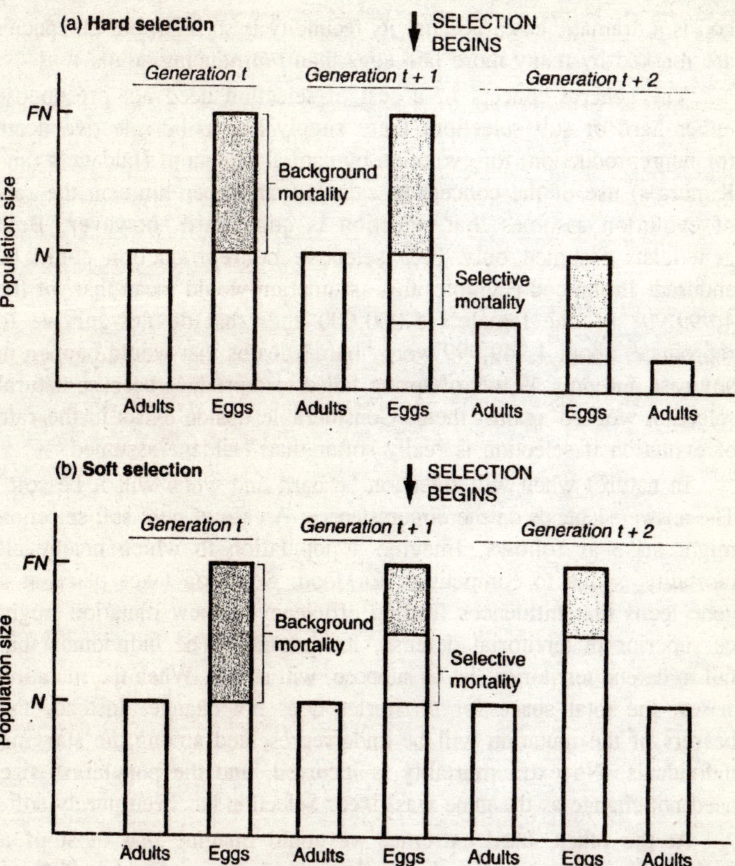

Fig. 1.4: A population with N adults, of average fecundity 2F per female, produces FN eggs.

frequency of 1/2,500,000 to 1 (i.e., 2,500,000/2,500,000) in a single generation while the population size stays constant. Such a scenario is highly unlikely to happen: the intensity of selection is unrealistic and much of the mortality suffered by a cod population will inevitably be nonselective. The point is theoretically interesting, however, because it shows that rapid evolution can take place under natural selection provided that selection is soft. In such a case, the selective mortality is substituted for the natural mortality of the population. The population size then need not be decreased by the action of natural selection.

Hard and soft selection are extremes along a continuum. In practice, natural selection could be any mixture of hard and soft selection. (The cod is a dramatic case, because its fecundity is so high, but all species are marked by many more laid eggs than reproducing adults.)

The general concept of a cost of selection need not presuppose either hard or soft selection; there simply has to be selective death (or nonreproduction) for evolution by natural selection. Haldane's (and Kimura's) use of the concept to calculate an upper limit on the rate of evolution assumes that selection is quite hard, however. Both geneticists assumed only 10% selective nonreproduction could be endured. In the cod example, this assumption would mean that, of the 4,999,998 of one female's 5,000,000 eggs that do not survive to reproduce, about 4,949,999 were "hard" deaths that would happen in any case and only 49,999 offspring failed to reproduce because natural selection worked against them. Considerable upside exists in the rate of evolution if selection is really softer than Haldane assumed.

In nature) when will selection be hard and when will it be soft? The answer depends on the circumstances. A case of pure soft selection might arise as follows. Imagine a population in which nearly all mortality is due to competition for food. Selection takes place at a gene locus that influences feeding efficiency—a new mutation might be superior in territorial defense, for instance. The individuals that fail to defend territories, let us suppose, will starve. When the mutation arises, the total space for territories does not change. Instead, the bearers of the mutation will be underrepresented among the starving individuals. No extra mortality is incurred, and the population size need not change as the gene was fixed. Selection has been purely soft.

At the other, hard extreme, we could imagine that most of a species's mortality happens during the juvenile phase. Perhaps 99% of those born die before reproducing. In the absence of selection, the remaining 1% breed and manage to maintain the population. Now suppose that an advantageous mutation arises and influences the relative chance of adult survival. The mutation has the effect of reducing the survival of the rest of the adult population (for instance, because the mutant individual takes more food). Extra mortality is now being added to the juvenile mortality; and the population will go into decline. Selection would have been soft if the selective mortality could have been shifted into the juvenile phase, but in this case the mutation influences only the adult and the juvenile mortality happens for some inevitable ecological reason. Selection, therefore, is hard.

The important point is that Haldane's argument assumed that selection is hard. In the case of hard selection, the limit on the amount

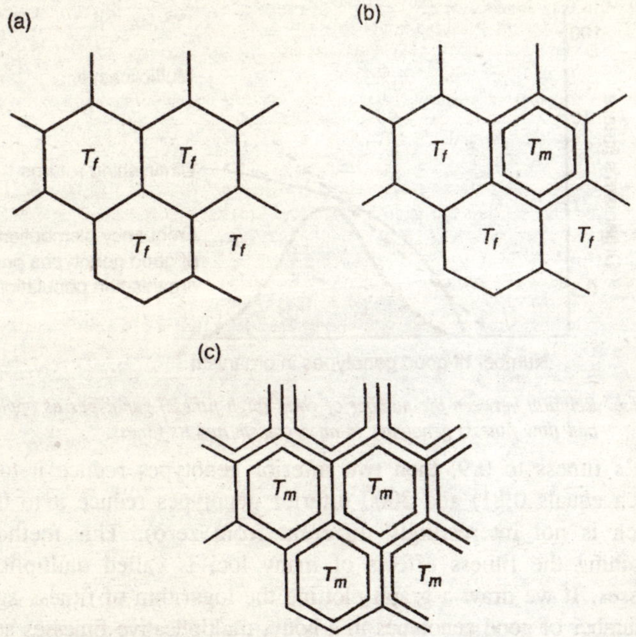

Fig. 1.5. Theoretical examples of pure soft selection. (a) The individuals (genotype T_f) divide up the available space and only those with territories breed; everyone else dies without reproducing. (b) A mutation (T_m) arises, whose bearers are superior at holding territories, the double border symbolizes this superior ability. (c) The mutation spreads, but as it does so the population size remains constant.

of selective mortality that can be tolerated will be lower than under soft selection. In some natural cases, selection may be hard, but it is not a requirement. To the extent that selection is soft, the neutralist argument is weakened.

Action of Natural Selection

The impossibly high figures for segregational and substitutional loads arose because an estimate for the load at one genetic locus was raised to the power of the number of loci (or of polymorphic loci) in the whole genome. In the calculation of segregational load in fruitflies, the average fitness reduction per locus (1 – Hs), was raised to the power 3000, representing the 3000 loci that may be polymorphic in a fruitfly population. This calculation assumes that selection operates independently at each locus, so that the effect on a fly's fitness of a homozygote is independent of whether the other loci are homozygous or heterozygous. We can then multiply the fitness contributions of all genotypes to find the fly's total fitness. If one inferior genotype reduces

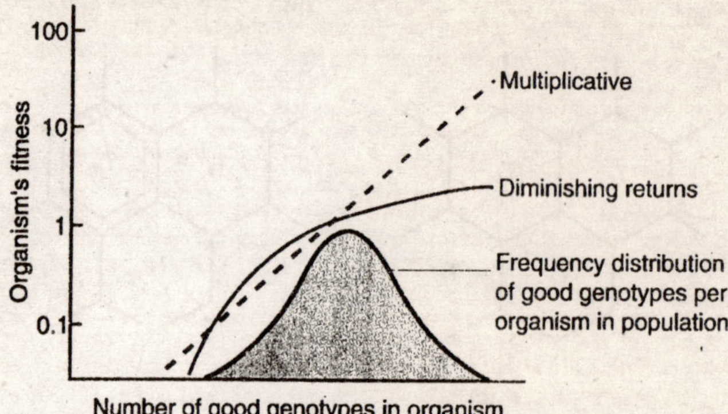

Fig. 1.6. Relation between the number of good (high fitness) genotypes as opposed to bad (low fitness) genotypes in an organism and its fitness.

a fly's fitness to 0.9, then two inferior genotypes reduce it to 0.9^2 (which equals 0.81) and 3000 inferior genotypes reduce it to 0.9^{3000} (which is not interestingly different from zero). This method of combining the fitness effects of many loci is called multiplicative fitnesses. If we draw a graph plotting the logarithm of fitness against the number of good genotypes in a body, multiplicative fitnesses appear as a straight line.

But does every locus independently influence an individual's fitness? Imagine a normally functional organism (a fruitfly, for example, or a human being) having good genotypes at many loci. If we assume a heterozygous advantage, the good genotypes would be heterozygotes (the argument that follows applies to all kinds of selection). Now imagine working down the genome, converting good genotypes into inferior genotypes. How will the fitness of the organism decline? The question is clearly an empirical one, but it cannot be answered in general with factual evidence. We can, however, reason about it in the abstract.

The idea that all, or even most, loci independently affect an organism's fitness is arguably biologically naive. Suppose the development of the right eye is controlled by one locus and the left eye by another: an organism with good genotypes at both loci has two eyes, and a bad genotype at either locus causes the loss of the appropriate eye. What will the fitnesses be for organisms with two good genotypes, one bad and one good genotype, or two bad genotypes? The organisms with two eyes clearly have the highest fitness. Organisms with one bad eye will have lower fitness, but perhaps not by a dramatic amount. A one-eyed organism can compensate for its defect by various means: it can turn its head, scan the landscape repeatedly, and spend slightly longer in looking around than would a two-eyed organism. Most of the time it will manage as well as a two-eyed organism, and its fitness on average will not be much reduced. On the other hand, the organism with two bad genotypes is blind and its fitness will crash. In this case, the fitnesses are not multiplicative. The fitness of a doubly defective organism is much lower than predicted by combining the fitnesses of organisms with the two kinds of single defect.

In fact, the development of right and left eyes is presumably not controlled by two genetic loci, but the argument illustrates what may be a general point. The relation between the number of good genotypes in an organism and the log of its fitness may not be a straight line but may instead demonstrate the decreasing returns pattern. The reason to think that the argument may apply in general is that organisms possess back-up systems and compensation mechanisms. If we introduce a small number of errors or defects into the development of an organism, it will be able to make up for them. As the number of defects increases, however, they eventually overwhelm the back-up mechanisms and fitness declines more precipitously.

If the fitness relation has the decreasing returns shape, it is easier for natural selection to maintain variation at a large number of loci (by heterozygous advantage or any other selective means). The individuals with multiple homozygotes have very low fitness and are disproportionately likely to die, as compared with the multiplicative fitness case. When a multiple homozygote dies, it takes all of its homozygotes with it; it may have enough homozygotes to kill it many times over, but it only needs to die once to eliminate them all. Thus, one selective death kills more than one inferior genotype. Natural selection is more efficient at purging homozygotes when fitness decreases more rapidly as the number of inferior genotypes (homozygotes, in this case) in an organism increases.

The pattern you expect will depend partly on how "atomistic" or "holistic" your view of biological organisms is. If organisms are atomistically organized into many self-contained subunits, the pattern would be more likely to be the straight line; if they are holistically organized, with multiple overlapping subsystems that can compensate for one another, the pattern would be more likely to be a curve. It is probable that extra good genotypes provide some "decreasing return" and therefore some curve in the fitness pattern, but whether this level is enough for selection to explain the observed levels of molecular variation remains an open question.

The main points of this section are that the absurd figures for genetic loads assumed that the fitness effects of different loci are independent. Other kinds of fitness relation make it easier for selection to maintain variation, as they require much lower genetic loads. An analogous argument can be made to explain how selection could drive evolutionary change, rather than maintain variation, at multiple loci, without running up an impossible cost or substitutional load.

A final point to note is that natural selection can, in theory, maintain polymorphism by frequency dependence of fitnesses as well as by heterozygous advantage. A genetic load is unavoidable if heterozygous advantage exists, but it is possible for a polymorphism to be maintained by frequency dependence without any load at all. (This statement is true for a sex ratio, for example; no selective death is needed at equilibrium to maintain the 50:50 sex ratio.)

We have now completed the three selectionist answers to the problem of genetic load. The paradoxical genetic loads were apparently implied by the discovery, in the 1960s, of high levels of molecular variation and high rates of molecular evolution. As we have seen, the problem exists only if selection is hard, fitnesses are multiplicative, and heterozygous advantage is the only selective regime—none of which need be true.

Levels of Polymorphism

Kimura's argument was that neutral drift can explain the high rate of molecular evolution and high levels of polymorphism, whereas natural selection cannot. His case against natural selection made assumptions about genetic loads that are unrealistic, however, and it is therefore unconvincing. Although the negative case against an explanation by natural selection is unconvincing, the positive case for it is no more powerful than for neutral drift. Neutral drift, we saw, can explain the observed evolutionary rates and heterozygosities by inserting more or less conjectural values for N and u into the equations for genetic drift. The actual values of N and u are unknown, and a neutralist could "explain" almost any evolutionary rate or heterozygosity by positing suitable values for N and u.

When natural selection is considered positively, as a possible explanation for molecular evolution, it turns out to occupy a similar position. Selectionists also do not generally know the value of the variables that control the rate of evolution. The probability of fixation of a selectively advantageous mutation with selective advantage s is approximately 2s, which makes the rate of evolution 2sm where m is the rate of selectively advantageous mutations. However, m remains just as much of an unknown as u, and not much can be said about s apart from that it is probably low (1% or less). Like the neutralist, the selectionist can explain almost any observation by inventing suitable values for unknown numbers.

The objections to selective explanation for molecular variation and evolution were successfully countered, as we saw in the previous section, but it does not mean that selection is the explanation for the facts. Indeed, few biologists now think that heterozygous advantage explains more than a few polymorphisms. The main reason is empirical: examples have accumulated too slowly. The main example, sickle cell anemia, was one of the first polymorphisms to be understood in selective terms and it was natural to assume (tentatively) that the process might be general. Not much evidence has since been found of heterozygous advantage at other polymorphic genes; in fact, 30 years of research between the discovery of heterozygous advantage in sickle cell anemia and a review by Endler in 1986 revealed only another six examples, which represents a tiny percentage of the polymorphic genes that have been investigated.

Frequency dependence is another possible selective explanation. Some population geneticists, such as Clarke, have argued that it is a general cause of polymorphism. The possibility has not been widely accepted, but it cannot be ruled out either. No general arguments have been set forth against this theory, except perhaps a vague intuition that not enough biological interactions have the correct characteristics to be negatively frequency dependent and to maintain 3000 polymorphisms in a fruitfly population. Some widespread biological processes such as the interaction of predators and prey and of parasites and their hosts produce frequency-dependent relations, although not necessarily stable polymorphisms. The importance of frequency-dependent selection in nature remains an open question. The same question plagues other kinds of selection, in which polymorphism is maintained by changes in the fitness of genotypes in time and space. In short, the observations about rates of molecular evolution and levels of polymorphism do not allow us to draw decisive conclusions.

Process Controlled by Natural Selection

The rate of molecular evolution can be measured for any pair of species. Each pair of species has its own number of molecular differences and time to their common ancestor. We can plot the point defined by these two numbers for many pairs of species on a graph, for hemoglobin.

The striking property of the graph is that the points for the different species pairs fall on a straight line. Molecular evolution appears to have an approximately constant rate per unit time—that is, it shows a *molecular clock*. Evolutionary change at the molecular level ticks over at a roughly constant rate, and the amount of molecular change between two species measures how long ago they shared a common ancestor.

A graph requires a knowledge of the time between the common ancestor and the evolution of each species pair. These times are estimated from the fossil record and are uncertain; the results are not universally trusted. However, we can also test the constancy of molecular evolution by another method, which does not require absolute dates, and this test also suggests that molecular evolution is fairly clock-like. An empirical controversy has arisen as to the constancy of the molecular clock; the statistical details of this issue are complex, and we will not discuss them here. It can reasonably be concluded at present that the rate of molecular evolution is constant enough to require explanation.

What does a constant rate imply about the roles of natural selection and neutral drift in molecular evolution? Kimura reasoned that constant rates are more easily explained by neutral drift than selection. Neutral drift has the property of a random process, and its rate will show the variability characteristic of a random process. Neutral mutations crop up at random intervals, but if they are observed over a sufficiently long time period the rate of change will appear to be approximately constant. Thus, neutral drift will drive evolution at a fairly constant rate. Natural selection, Kimura argued, does not produce such constant change. Under selection, the rate of evolution is influenced by

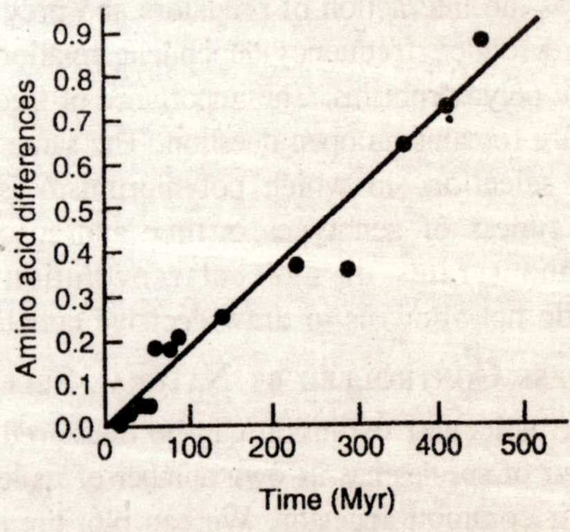

Fig. 1.7. The rate of evolution of hemoglobin.

environmental change as well as the mutation rate. It would require a surprisingly steady rate of environmental change, occurring over hundreds of millions of years, in organisms as diverse as snails, mice, sharks, and trees to produce the constant rate of change.

Moreover, if we examine characters (e.g., any adaptive morphological characters) that have undoubtedly evolved by natural selection, they do not appear to evolve at constant rates. Kimura's discussion of the evolution of the vertebrate wing illustrates the argument. A long period preceded the evolution of the wing, during which the vertebrate limb remained relatively constant (in the form of the tetrapod limb of amphibians and reptiles); then came a shorter period when the wing originated and evolved; finally, a more or less finished wing form was fine-tuned over a long period. The wing undoubtedly evolved under the influence of natural selection, and the rate of change probably underwent large fluctuations between fast and slow evolution. The argument suggests that rates of molecular evolution are strikingly constant; it is also Kimura's reason for confining the neutral theory to molecules, and not applying it to the gross phenotypes of organisms. Molecular evolution does appear to proceed at a fairly constant rate, as would be expected for a random process. However, the rate of morphological evolution has a different pattern, and was probably produced by the nonrandom process of selection.

Molecular evolution in "living fossils" provides a striking example both of the constant rate of molecular evolution and of the independence between molecular and morphological evolution. The Port Jackson shark *Heterodontus portus-jacksoni* is a living fossil—a species that closely resembles its fossil ancestors (fossils over 300 million years old are very similar to the Port Jackson shark). Its molecules have been evolving very differently from its morphology. Hemoglobin duplicated into α and β forms before the ancestor of mammals and sharks, at the beginning of the chordate radiation. We can count the amino acid differences between α- and β-globin as a measure of the rate of molecular evolution in the lineages leading to the modern species. Thus, the rates of molecular evolution in the two lineages are roughly equal. In contrast, the two lineages show large differences in their rates of morphological evolution. The Port Jackson shark lineage has hardly changed at all, but humans have evolved from fish-like ancestors, and passed through amphibian, reptilian, and several mammalian stages.

Moreover, human β-globin is as different from human α-globin as it is from carp α-globin. This disparity has evolved despite the fact that human α- and β-globin have shared much more similar external selective pressures, being locked in the same kind of organisms throughout evolution, than have human β-globin and carp α-globin. The result suggests that the α- and β-globin molecules have been accumulating changes independently, at roughly constant rates, regardless of the external selective circumstances of the molecule. This evidence, in turn, suggests that most of the evolutionary changes in the globin molecule have been neutral shifts among equivalent forms, of equal adaptive utility. While the rates of morphological change vary greatly among the various evolutionary lineages of vertebrates, the rates of molecular evolution all seem to have been more similar.

Protein Molecular Clock

Wilson et al. (1977) provided the first powerful evidence that the rate of protein evolution is constant relative to absolute time, rather than the number of generations. Their arguments are based on comparisons between pairs of species. In each pair, one of the species had a long generation time (elephant or sperm whale, for example) and the other had a short generation time (mouse or rabbit, for example). Wilson et al. found 12 proteins that had been sequenced in one or more pairs of a long and a short generation time species, and they were able to make 25 species-pair comparisons in all. Each comparison used the calculations of the relative rate test, and the rate of evolution was estimated for the lineage leading to the short and to the long generation time species. It is a reasonable assumption that the generation times of the stream of species leading up to the modern sperm whale were, on average, longer than those of the species leading to the modern mouse.

If generation times influence the rate of evolution, the amount of change in the long generation time lineage should be less (Wilson et

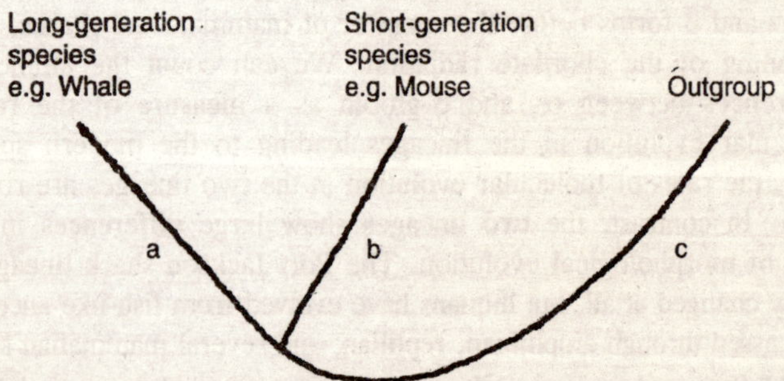

Fig. 1.8. Wilson et al.'s method to test for a generation time effect on the rate of protein evolution.

al. reckoned it would be about $^1/_5$ to $^1/_8$ of the amount in the short generation time lineage, if the molecular clock keeps time relative to numbers of generations). In fact, the evidence suggested that the rate of evolution was about the same for the two lineages. Thus, the rate of evolution (and molecular clock) ticks at a rate proportional to absolute time, in years, rather than number of generations.

Notice that this result applies to the amino acid sequence of proteins. We will see later that silent changes in the DNA—changes that have no effect on the amino acid—may tell a different story. Some subsequent work hints at small generation time influences on the rate of protein evolution, but the influences, if they exist at all, are always much smaller for protein evolution than for silent changes in the DNA. For proteins, Wilson et al.'s conclusion that the molecular clock runs relative to years and not numbers of generations is reasonably well substantiated.

According to the neutral theory, the rate of evolution is equal to the neutral mutation rate. The rate will, therefore, be proportional to the time scale of the process-generating mutations. This relation is normally thought to imply generation time. Most mutations probably arise as copy errors during mitotic or meiotic replication. The mutation rate should therefore be proportional to the number of DNA replications in the germ line. Fewer such replications are produced per unit of absolute time in species with longer generations, because the number of mitotic divisions in the germ line is much less than proportional to longevity. For visible mutation, mutation rates per generation are not noticeably longer in species with longer generation times. If (as seems reasonable) neutral mutations are like other mutations, neutral evolution should occur more slowly in species with longer generation times. Selectionists can, therefore, retort that neutralism, far from predicting the molecular clock, actually does not predict it. The neutralist's clock would be quite different from the clock that has been found, and would tick in proportion to numbers of generations, not of years.

Selection probably provides a better explanation than neutral drift for why the clock depends on absolute time. Under neutral evolution, the rate of evolution exactly equals the mutation rate; as a result, neutralism strongly predicts a relation between number of generations and evolutionary changes. Under selection, the rate of evolution can be controlled by the pace of environmental change, which could depend on absolute time.

Suppose, for instance, that two kinds of mammal, such as a mouse (generation time approximately 0.33 year) and an elephant (generation time approximately 33 years), were both separately evolving in relation to changes in parasites (e.g., bacteria) with short generation times. After 0.33 year, the bacterial population will have accumulated 0.33 year's worth of change, which acts as an agent of selection on the mouse population. On the other hand, after 33 years, the bacteria that parasitize elephants will have accumulated 100 times as much change, which will select 100 times as strongly on the elephants. The larger elephant generation time will, to some extent, be compensated by the stronger selection it will experience. The resulting rates of evolution in mice and elephants might be quite similar. This argument has been simplified, of course, but it suggests that natural selection can explain why the molecular clock keeps absolute) rather than generational, time.

DNA Sequences

As an increasing number of DNA sequences have been analyzed, it has become possible to test for the constancy of molecular evolution in the DNA sequences. The genetic code contains 64 codons, of which 61 code for 20 amino acids; the three-fold redundancy in the code means that not all base changes in the DNA cause amino acid changes in the protein. As an almost inevitable theoretical consequence) there cannot be an evolutionary molecular clock for both base changes and amino acid changes. We have seen that the clock for proteins keeps reasonably good time, although it ticks over according to absolute, not generational, time. What happens in the DNA?

The evidence suggests the presence of a generation time effect. We can distinguish the results for "silent" and "replacement" base substitutions. Silent (sometimes also called "synonymous") base changes do not result in an amino acid change in the protein; the protein is the same after a silent mutation as before. In contrast, replacement (sometimes also called "meaningful") base changes do alter the amino acid. It seems likely that silent mutations will more often be neutral than will replacement mutations.

Rodents, such as mice and rats, have shorter generation times than primates and artiodactyls (such as cows). For both silent substitutions and replacement substitutions, evolution is faster in rodents than in artiodactyls, and faster in artiodactyls than in primates. The effect is particularly marked in silent sites. Silent substitutions occur more rapidly in species with shorter generation times, as the neutral theory predicted. Perhaps the silent parts of the DNA do, in fact, evolve neutrally. (We will have more to say about the rate of evolution in DNA sequences in later sections; here we are concerned only with whether they show a generation time effect.) In conclusion, the absence of a generation time effect in the molecular clock for proteins suggests that neutral drift is not as important in their evolution as Kimura originally thought. For silent base changes in DNA, however, the clock depends on generation time and the neutral explanation is much more plausible.

Functionally Constrained Parts of Proteins

A protein contains functionally more important regions (such as the active site of an enzyme) and less important regions. It has consistently been found that the rate of evolution in the functionally more important parts of proteins is slower. Insulin, for example, is formed from a proinsulin molecule by excising a central region. The central region is then discarded, so its sequence is probably less crucial than that of the outlying parts that form the final insulin protein. The central part evolves six times more rapidly than the outlying parts.

The same result has been found by comparing evolutionary rates in the active sites and in other regions of enzymes (or of proteins generally). The surface of a hemoglobin molecule, for example, may be functionally less important than the heme pocket, which contains the active site. The evolutionary rate is about 10 times faster in the surface region.

A similar tendency may underlie differences in the rates of evolution of whole genes, or proteins. One generalization is that "housekeeping" genes, which control the basic metabolic processes of the cell, evolve slowly. In contrast, genes like the globins, fibrinopeptides, and immunoglobulins, which have more specialized

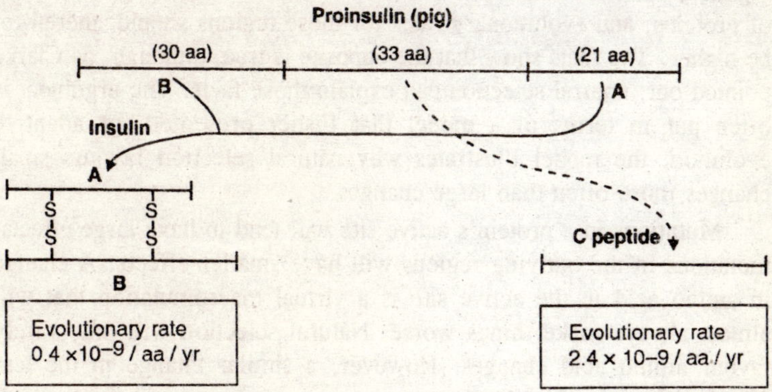

Fig. 1.9. The insulin molecule is made by snipping the center out of a larger proinsulin molecule.

functions and only operate in specific cell types, evolve more rapidly. This pattern is less clear-cut than the pattern seen within a gene for insulin and for hemoglobin, but it is possible that differing degrees of functional constraint are related to rates of evolution for different genes as a whole. A basic housekeeping gene may be more difficult to change during evolution than a gene with a more localized function.

Natural Selection and Neutral Drift

The neutral explanation for the relation between evolutionary rate and functional constraint follows. In the active site of an enzyme, an amino acid change will probably modify the enzyme's activity. Because the enzyme is relatively well adapted, the change is likely to be for the worse and may well spoil the enzyme's function. In the other parts of the molecule, the identity of an amino acid occupying a site may matter less, and a change is more likely to be neutral. The proportion of mutations that are neutral will be lower for the functionally constrained regions. As a result, if the total mutation rate is similar throughout the enzyme, the number of neutral mutations will be lower in the active site. The evolutionary rate will be lower as well. Geneticists have found a good fit between facts and prediction, and the relationship has been confirmed repeatedly. We will examine further examples involving DNA later.

What is the selective explanation for this trend? It has been argued that, if all evolution occurs by natural selection, there should either be no relation between functional constraint and evolutionary rate, or the opposite relation from what is found. Natural selection would arguably be more important in the functionally more important parts of proteins, and evolutionary rates for those regions should, therefore, be higher. The facts show that the opposite is true, although, as Clarke pointed out, natural selection can explain those facts. The argument is often put in terms of a model that Fisher presented for adaptive evolution; the model illustrates why natural selection favours small changes more often than large changes.

Mutations in a protein's active site will tend to have large effects; mutations in the outlying regions will have smaller effects. A change in amino acid in the active site is a virtual macromutation that will almost always make things worse. Natural selection will only rarely favour amino acid changes. However, a similar change in the less functionally constrained parts may have more chance of being a small fine-tuning improvement that would be favoured by natural selection. Selection will then more often favour changes in the less constrained regions of molecules, because those parts offer more opportunities for fine-tuning. An apt analogy might be a radio, in which the more and less constrained regions of a protein are like two tuning knobs—one for fine adjustments and the other for larger movements. Once a radio station has been located, we make more use of the fine adjustment knob, to keep track with the signal. Likewise, changes outside the active site are more frequent in evolution.

Kimura (1983) rejected this argument as "completely false." It ignores the probability that a mutation, even if it is advantageous, eventually becomes fixed. The chance of fixation of a favourable mutation is, as we have seen, approximately 2s and, as a consequence, is lower for a mutation of smaller s. The fine-tuning mutants will, Kimura reasons, have lower selective advantage, and the larger proportion of selectively advantageous small mutations will be canceled by their higher chance of random loss.

Kimura's reasoning does not work for all cases. His argument works for a unique mutation. If a favourable mutation occurs recurrently, however, it will eventually be fixed even if its chance of survival is low each time. In that instance, the recurrent, advantageous mutations in unconstrained regions will eventually be fixed, while the recurrent disruptive changes in active sites will be rejected. The rate of evolution will be higher in the functionally less constrained parts of molecules, as is observed. Thus, the selective explanation is quite plausible, and both theories can explain the well-documented relation between functional constraint and evolutionary rate.

SILENT SITES IN DNA

The same relationships between functional constraint and evolutionary rate have been found, and the same explanations defended, for DNA as for proteins. Two properties of DNA sequences are particularly interesting: the relation between silent and replacement changes in the third codon position, and the evolutionary rate of *pseudogenes*.

A pseudogene is a region of a DNA molecule that clearly resembles the sequence of a known gene, but differs from it in some crucial respect and probably has no function. Some pseudogenes, for example, cannot he transcribed, because they lack promotors and introns. Pseudogencs, once formed, are probably under little or no constraint, and mutations will accumulate by neutral drift at the rate at which they arise. They will show pure neutral evolution in the pan-neutralist sense that all mutations are neutral.

The neutral theory predicts that pseudogenes should evolve rapidly—and they do. It is tempting to use these rates of pseudogene evolution as direct estimates of the mutation rate. For instance, taking an approximate figure of one change per site per 10^9 years, we could multiply the rates for the whole genome, with a haploid content of about 3×10^9 bases in humans, to suggest that about three mutations occur on average per year (or more like 60 per generation in humans). Other pseudogenes evolve at other rates, and the evidence as a whole suggests that about 100 mutations on average appear between a parent and its offspring. However, as we will see below, some evidence also exists for constraints on pseudogene evolution; if some mutations are selected against, then the total mutation rate must be higher than the evolutionary rate.

A selectionist could argue that the fast evolution of pseudogenes is due to particularly frequent fine-tuning in these molecules, but to date no one has advanced this theory. Selectionists will be more tempted to concede pseudogenes as an exception, but an exception that tells us nothing about the reason for faster evolution in other cases of reduced functional constraint. As a result, it is widely accepted that the rapid evolution of pseudogenes is due to neutral drift.

Silent base changes, which do not alter the amino acid, should be less constrained than replacement sites. Kimura had predicted, before DNA sequences were available, that silent changes would evolve more rapidly. It has now been confirmed that Kimur'a was correct; evolution in silent sites runs roughly five times as rapidly as evolution in replacement sites. A few exceptions are known, including the antigen recognition site of the HLA genes. The debate over whether the usual pattern of rapid change in silent sites involves solely neutral drift or frequent selected fine-tuning has not been conclusively settled, but most people accept that neutral drift is the main force, if only because it is difficult to imagine why silent changes should be selected for so often.

If silent codon positions were selectively unconstrained, they should show pan-neutral evolution. They should then evolve in a different manner from proteins. Different proteins, as we have seen, evolve at different rates, which neutralists explain by different degrees of constraint (and selectionists by different intensities of selection). If the silent positions in the DNA encoding those same proteins are unconstrained, they should not show different evolutionary rates. Instead) they should evolve at the rate of a "universal" molecular clock (i.e., at their total mutation rate). This idea can be tested by comparing the

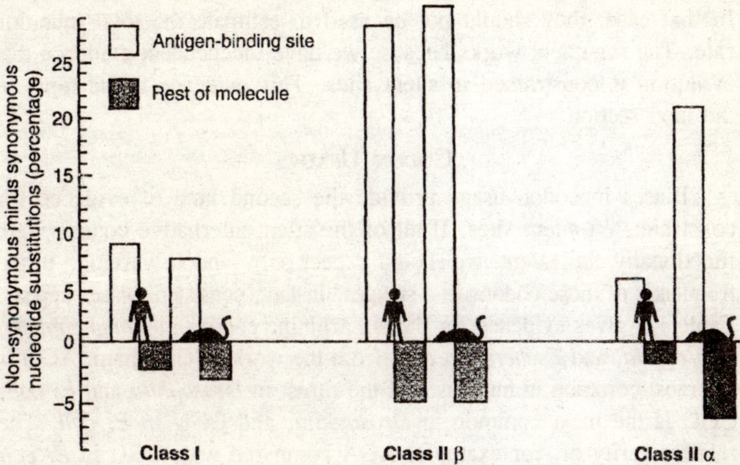

Fig. 1.10. Evolutionary rates of silent and amino acid replacing base changes in the human (HLA) and mice (H-2) major histocompatibility loci.

rates of change in the silent sites of different genes to see whether evolutionary rates remain constant in silent sites. Initially, Miyata and others suggested that the silent rates were constant, but more recent evidence has not supported them. Evolutionary rates are less variable in silent sites than in replacement sites, suggesting a lower degree of constraint, but they vary nonetheless. Also, evidence showing that evolutionary rates for a single gene can vary between different lineages, also contradicts the idea of pan-neutral evolution.

Other, more direct, evidence gathered about selective constraints in silent base sites involves two kinds. The first concerns the relative rates of evolution in pseudogenes and silent positions. It was initially thought that pseudogenes evolve twice as fast as silent codon positions, but more recent evidence suggests that evolution occurs just as rapidly in silent codon positions as it does in pseudogenes. We can provisionally conclude that pseudogenes and silent sites evolve at about the same rate, although future evidence may suggest otherwise.

The question of these evolution rates is interesting, because when it was thought that pseudogenes evolve faster than any other DNA, it was tempting to argue that their evolution was completely unconstrained. If true, their rate of evolution would then be a direct estimate of the total mutation rate. We now think that pseudogenes evolve at the same rate as silent base changes, and it is argued that evolution is probably constrained even in pseudogenes to some extent.

In that case, they should not be used to estimate the total mutation rate. The argument works because we have independent evidence that evolution is constrained in silent sites. This evidence is the topic of the next section.

Codon Usages

Biases in codon usage provide the second kind of evidence for constraints on silent sites. If all of the silent alternative codons were functionally equivalent, we should expect only random variation in the frequency of those codons in a species. In fact, consistent biases appear. Table 1.3 gives evidence for the six arginine codons in *Homo sapiens*, *Drosophila*, and *Escherichia coli*, from the work of Grantham. AGG is the most common in humans, but the rarest in *Drosophila* and *E. coli*; CGC is the most common in *Drosophila*, and CGU in *E. coli*. The relative rarity of, for example, AGA compared with CGU in *E. coli* suggests that mutations from CGU to AGA tend to be selected against. The changes are not all neutral. Pan-neutralism is not valid in this case. (Notice again that we are concerned only with the possibility that some codons, out of a synonymous set, are deleterious, and selected against; the nonrandom frequency distributions could then be explained. This explanation does not argue against Kimura's neutral theory. Logically, the nonrandom frequency distributions could arise because of positive selection for certain codons. The neutral theory would then be wrong, but that is another matter.)

Table 1.3. Frequencies of six arginine codons in the DNA of three species.

Codon	*Human*	*Drosophila*	*E. coli*
Arginine			
AGA	22%	10%	1%
AGG	23%	6%	1%
CGA	10%	8%	4%
CGC	22%	49%	39%
CGG	14%	9%	4%
CGU	9%	18%	49%
Total number of arginine codons	2403	506	149
Total number of genes	195	46	149

How could selection discriminate among silent codons? The mechanism is not known, although two suggestions have been put forth. Both possibilities are theoretical only—it has not been shown that natural selection does operate on silent codon changes. The first suggestion is that the nucleotide sequence controls the secondary structure of the DNA molecule; changes in nucleotides might then influence the molecular shape, which could make a difference to the organism's fitness. Silent substitutions would be as likely to influence structure as replacement substitutions, and selection would work on both for the same reason.

The second factor is transfer RNA. The different silent codons use, to some extent, different tRNA molecules. It has been observed, in yeast and in *E. coli*, that the frequency of use of the codons in a set of synonyms is correlated with the tRNA abundances in the cell. We should expect the relation if the frequencies of codons were set by some other factor; the tRNA abundances would then adjust (by natural selection) to the quantity needed. The relation could also arise if the tRNA abundances were fixed by some other constraining factor. In that case, the mutations among the silent codons would not be neutral, as mutations to codons whose tRNA was in short supply would be selected against relative to those with abundant tRNA.

Directional mutation pressure is another possible explanation for biases in codon usage. A tendency for A and T bases to mutate more often to G or C than do G and C to A or T would result in a build-up of GC bases. Codon biases would result automatically. It has, in fact, been suggested that GC is more stable than AT, and favoured by mutation pressure. The task is then to apply these chemical mechanisms to explain the differences between species in their codon biases.

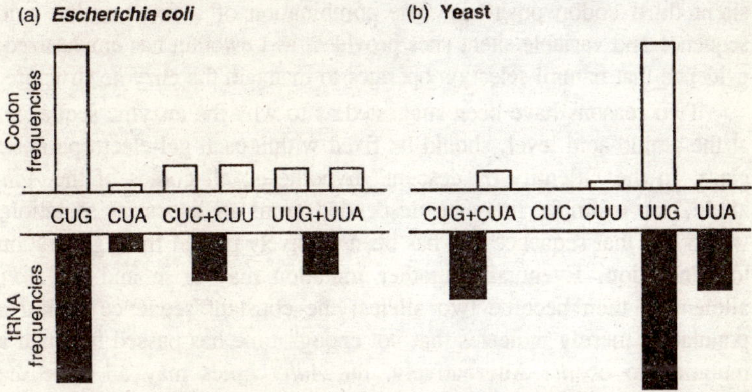

Fig. 1.11. Relative frequencies of codons match tRNA abundances. (a) the above column are the relative frequencies of six leucine codons in E. coli; the below column are the relative frequencies of the corresponding tRNA molecules in the cell. (b) Same relation, but in yeast.

In summary, the functionally more important parts of proteins (such as the active sites of enzymes) and DNA (such as replacement rather than silent sites) evolve more slowly than the less important parts. Two explanations have been suggested for the trend. The first theory is that it may arise because a higher proportion of mutations are neutral in the less important parts; this theory is the generally accepted explanation for the rapid rate of evolution for silent base changes. However, even these sites are not completely unconstrained—different taxonomic groups show different biases in codon usage and have different rates of evolution in silent sites. The second explanation for the trend is that natural selection works more readily with small changes, and a change by one amino acid is more likely to be a small change if it is found in a less important part of a molecule than if it is located in a more important part. Biologists do not agree as to whether selection or drift is the reason why the functionally important parts of proteins evolve slowly.

Role of Natural Selection on Protein Structure

When we considered the evidence for biological variation, we noticed that many DNA sequence variants can be uncovered if a single gel electrophoretic class of a protein is sequenced at the DNA level. This observation has important implications for molecular evolution. At the alcohol dehydrogenase (*Adh*) locus in the fruitfly, two electrophoretic classes (fast and slow) are present. When Kreitman sequenced proteins from the two classes, he found that the amino acid sequence of each was uniform, but many variants were observed in silent third codon positions. The combination of a fixed amino acid sequence and variable silent sites provides, as Lewontin has emphasized, evidence that natural selection operates to maintain the enzyme structure.

Two reasons have been suggested as to why the enzyme sequence, at the amino acid level, should be fixed within each gel electrophoretic class. In the "identity by descent" hypothesis, all copies of the Adh_f allele, for example, may be descended from an ancestral mutation, which had that sequence and has been passively passed from generation to generation. Eventually another mutation may arise and the Adh_f allele will then become two alleles; the constant sequence within a population merely indicates that not enough time has passed for such a mutation to occur. Alternatively, the Adh_f copies may all have the same sequence because the sequence is maintained by natural selection; when a mutation arises) selection removes it. The observed variability distinguishes between these two hypotheses. The variability in the silent base sites means that there has been time for mutations to arise in the molecule. If mutations have arisen in silent sites, they will surely have occurred in replacement sites as well. Therefore, we can reason that the identity in amino acid sequence is unlikely to be identity by descent. Mutations in replacement sites have presumably not been retained because natural selection eliminated them.

If the *Adh*$_f$ allele has been found to be fixed for one DNA sequence at all sites—both replacement and silent—we could not know whether the uniformity was due to selection or identity by descent. We would be in the same position as when we had only the gel electrophoretic evidence. Perhaps the fixity meant simply that no mutations had occurred. The DNA sequences thus provide evidence for selection that could not have been obtained with amino acid sequences alone.

The absence of amino acid sequence variation within the *Adh*$_f$ (and *Adh*$_s$) allelic class is particularly striking because 30% of the enzyme consists of isoleucine and valine, which are biochemically very similar (and electrophoretically indistinguishable). A neutralist might have predicted that some of the valines could be changed to isoleucines, or vice versa. The only amino acid sequence variant is the one that causes the *AdH*$_f$/*AdH*$_s$ polymorphism. Because abundant evidence exists that natural selection operates on that change, Lewontin concluded that there are actually no neutral amino acid sites in the 255 amino acid alcohol dehydrogenase enzyme of the fruitfly. Interestingly, that conclusion means that we could almost construct for alcohol dehydrogenase at the amino acid level. Natural selection is powerfully maintaining the amino acid sequence, while the silent base changes are probably neutral.

HETEROZYGOSITIES

For the fourth kind of test we return to the two phenomena we discussed in the first test: the degree of polymorphism (or heterozygosity) within species and the rate of evolutionary change between species. In this test, we will look not at their absolute rates) but at the relation between the two. The neutral theory predicts the two rates will be correlated. The same factor—the neutral mutation rate—explains both polymorphism and the rate of change, in the neutral theory. All polymorphism is "transient," and both polymorphism and the substitution of one gene for another are manifestations of the same process—the neutral drift of genes up or down in frequency. If we look at different genes (or different parts of one gene), we find some have higher heterozygosities than others, which the neutral theory explains by their having a higher neutral mutation rate. The same factor also produces a higher rate of change. We should, therefore, find a positive correlation between the rate of evolutionary change and the heterozygosity of different regions of the DNA.

If natural selection dominates molecular evolution, a positive correlation would not necessarily be expected. Natural selection can produce transient polymorphism. It takes time for selection to substitute one allele for another (although not as long as it takes with drift) and the population will be polymorphic while the substitution proceeds. The process has been seen in action in the peppered moth and some other examples, but it is generally supposed to be rare. If approximately 10,000 gene loci in a population are polymorphic, it is unlikely that substitutional events of the peppered moth kind are occurring at more than a minority of them in an average population.

This intuition could be wrong, but it seems more likely that most selected polymorphisms are not transient but stable, being due to frequency dependence, heterozygous advantage, or some other selective process. If selected polymorphisms are not transient, then different processes control evolutionary change and polymorphism. Occasional rounds of directional selection cause change, but frequency dependence (or whatever) causes polymorphism. If we look at different genes (or regions of one gene), the two can be correlated only by coincidence. Genes with high evolutionary rates will be those that experience strong and persistent directional selection; we have no particular reason to suppose they will also be the ones at which selection favours a high heterozygosity. Thus, there is now less reason to expect a positive correlation between the evolutionary rate and polymorphism.

What do the facts show? The relation was first examined in gel electrophoretic data for proteins, and broadly supported the neutral theory. That test was relatively coarse, however, and an expanding program of research is now making better controlled comparisons in the DNA.

The first important study of this kind was performed by McDonald and Kreitman in 1991. They did not compare different genes, but silent and replacement events within one gene, alcohol dehydrogenase (*Adh*) in two closely related species of fruitfly (*Drosophila*). McDonald and Kreitman developed a more exact version of the argument given above. For the two species, they measured the number of (fixed) differences for replacement and for silent nucleotide changes; they divided one by the other to give a ratio. They also measured the polymorphism within

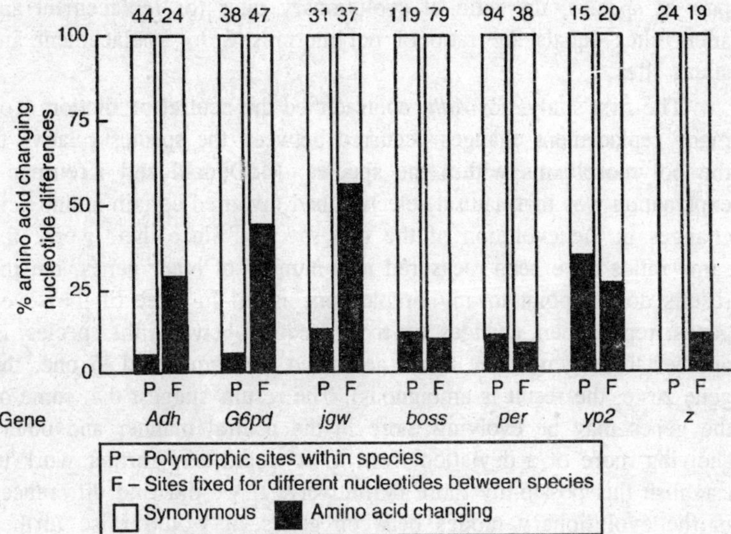

Fig. 1.12. Ratio of replacement to silent changes, for differences among species and for polymorphisms within a species, in seven genes in fruitflies.

each species for replacement and silent sites and expressed them as a ratio. The two ratios should be equal if all the changes are neutral, for the following reason. The rate of neutral evolution equals the mutation rate, and the ratio of species differences for replacement sites/species differences for silent sites will, therefore, be the ratio of the neutral replacement change mutation rate/neutral silent change mutation rate. (The neutral replacement change mutation rate will arguably be the lower of the two, because it is less likely that any one replacement mutation will be neutral than any one silent mutation.) The amount of polymorphism under neutral drift is given by a more complicated formula, but if the formula for the heterozygosity for neutral replacement changes is divided by that for the heterozygosity for silent changes, everything cancels except the respective mutation rates. The predicted replacement:silent ratio for the heterozygosities within species is then the same as for the number of differences between species.

The virtue of the test carried out by McDonald and Kreitman is that it is so well controlled. Many factors can influence both polymorphism and evolutionary rates, but all those factors should operate in the same way if we concentrate on a comparison for a single gene. The factors cancel out in the comparison and can be ignored. We can, therefore, test the neutral theory by seeing whether, in one gene in a pair of species, the ratio of evolutionary rates for replacement and silent sites equals the ratio of polymorphisms for replacement and silent sites.

The first study, for *Adh*, contradicted the neutral prediction. Too many replacement changes occurred between the species relative to the polymorphisms within the species. McDonald and Kreitman's explanation was that natural selection had favoured certain amino acid changes in the evolution of the two species. Since their work, the same ratios have been measured in a number of other genes, but the results do not point to any simple conclusion. In three of the seven genes, replacement changes are more frequent between the species; in another three genes, the ratios are about the same (and in one, the gene zeste, the result is ambiguous). The results suggest that some of the genes may be evolving more in the neutral manner and others showing more of a deviation from it. It would take further work to establish this possibility more definitively. If we did find differences in the evolutionary modes between genes, it would raise further questions. For now, the main point is that the relation between evolutionary rates and levels of polymorphism offers a promising line of research. At this time, the evidence does not point to any simple conclusion—but that may be because no simple conclusion exists.

Analysis of DNA Sequences

The controversy between the selective and the neutral theories has moved through several stages. The first stage was concerned with the amino acid sequences of proteins. Kimura suggested that the high rate, and constancy, of protein evolution, and high levels of heterozygosity, could not be explained by natural selection. It seems likely, however, that natural selection could explain these observations. Moreover, further evidence suggested that the levels of heterozygosity are too low for the neutral theory; evidence has also suggested that the molecular clock may be more erratic than the neutral theory predicts for proteins and that it ticks according to generational, rather than absolute, time.

More recently, the availability of base sequences for DNA has added another dimension to these tests. DNA contains both silent and replacement base positions. It seemed likely, or at least possible, that silent base changes—which do not alter the amino acid—would be neutral. Silent changes should have a higher rate of evolution than changes at replacement sites, and it has been confirmed that they do (with a few interesting exceptions, such as the HLA genes). Moreover, the evolutionary rate for silent sites should be proportional to generation time; the evidence suggests it is. Silent sites are not pan-neutral in their evolution, however. Constraints have also been suggested by the different patterns silent sites show in different taxonomic groups. The DNA sequence has also provided strong evidence that selection maintains the amino acid sequence of enzymes. New, more subtle tests have recently been developed for the pattern of heterozygosities and evolutionary rates in different regions of the same gene.

Thus, the controversy is not settled. However, it is unlikely that neutral drift is as important, and selection as unimportant, in protein evolution as Kimura originally suggested. His case against selection has been rebutted, and the facts do not comfortably fit the neutral predictions. As evidence for DNA sequences has accumulated, the controversy has increasingly moved away from simple generalizations about all molecular evolution, and toward understanding which kinds, or regions, of molecules evolve in which mode. For silent sites in the DNA, the neutral theory is widely accepted; the theory made clear predictions, many years before the facts were discovered) and the predictions have turned out correct. Ample room remains for disagreement about which molecular evolutionary changes were driven by selection, and which by neutral drift.

2

HISTORICAL BACKGROUND

Molecular evolution encompasses two areas of study: (i) the evolution of macromolecules and (ii) the reconstruction of the evolutionary history of genes and organisms. The first area includes the rates and patterns of change in the genetic material (e.g., DNA sequences) and its encoded products (e.g., proteins) during evolutionary time and the mechanisms responsible for such changes. The second area, also known as "molecular phylogeny" or "molecular phylogenetics", deals with the evolutionary history of organisms and macromolecules, as inferred from molecular data.

It might appear that the two areas of study constitute independent fields of inquiry, for the object of the first is to elucidate the causes and effects of evolutionary changes in molecules, while the second uses molecules merely as a tool to reconstruct the evolutionary history of organisms and their genetic constituents. In practice, however, the two discipline are intimately interrelated, and progress in one area facilitates studies in the other. For instance, phylogenetic knowledge is essential for determining the order of changes in the molecular characters under study, and knowing the order of such changes is usually the first step in inferring their cause. Conversely, knowledge of the pattern and rate of change of a given molecule is crucial for attempts to reconstruct the evolutionary history of a group of organisms.

Traditionally, a third area of study, prebiotic evolution or the "origin of life," is also included within the framework of molecular evolution.. This subject, however, involves a great deal of speculation and is less amenable to quantitative tr' atment. Moreover, the rules that govern the process of information transfer in prebiotic systems (i.e., systems devoid of replicable genes) are not known at the present time.

The study of molecular evolution has its roots in two disparate disciplines: population genetics and molecular biology. Population genetics provides the theoretical foundation for the study of evolutionary processes, while molecular biology provides the empirical data. Thus, to understand molecular evolution it is essential to acquire some basic knowledge of both molecular biology and the theory of population genetics.

Although as a discipline molecular evolution is still at a young stage of development, the pursuit of this subject actually began at the turn of the century. Studies in immunochemistry at the end of the last century and the beginning of this showed that serological cross-reactions were stronger for more closely related organisms than for less related ones. Seeing the evolutionary implications of this finding, Nuttall (1904) conducted precipitin tests of serum proteins to infer the phylogenetic relationships among various groups of animals. He determined, for example, that man's closest relative were the apes, followed, in order of relatedness, by Old World monkeys, New World monkeys, and prosimians. However, Nuttall was ahead of his time and intense research in molecular evolution did not start until the mid 1950s when the methods of protein sequencing, tryptic fragment pattern analysis, and starch-gel electrophoresis as well as better immunological techniques were introduced into evolutionary studies. These techniques stimulated much interest in the molecular phylogeny of humans, apes, and other primates and made a great impact on the study of molecular phylogeny in general.

The first complete sequence of a protein (insulin) was determined in 1952 by F. Sanger and colleagues, and by the mid-1950s a substantial amount of protein sequence information had been accumulated. The data revealed that amino acid substitutions occurred nonrandomly among different regions of a protein; for example, among the cattle, sheep, pig, horse, and whale insulins, changes were restricted to positions 8-10 of the A chain. The data also revealed that the majority of amino acid substitutions found in the same protein from different species did not seem to have significant effect on its biological activities. Yet a small number of amino acid differences may account for the large differences in biological activities of two different but related proteins; for example, cattle vasopressin and oxytocin differ by only two amino acids but have very different biological properties.

The 1960s were a period of rapid progress and heated controversies in molecular evolution. Considerable sequence data for hemoglobins and cytochromes c had become available in the early 1960s and comparative studies of the data suggested that the rate of amino acid substitution in each of these proteins was approximately the same among different mammalian lineages. Zuckerkandl and Pauling (1965) therefore proposed that for any given protein the rate of molecular evolution is approximately constant in all evolutionary lineages or, in other words, that there is a molecular clock. This proposal immediately stimulated a great deal of interest in the use of macromolecules for evolutionary studies. Indeed, if proteins evolve at constant rates, they can be used to estimate the dates of species divergence and to reconstruct phylogenetic relationships among organisms. This proposal has had a tremendous impact on the development of molecular evolution.

The molecular clock hypothesis, however, also provoked a great deal of controversy because the concept of rate constancy is diametrically opposed to the erratic tempo of evolution at the morphological and physiological levels. The rates of morphological evolution in different evolutionary lineages and at different times in the same lineage are strikingly variable.

The controversy became particularly heated after the rate-constancy assumption was used as a basis for estimating the divergence times between man and the African apes, leading to values much younger than traditional dates, and after it was taken as strong support for the neutral mutation hypothesis. Another controversy, stimulated by both protein sequence and DNA hybridization data, arose over whether the length of generation time has any significant effect on the rate of molecular evolution. In the 1970s statistical analyses of protein sequence data showed that amino acid substitution usually did not follow a simple Poisson process. Nevertheless, the rate of substitution was roughly constant over time and the approximate constancy seemed more striking than the relatively minor deviations therefrom for a discussion of why deviations from the Poisson process may not be used to exclude the constancy of evolutionary rate. Further support for the hypothesis came from quantitative immunological comparisons of proteins. Therefore, by the late 1970s the molecular clock hypothesis was accepted by many molecular evolutionists. However, some others, particularly Goodman (1976, 1981) continued to challenge it.

The 1960s also witnessed a revolution in population genetics. The availability of protein sequence data removed the species barriers in population genetic studies—traditional genetic analysis had been restricted to groups that could be crossed. For the first time, molecular studies had provided adequate data for examining the theory of gene substitution. Extrapolating the rate of nucleotide substitution estimated from several protein sequences to the entire mammalian genome, Kimura (1968a) arrived at a genomic rate that he thought was much too high to be compatible with Haldane's (1957) principle of the cost of natural selection. He therefore proposed that the majority of molecular changes in evolution are due to random drift of neutral or nearly neutral mutations. ("Nearly neutral" means that the selective differences between alleles are of the order of the reciprocal of the species population number or less, so that the effect of selection on the change in allele frequencies is similar to or weaker than that of random drift.) The same hypothesis was proposed independently by King and Jukes (1969). At the same time, the discovery that natural populations contain large amounts of protein polymorphism detectable by electrophoresis also raised many new issues, especially with respect to the mechanism of maintenance of high levels of polymorphism in populations. In the classical view, evolution and polymorphism were regarded as quite different processes, the former being dynamic whereas the latter is static. Kimura (1968a,b) proposed instead that most of the polymorphism is due to neutral or nearly neutral mutations and is not maintained permanently in the population, that is to say, polymorphism is a transient phase of molecular evolution. The hypothesis is proposed by Kimura and King and Jukes has been known as the neutral mutation hypothesis or the neutral theory of molecular evolution.

Under this hypothesis, evolution at the molecular level is largely driven not by natural selection but by mutation and random drift. Over a long time period the average rate of substitution is equal simply to the mutation rate of neutral mutation, if both are measured in the same units, e.g., years or generations. This provided a theoretical basis of rate constancy from the neutralist view; that is, if the majority of amino acid or nucleotide substitution in evolution are indeed due to neutral or nearly neutral mutations and if the rate of such mutation per unit time is constant over time, then the rate of evolution will be approximately constant.

The neutral mutation hypothesis provoked a great controversy, for it contradicted the traditional view that natural selection is responsible for both within-species polymorphisms and between-species differences. A number of issues quickly arose. First, there were challenges to the concept that gene substitution creates a cost to the population, and there was divergence of opinion as to how to calculate the cost of natural selection when a large number of loci are involved.

Second, after the rate-constancy assumption was used to support the neutral mutation hypothesis, the validity of the assumption became a major issue and the rejection of the assumption (i.e., the molecular clock hypothesis) has sometimes been mistaken as a rejection of the neutral mutation hypothesis. After all, the mutation rate per unit time and other factors need not have remained constant.

Third, there was a debate on whether certain types of mutation are really selectively neutral. For example, synonymous mutations, which cause no amino acid changes, were then considered by neutralists as good candidates for neutral mutations. This view was challenged by Clarke (1970a) and Richmond (1970), who argued that different synonymous codons for an amino acid may have different fitnesses because the tRNAs recognizing different codons may have different concentrations in the cell and because a tRNA recognizing multiple synonymous codons may have different binding affinities for different codons.

Fourth, studies on the pattern and rate of amino acid substitution in protein evolution revealed that amino acids that are similar in physicochemical properties are interchanged more frequently than dissimilar ones and that functionally less important molecules or parts of a molecule evolve faster than more important ones. These observations suggest that mutations with smaller effects on protein structure or function have greater changes of fixation (i.e., spreading through the species) than those with more radical effects. Clarke (1970a, 1971) submitted that these observations are consistent with the selectionist view because mutations with smaller effects would have higher changes of being beneficial. Kimura and Ohta (1974), however, reasoned that mutations, if their effects are very small, are likely to have very small selective advantages and therefore have correspondingly small probabilities of fixation. They as well as Jukes and King (1971) argued that the above observations can be much more simply explained under the neutralist view, namely that the smaller the effect a mutation has, the higher the probability will be that it is selectively neutral. This latter argument was apparently more acceptable to biochemists and molecular biologists. For this reason and because of the substantial support gained in the 1970s for the rate-constancy assumption, the neutral mutation hypothesis had gained considerable acceptance before 1980.

Fifth, Ohta's (1973, 1974) suggestion of the importance of slightly deleterious mutations in molecular polymorphism and evolution has attracted much attention—given the fact that most mutations to be frequent. However, whether the slightly deleterious hypothesis can explain the patterns of molecular polymorphism and evolution is a subject of debate.

The controversy over the neutral mutation hypothesis has had two strong impact on molecular evolution and population genetics. First, it has led to the general recognition that the effect of random drift cannot be neglected when considering the evolutionary dynamics of molecular changes. Second, the fusion of molecular evolution and population genetics, which began with the availability of protein sequence and electrophoretic data, was much accelerated by the introduction of the concept that molecular evolution and polymorphism are two facets of the same phenomenon. Thus, an adequate theory of evolution at the molecular level must be consistent with both. Furthermore, because it is simple and testable, the neutral hypothesis has been of great value as a starting point of analysis, i.e., it can serve as a null hypothesis.

The 1960s and 1970s also were years of tremendous progress in molecular phylogenetics. The accumulation of protein sequences, which were more informative and easier to analyze than other types of molecular data then available, provided for the first time adequate data for studying long-term evolution such as evolutionary relationships among orders or beyond. These data stimulated many studies on the reconstruction of phylogenetic trees and on the development of tree-making methods. In particular, the finding that the tree reconstructed from the sequences of a *single* protein (cytochrome c) was similar to the species tree then known for many vertebrate and invertebrate species revealed the power of molecular phylogenetics. Moreover, the data also stimulated much interest in the evolutionary relationships among protein sequences and in the methodology of sequence alignment. On the other hand, although the method of protein electrophoresis is less accurate than that of protein sequencing, it is far less time-consuming and was therefore extensively used in the study of phylogenetic relationships among relatively closely related populations or species. The application of this method prompted the development of measures of genetic distance and methods of tree-making; for example, Nei's (1972) genetic distance has greatly facilitated the study of evolutionary relationships among populations or closely related species. In addition, immunological techniques such as the method of microcomplement fixation were also extensively used in phylogenetic studies. The DNA hybridization method was also extensively used, particularly in the study of primate evolution and bird systematics.

The rapid progress in the study of molecular evolution has been greatly facilitated by the development of high-speed computers. Most of the procedures used in molecular evolution and population genetic analysis were computation-intensive and could not have been routinely done without computers. Thanks to their ever-increasing speed and accessibility, analysis methods became more and more sophisticated. Only recently has it been possible to use some of the more rigorous methods such as maximum likelihood for the construction of trees.

This is a brief history of molecular evolution in "the pre-DNA era." The advent of various DNA techniques such as restriction analysis, gene cloning, polymerase chain reaction, and DNA sequencing techniques since the 1970s has brought an explosion of knowledge of molecular biology and a new era of exciting development in the study of molecular evolution. A better understanding of gene structure and genomic organization and the availability of a large body of molecular data have allowed a much more extensive and closer examination of long-standing issues and have raised numerous new ones.

3

MOLECULAR ANCESTRY

Modern efforts to overcome some of the usual phylogenetic problems include the use of molecular information rather than exclusive reliance on morphological information. On the molecular level, we can get information by comparing sequences of nucleotides in various DNA and RNA molecules, as well as by comparing sequences of amino acids (and their molecular configurations) in different proteins. These techniques are obviously restricted to recent organisms from which we can extract such compounds, and the information gathered so far has generally been limited to only some proteins and some nucleic acid sequences, although in ever-increasing number.

Nevertheless, the advantage of a molecular approach offsets many of its sampling limitations, since evolutionary changes on the level of amino acid and nucleotide substitutions can be measured and compared between existing organisms no matter how greatly they differ in other phenotypic features. That is, we can compare differences between sampled molecules from different organisms on a unit scale of amino acids or nucleotides when the organisms have no simple, comparable units of morphology, behaviour, ecology, physiology, and so on. Molecular comparisons thus transcend barriers between organisms whose relationships cannot be evaluated by traditional experimental techniques such as genetic exchange. For example, we can compare an amino acid sequence in a particular protein among humans, pigs, horses, rats, chickens, alfalfa plants, yeast, and bacteria.

Although most sequence relationships are still unknown, this close tie to molecular biology has thus transformed evolution from a "theoretical" explanation of historical events to an observable and

Bacteria PLFDFAYQGFARG-LEEDAEGLRAFAAMHKELIVASSYSKNFGLYNERVG
Yeast ALFDTAYQGFATGDLDKDAYAVRXXLSTVSPVFVCQSFAKNAGMYGERVG
Alfalfa PFFDSAYQGFASGSLDADAQPVRLFVADGGELLVAQSYAKNMGLYGERVG
Chicken PFFDSAYQGFASGSLDKDAWAVRYFVSEGFELFCAQSFSKNFGLYNERVG
Rat PFFDSAYQGFASGDLEKDAWAIRYFVSEGFELFCPQSFSKNFGLYNERVG
Horse PFFDSAYQGFASGNLDRDAWAVRYFVSEGFELFCAQSFSKNFGLYNERVG
Pig PFFDSAYQGFASGNLEKDAWAIRYFVSEGFELFCAQSFSKNFGLYNERVG
Human PFFDSAYQGFASGNLERDAWAIRYFVSEGFEFFCAQSFSKNFGLYNERVG

Fig. 3.1. A comparison of eight organisms for a 50-amino-acid-long sequence of the enzyme aspartate transaminase.

unbreakable link between all living forms. That is, organisms, past and present, can all be bound together through the genetics of evolution by innumerable threads of inherited molecular sequences and their variations.

Immunological Techniques

The earliest, and still among the simplest, comparative molecular methods uses immunological techniques in which antibodies produced in a particular host (usually a rabbit) against proteins (antigens) of one species are measured as to their activity against proteins of other species. For example, if antibodies against species A precipitate much of the protein in species C but little of the protein in species B, then the researchers presume that the A and C proteins have more similar molecular configurations (antigenic components) and are therefore more evolutionarily alike (smaller *antigenic distance*) than those of A and B.

Using a variety of antibodies, researchers can then analyze immunological data by various mathematical rules (algorithms) to construct a phylogenetic tree that best correlates the antigenic distance between species with the length of time since they shared a common ancestor. Thus, if species A and C are antigenically closer to each other than to species B, species B is presumed to have broken off earlier from the common stem that all three originally shared. Successive comparisons are then made between all possible combinations of species until the entire phylogenetic tree is obtained.

In contrast to traditional taxonomy of the time, this technique showed that humans (*Homo*), gorillas (*Gorilla*), and chimpanzees (*Pan*) are antigenically closer to each other than to the Asian orangutan (*Pongo*). Using this method, we can therefore separate the first three genera from the orangutans and place them in a separate group.

Other immunological techniques, such as the *microcomplement fixation* that Sarich and Wilson used, involve measuring antibodies produced against specific proteins found in blood serum (albumin and

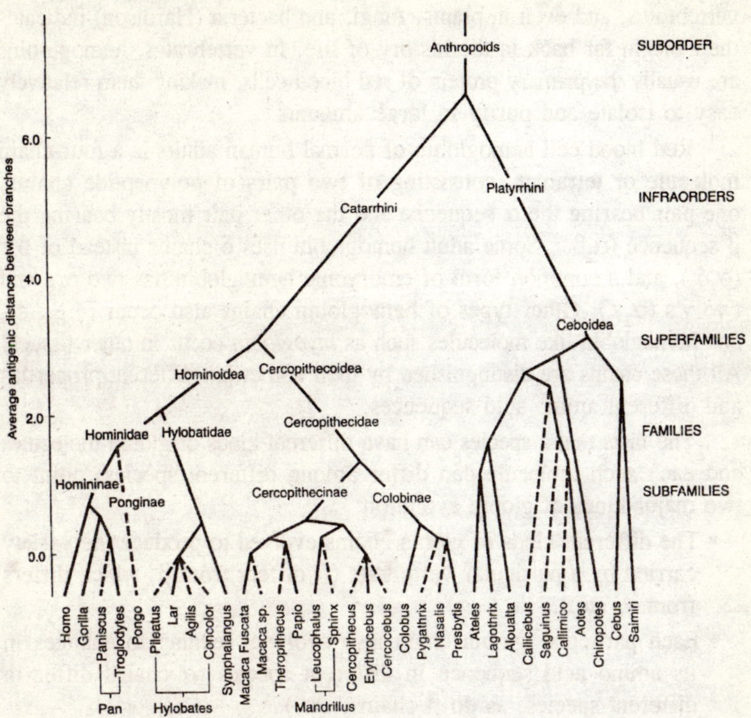

Fig. 3.2. A phylogenetic tree for the anthropoid suborder of primates based on immunological distances derived from blood serum antigens.

transferrins) or enzymes such as lysozyme. Antigenic distances, which researchers can detect by measuring the amount of antigen-antibody reactions provide data that generally support the phylogenies we can obtain by other taxonomic methods, although some differences occur.

Amino Acid Sequences

A more direct, and now more popular, approach to molecular phylogeny is actually sequencing the amino acids in proteins by biochemical methods and comparing such sequences for the same protein in different species. Among the first proteins to yield its amino acid sequence was *hemoglobin* (Ingram), and it probably still remains one of the most investigated of all proteins. The basic unit of hemoglobin consists of an iron-containing porphyrin (heme) that can reversibly bind oxygen attached to a globin polypeptide chain that is usually no less than 140 amino acids long. The demonstration that hemoglobin-like molecules appear in a wide range of organisms, from invertebrates to vertebrates, and even in plants, fungi, and bacteria (Hardison) indicates their origin far back in the history of life. In vertebrates, hemoglobins are usually the primary protein of red blood cells, making them relatively easy to isolate and purify in large amounts.

Red blood cell hemoglobin of normal human adults is a four-chain molecule or tetramer, consisting of two pairs of polypeptide chains, one pair bearing the α sequence and the other pair mostly bearing the β sequence ($\alpha_2\beta_2$). Some adult hemoglobin uses δ chains instead of β's ($\alpha_2\delta_2$), and a common form of embryonic hemoglobin has two α's and two γ's ($\alpha_2\gamma_2$). Other types of hemoglobin chains also occur (e.g., ε), and hemoglobin-like molecules such as *myoglobin* occur in other tissues. All these chains are distinguished by their somewhat different properties and different amino acid sequences.

The facts that a species can have different kinds of globin molecules and each such molecule can differ among different species, point to two major kinds of globin evolution:

- The different kinds of globin chains evolved to produce the variety carried by a particular vertebrate (α differs from β, which differs from γ, etc.).
- Each particular globin chain also evolved, leading to changes in its amino acid sequence in different species (α chains differ in different species, as do β chains, etc.).

They are obviously all variations on a single homologous theme: the chains are all about the same length; they possess identical amino acids at a significant number of positions; and all functioning human globin genes share a similar exon-intron structure—three exons separated by two introns. The three-dimensional structures that the amino acid sequences dictate are also similar, leading to their similar physiological functions. In addition, the genes for the β, γ, and δ chains are closely linked (chromosome 11), although the gene for the α chain lies on a different chromosome (number 16).

How to explain these events? These observations, especially the sequence similarities, strongly suggest that rather than arising from different genes that accidentally converged in sequence and function, all the different globin chains arose as *gene duplications* of an original globin-type gene. Once such duplicated globin genes appeared, each could then undergo its own subsequent evolutionary changes. The temporal order in which the duplications occurred can be discerned from amino acid differences on the basis that the greater the amino acid differences between any two chains, the further back in time was their common ancestor. Thus, we can order a number of events from the following three observations:

1. The myoglobin chain differs most from all others because it has distinctive amino acids at more than 100 sites.
2. The α chain differs from β at 77 sites.
3. The β chain differs from γ at 39 sites but differs from δ at only 10 sites.

These findings mean that the gene for myoglobin must have formed from a very early duplication, which was then followed by a later duplication that separated the α and β genes. Because they differ least, the separation between the β and δ chains derives from a fairly recent duplication.

The genetic phylogeny of the five globins in terms of the numbers of nucleotides necessary to account for the amino acid differences, along with the chronological periods in which evolutionists presume each duplication to have occurred. We can see that the duplication events led to the early coexistence of myoglobin with an α-like chain, the former probably assuming (or maintaining) an intracellular function and the latter probably assuming a circulatory function.

When a duplication of the α-like gene further evolved into a β-like gene, the advantage of having two pairs of different chains in a tetramer hemoglobin molecule must have been sufficiently great to

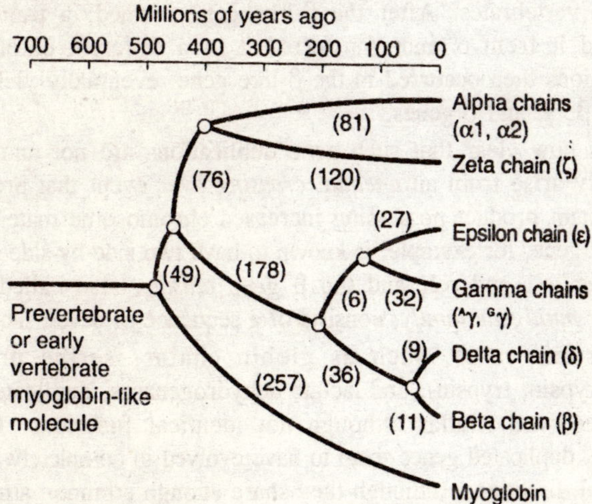

Fig. 3.3. Phylogenetic relationships between globin-type proteins found in humans, showing the estimated times at which they originally diverged from each other.

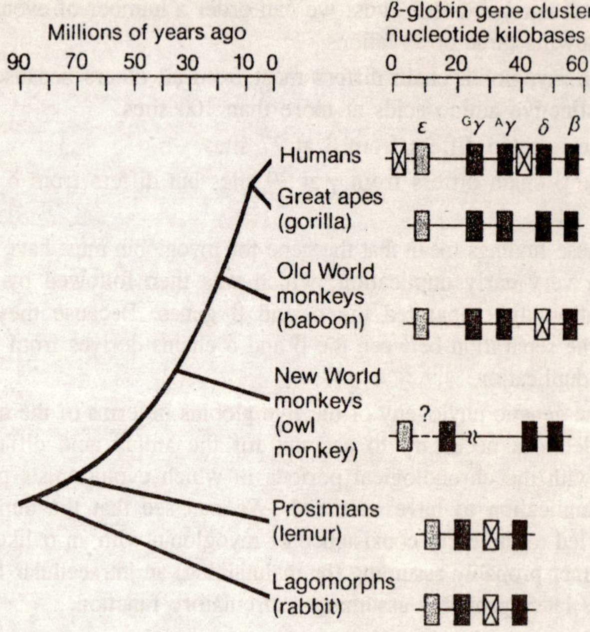

Fig. 3.4. The clustered organization of the β-globin-type genes in five different primates and in the rabbit, along with a proposed evolutionary tree.

account for preserving tetramer organization in the circulating blood of most vertebrates. After the β-like gene formed, a translocation separated it from α and transferred it to a different chromosome. Duplications then occurred in the β-like gene, eventually yielding the modern β, γ, and δ genes.

It is now clear that such gene duplications are not unusual and can easily arise from an *unequal crossing over* event that produces a recombinant product possessing increased chromosome material. The human α gene, for example, is known to have two side-by-side (tandem) duplicates, α1 and α2, and the β *gene cluster*, also called a *gene family* or *multigene family*, consists of a sequence of seven such genes.

In some cases, such as globin chains, serine proteases (chymotrypsin, trypsin), and lactate dehydrogenases, duplicated genes have preserved similar although not identical functions. In other instances, duplicated genes seem to have evolved in completely different functional directions, although they share enough common amino acid sequences to indicate their common origin. Thus the vertebrate nerve growth factor protein that enhances the outgrowth of neural cells from sympathetic and sensory ganglia has amino acid sequences similar to insulin and may share some functional similarities as well. Also, the α-lactalbumin protein, which is part of an enzyme used in the synthesis of lactose in mammalian milk, has an amino acid sequence remarkably similar lo that of the lysozyme enzyme found in tears that degrades the mucopolysaccharides of bacterial cell walls. This similarity is further reflected in evidence that both proteins are the products of tissues (mammary, tear duct) that were at one time probably sebaceous glands, and they both use sugar molecules as their substrates.

Duplications also exist within genes themselves, as shown by the presence of three homologous amino acid regions (polypeptide domains) within the γ heavy chain of the immunoglobulin G antibody. In the human haptoglobin α-2 protein of blood serum, a segment of 59 amino acids (positions 13 to 71) almost exactly repeats an adjacent segment (positions 72 to 130). Researchers have also found repeated amino acid sequences within the ferredoxin protein (used as an electron carrier in various biochemical processes) in the glutamate dehydrogenase enzyme, and various other proteins.

In spite of these and other fairly conspicuous examples, ancestral homologies based on similarities between amino acid sequences are not always obvious. Researchers have therefore devised various tests for detecting amino acid sequence similarity with some more sensitive in distinguishing homologous relationships than others. In general, the most sensitive of these tests depend on comparing fairly long amino acid sequences, since comparisons between short sequences can be misleading; for example, similar sequences only three or four amino acids long may often arise in nonrelated proteins. Workers in the field have also proposed methods to overcome difficulties caused by gaps or insertions.

Geneticists have used the second aspect of protein evolution, comparisons between their amino acid sequences in different species, to produce phylogenies involving a large number of organisms. For such comparisons, Fitch has suggested that we should use different terms for duplicate genes in a species (e.g., α, β, and γ hemoglobins in humans), although homologous to each other in the sense that they derive from a common ancestor, and for homologies for the same gene in different species (e.g., α hemoglobins in horses). He called the former *paralogous*, and the latter *orthologous*. It is therefore comparisons between orthologous genes that are valid for determining the phylogenetic relationships of the species that carry them.

One basic technique for detecting orthologous amino acid sequences does not essentially differ from that used in determining phylogenies based on immunological distances, except that it can furnish more precise phylogenetic positioning because it begins with an alignment of amino acid sequences from the same known protein in different species. Once aligned, the minimum number of mutations necessary to transform ore amino acid in one sequence to that of a different amino acid in the same position in the other sequence can then provide an estimate of mutational distances.

This method of choosing an hypothesis that minimizes explanations to account for an observation invokes the principle of *parsimony* and is widely used in science. Parsimony has the advantage that a true hypothesis generally involves fewer assumptions than a false hypothesis; for example, we can better explain a phenylalanine codon (UUU) as deriving from a single nucleotide substitution in a serine codon (UCU → UUU) than as deriving from a triple nucleotide substitution in a glutamic acid codon (GAA → UUU). Note that "unparsimonious" explanations are not necessarily false, but in the *absence of further information*, choices are made on the basis of parsimony.

For examine, if the minimum, or most parsimonious mutational distance for a particular protein comparison between species A and B is 25, between A and C is 20, and between B and C is 30, then the two most closely related species are obviously A and C. If we assign legs x, y, and z to represent the numbers of mutations responsible for their divergence, we can portray the phylogenetic relationship among the three species as follows:

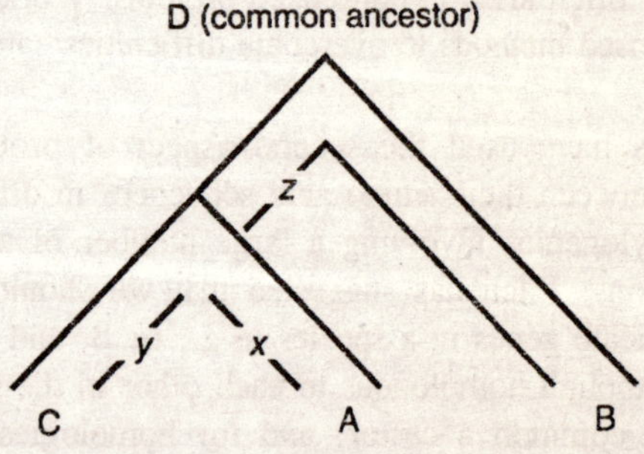

We can determine the lengths of these legs by noting that the A-B distance (25) is less than the C-B distance (30), hence the leg x must be 5 mutations less than y. Since x + y = 20 and y - x = 5, we can subtract one equation from the other and solve for x:

$$\begin{aligned} x + y &= 20 \\ \underline{-\ (-x + y} &\underline{= 5)} \\ 2x &= 15 \\ x = 7.5,\ y &= 12.5 \end{aligned}$$

The leg z must therefore equal the A-B distance minus x (or the C-B distance minus y); that is, z = 25 - 7.5 (or 30 - 12.5) = 175.

These mutational distance therefore yield

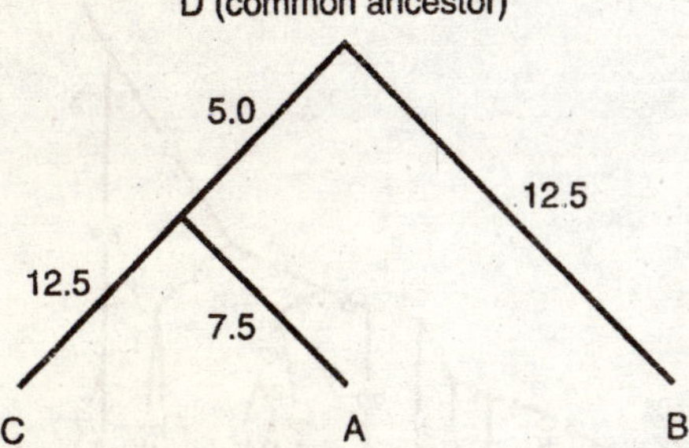

Note that an important condition of the parsimony method is that the observed mutational distance between two species must be less than or equal to the sum of the distances from their common ancestor to each species in the projected phylogenetic tree. In this example, the observed distance between A and B is 25, and the distances from their hypothesized common ancestor at the triangular apex (e.g., D) to these two species must total at least 25, if not more. The sum of the (D-A) + (D-B) legs may be more than 25 because undiscovered *back mutations* (e.g., adenine → guanine → adenine) may exist in these two legs, or *parallel mutations* may have occurred in each leg (e.g., cytosine → adenine; guanine → adenine), so that the difference observed between A and B is really an underestimate of the actual mutational distance. This triangle inequality condition therefore signifies the evolutionary presence of back and parallel mutations, and researchers often take it into account by proportionally augmenting the number of mutations in an ancestral leg in accord with its length. That is, not only do they increase, for example, the A-D-B distance numerically over the measured A-B distance, but they even further augment longer distances.

As geneticists use increased numbers of species to provide mutational data for a particular protein, they can establish a *phylogenetic tree* for these species by different numerical and algorithmic methods. Figure 3.5 shows one example for the sequence of α-

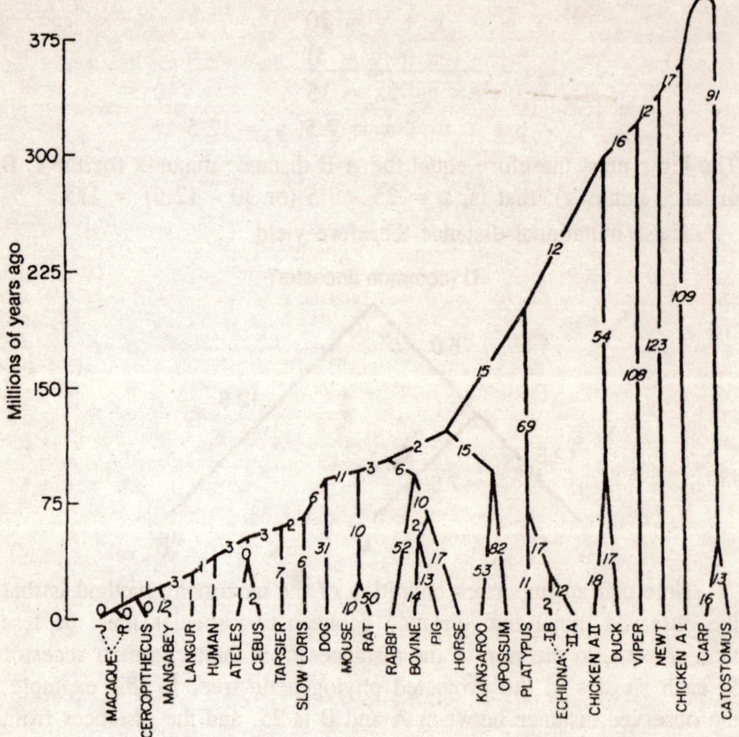

Fig. 3.5. A phylogeny of α-hemoglobin chains in a variety of vertebrate species, determined by calculating the number of nucleotide represents that account for the number of observed amino acid substitutions presumed to have occurred during each evolutionary interval.

hemoglobin chains in 29 different vertebrates involving a total of 630 nucleotide replacements.

Because hemoglobin phylogenies exclude invertebrates, plants, and fungi, researchers have used other, more ubiquitous proteins that they can compare among more widely varied groups of organisms. One such protein, cytochrome c, also contains heme but generally functions in aerobic organisms as part of the respiratory electron transport chain.

Not surprisingly, we can hardly expect phylogenies derived from studies of only one or two proteins to always reflect true evolutionary history nor discriminate accurately even between different species. That no differences show up between human and chimpanzee α-hemoglobins marks such lack of discrimination. Similarly, that camels and whales share the same cytochrome c amino acid sequence hardly

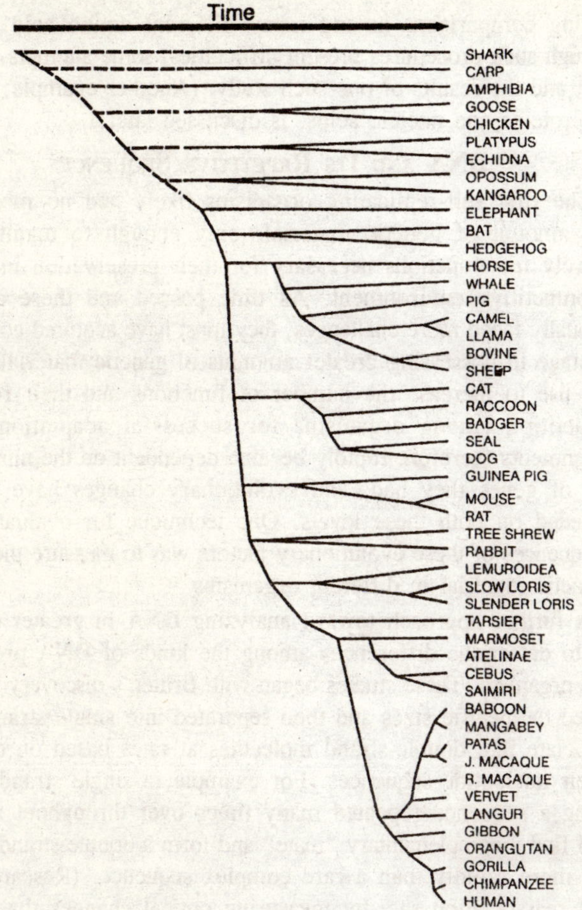

Fig. 3.6. A phylogenetic tree for 49 vertebrate taxa derived from aligning the amino acid sequences in up to seven different polypeptide chains: α-hemoglobin, β-hemoglobin, myoglobin, lens α-crystallin A, fibrinopeptide A, fibrinopeptide B, and cytochrome c.

reflects their obvious evolutionary divergence. (Note also, for example, the similar sequence in cows, pigs, and sheep.) Nevertheless, it is remarkable how closely the information obtained from this limited sample of only two proteins out of many thousands approximates the phylogenetic relationships obtained from extensive studies in comparative anatomy and paleontology.

We might expect a more faithful reflection of evolutionary history, therefore, if we could combine studies from many different proteins, allowing comparisons among many different amino acid positions. Although such procedures present difficulties, some attempts have been made, and the results of one such study. (Another example, involving both proteins and nucleic acids, is discussed later.)

DNA and Its Repetitive Sequences

The first self-replicating organisms likely had no more than a small amount of genetic material, only enough to maintain those relatively few functions necessary for their preservation in a mostly noncompetitive environment. As time passed and these organisms continually faced more challenges, they must have acquired considerable advantage in possessing greater amounts of genetic material that they could use to increase the number of functions and their regulation. Competition among organisms for successful adaptation to their environments therefore rapidly became dependent on the numbers and kinds of genes they had, and evolutionary changes have obviously proceeded on both these levels. One technique for evaluating some consequences of these evolutionary factors was to measure the amounts of genetic material in different organisms.

A further approach toward analyzing DNA in greater detail has been to determine differences among the kinds of DNA present in a single organism. These studies began with Britten's discovery that DNA sheared to specific sizes and then separated into single strands would reassociate into double-strand molecules at rates based on the nature of their nucleotide sequences. For example, a single strand of DNA bearing a sequence repeated many times over throughout a genome would find a complementary "mate" and form a double-strand molecule much more rapidly than a rare complex sequence. (Researchers can detect reassociation rate by measuring optical changes that occur in the transition from single- to double-stranded DNA.) By these means, Britten and co-workers were able to classify DNA as either *repetitive* or *unique*, referring to sequences that occurred frequently and those that occurred only as single copies.

The lengths of these repetitive sequences may vary from less than 100 to more than 2,000 nucleotides long. With the exception of *Drosophila*, the shorter-length repetitive sequences seem to be interspersed among sequences of nonrepetitive unique DNA, although the function of this arrangement is unknown. According to some authors, these interspersed sequences act as regulatory genes similar to prokaryotic operators and promoters, whereas others have suggested that they are involved in packaging long, transcribed strands of heterogeneous nuclear RNA (HnRNA) so they can be further processed into messenger RNA (mRNA). The fact that many eukaryotic mRNA molecules are shorter than their parental HnRNA molecules certainly indicates that extra DNA nucleotides in each gene are not translated into protein: some are at the gene termini, and others, known as *introns*, are distributed throughout eukaryotic protein-coding genes.

Among the long-length repetitive DNA sequences, some undoubtedly code for multigene families such as the many duplicate, tandemly arranged ribosomal RNA genes, transfer RNA genes, and histone genes. The function of the more repetitive but shorter nucleotide sequences in *satellite DNA* (recognized by its separation from the main portion of DNA after centrifugation) seems more obscure, although these have often been localized to distinctively staining chromosome sections and centromere regions (*heterochromatin*).

In the house mouse (*Mus musculus*), a satellite DNA that comprises about 10 percent of the genome interestingly shows no homology to the DNA of related rodents such as rat, field mouse, and hamster. In contrast, a number of satellite sequences are widely conserved in different species groups such as crustaceans (crabs) and insects (*Drosophila*). Such studies and others seem to make clear that at least some satellite DNA can arise quite rapidly during the evolution of a species by adding many copies of a new DNA sequence or by amplifying DNA sequences present in ancestral species. The different kinds of satellite DNA, their different amounts, and their possible different origins indicate that they probably have different functions, or perhaps no function at all. Various biologists have been tempted to consider some or many such sequences as forms of "selfish DNA."

Nucleic Acid Phylogenies Based on DNA-DNA Hybridizations

To estimate the extent of homology between nucleic acids of different sources, researchers can measure the degree to which homologous nucleotide sequences in different single strands pair up to form double-stranded sections. In one technique presently used, experimenters extract DNA from two organisms, X and Y, dissociate it into single strands, and then give it the opportunity to form X-Y hybrid double-strand DNA by incubating the different DNA molecules together at appropriate temperatures.

To let interspecific X-Y DNA separate from intraspecific X-X or Y-Y DNA, experimenters radioactively label the nonrepetitive DNA of one species, X, and use only relatively small amounts of it in the

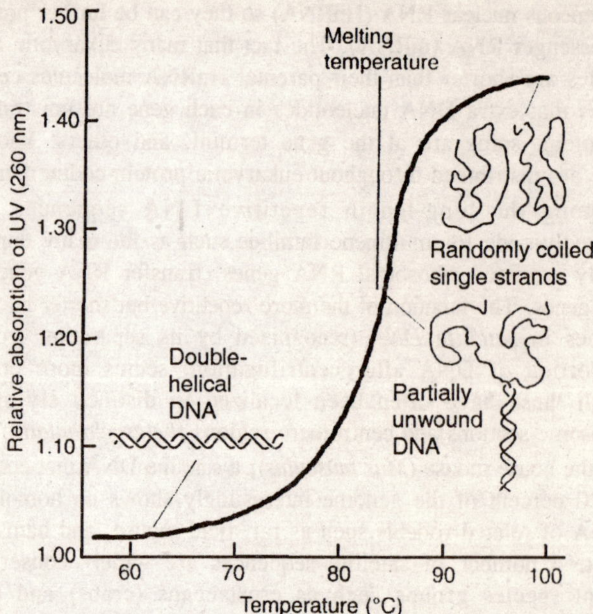

Fig. 3.7. Melting temperature curve for DNA of T4 bacteriophage showing the marked change in ultraviolet absorption that occurs at approximately 84°C when about half the DNA has changed from double-helix form to single-strand form.

incubation mixture. Because of its rarity, the DNA of X will therefore have very little chance of forming X-X double strands, and we can assume that all radioactively labeled double-strand DNA is X-Y. We can then extract this double-strand DNA (on hydroxyapatite crystals) and examine its properties. If the DNA from X and Y are perfectly homologous and have no nucleotide differences at all (i.e., they are of the same species), then the *melting temperature* at which the hybrid X-Y DNA dissociates into single strands will, of course, be the same as either X-X or Y-Y. However, should there be nucleotide differences between X and Y, then the X-Y hybrid will dissociate more easily because of nucleotide mismatching; that is, its stability reduces and its melting temperature lowers. Various experiments show that for each 1 percent difference in nucleotide composition between X and Y, the thermal stability of the X-Y hybrid DNA molecule lowers by about 1°C.

Sibley and Ahlquist point out that such techniques enable comparisons among perhaps a billion or more nucleotides simultaneously and can provide considerably more information than usually obtains

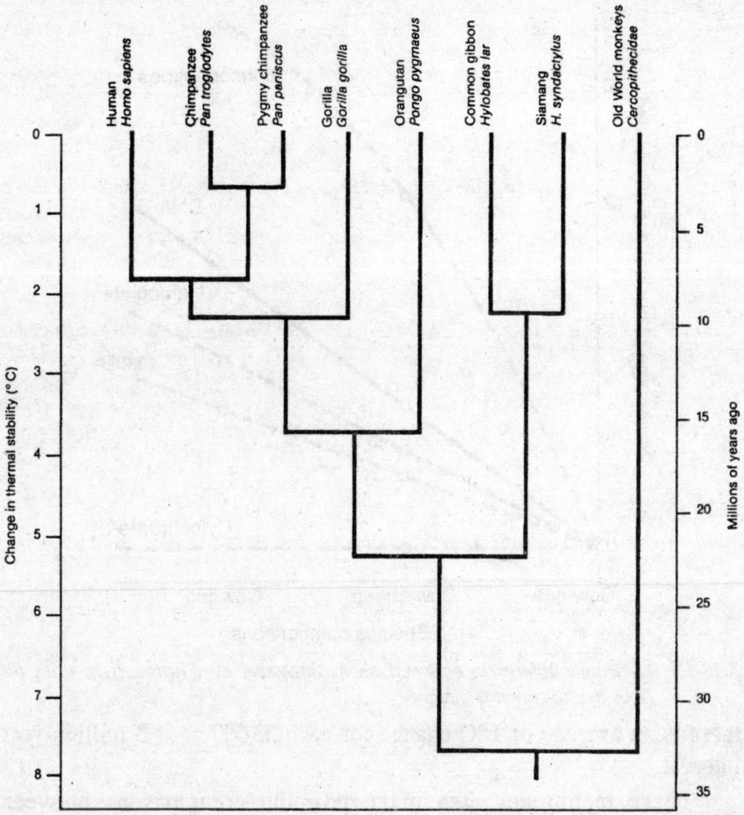

Fig. 3.8. A phylogenetic tree and dates of divergence for humans, apes, and Old World monkeys based on DNA-DNA hybridization studies.

from comparing a few characters at a time. For example, they have produced detailed bird order phylogenies that once seemed difficult or impossible to determine. They have also suggested that primate phylogenies based on such DNA-DNA comparisons can specify relationships that formerly seemed obscure, such as human-chimpanzee-gorilla. Moreover, they propose that since billions of nucleotides are involved in these determinations and changes accrue over millions of years, they can estimate average rates of change over time quite confidently.

Thus, assuming from paleontological evidence that the divergence between the lineages of Old World monkeys and apes-humans occurred about 33 million years ago, and observing an approximate 7.7°C change in thermal stability between these groups and their common ancestor,

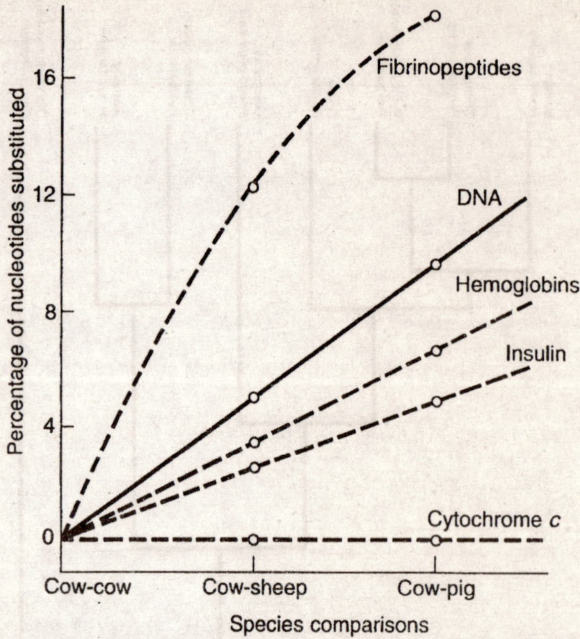

Fig. 3.9. Nucleotide differences observed among three species of artiodactyls, using the DNA hybridization technique.

there is an average of 1°C change for each 33/7.7 = 4.3 million-year interval.

These techniques also make possible comparisons between *nucleotide substitution rates* obtained from DNA-DNA hybridization data and substitution rates from amino acid changes in known proteins. When researchers undertake such comparisons, the rate of change in DNA seems more rapid than the rate of change from most proteins except for fibrinopeptides. Therefore considerable portions of these tested DNA sequences probably do not code for essential proteins such as cytochrome and insulin, and perhaps code for no proteins at all. Such sequences, perhaps largely "junk" DNA or "selfish" DNA can consequently accumulate changes more rapidly than can genes that code for stringently selected proteins.

However, despite the value of DNA-DNA hybridization techniques in detecting divergences between entire genomes, some critics object to them nevertheless. One repeated objection is that DNA hybridization experiments compress all divergence information into a single distance measurement, thereby losing information on specific nucleotide sequence changes that could provide a more statistically supported choice among the different possible phylogenetic trees. Therefore researchers have also explored DNA comparisons by other techniques, discussed as follows.

Nucleic Acid Phylogenies Based on Restriction Enzyme Sites

One approach to comparative DNA analysis is to use *restriction enzymes* that recognize specific short nucleotide sequences and cleave the molecule at these sites. For example, the enzyme *Eco*RI, isolated from *E. coli* bacteria, recognizes the GAATTC/CTTAAG hexanucleotide sequence in double-strand DNA (N represents non-specified nucleotides):

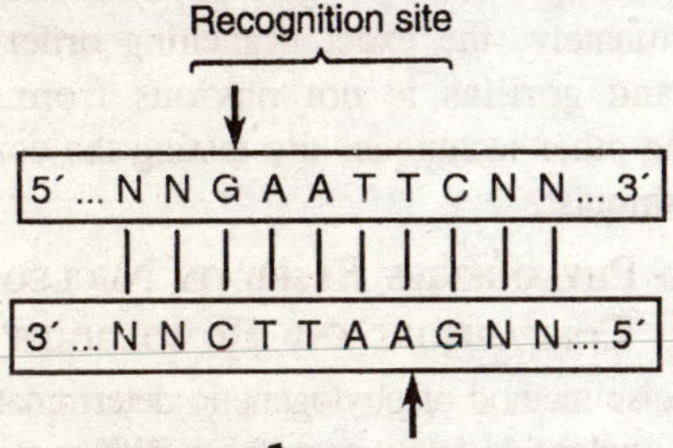

and produces cleavage products at the points between G and A on both strands indicated by the arrows:

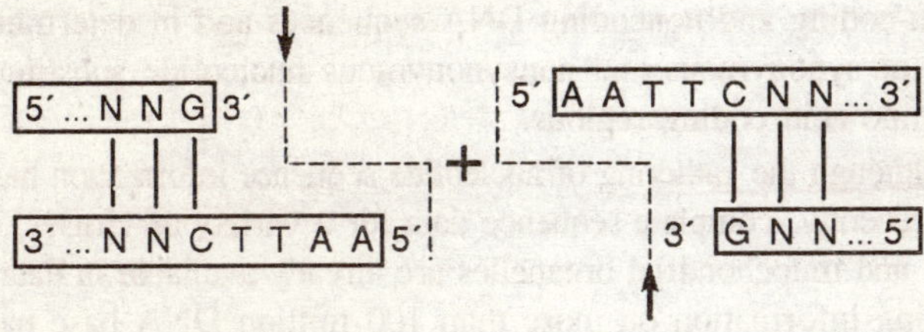

Since DNA molecules can differ from each other in nucleotide sequence, and therefore differ in the number and placement of sites recognized by *Eco*RI, each particular kind of DNA will have fragments of characteristic length when subjected to the enzyme. Also, since there are different kinds of restriction enzymes, many of which recognize target sites different from those recognized by other such enzymes, a DNA molecule subjected to a battery of different enzymes will produce cleavage products unique for that particular kind of DNA. Scoring differences in these inherited fragmentation patterns between individuals (*restriction-fragment-length polymorphisms*, also called RFLPs) then enables estimates of genetic variation in populations, with each unique pattern designated as a *haplotype*.

The distinct sites at which restriction enzymes fragment DNA also provides *restriction site maps* for sequences ranging from relatively small repetitive DNA sequences to mitochondrial DNA and the DNA of even larger chromosomes. In one example that Ferris and co-workers analyzed, mitochondrial DNA from humans and apes was subjected to 19 different restriction enzymes. These enzymes cleaved approximately 50 sites in each mitochondrial chromosome, allowing a detailed comparison of target site sequences among the five species. Thus, in accord with previously determined phylogenies, humans share more such sites with chimpanzees and gorillas than with orangutans and gibbons. Unfortunately, the exact branching order among humans, chimpanzees, and gorillas is not obvious from these data, and researchers using other techniques are testing the conclusions reached by Sibley and Ahlquist.

Nucleic Acid Phylogenies Based on Nucleotide Sequence Comparisons and Homologies

A more precise method of phylogenetic determination is to compare actually known nucleotide sequences from different organisms rather than to infer relationships from hybridization studies or restriction enzyme maps. This offers advantages in comparing changes between protein-coding and noncoding DNA sequences and in determining the extent of synonymous and nonsynonymous nucleotide substitutions in the amino acid coding regions.

Although the gathering of nucleotide sequence information has begun fairly recently, complete sequence data for a variety of viruses, cellular genes, and mitochondrial organelles are already available in data banks, including information on more than 100 million DNA base pairs and on molecules more than 2 million nucleotide base pairs long. Many laboratories are now coordinating efforts to sequence the complete DNA genomes of organisms ranging from prokaryotes to yeast, nematodes, *Drosophila*, and humans. Interestingly, the technology now even allows sequencing of some DNA from fossil organisms.

Figure 3.5 shows one phylogeny derived from characterizing DNA sequences in considerable detail for the β-globin gene clusters of various primates. Some other examples come from nucleic acid structures that are more widely distributed, such as 5S RNA, a component of the larger of the two ribosomal subunits that functions in ribosome binding of the various transfer RNA molecules that carry the different

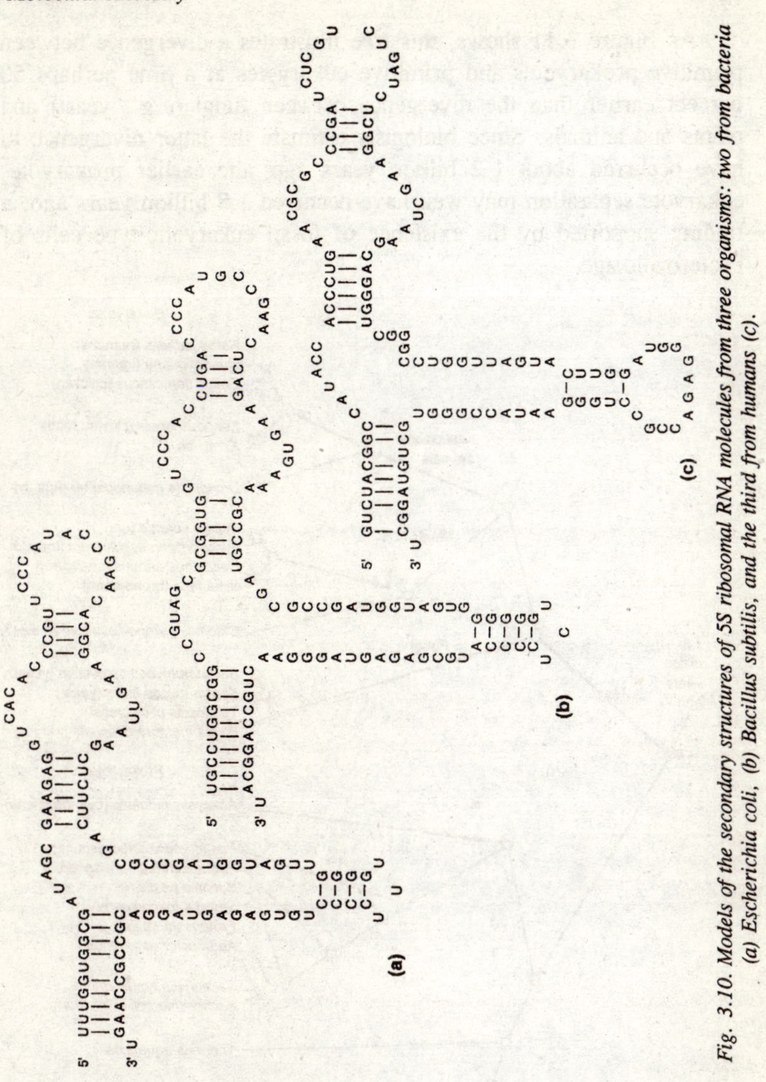

Fig. 3.10. Models of the secondary structures of 5S ribosomal RNA molecules from three organisms: two from bacteria (a) Escherichia coli, (b) Bacillus subtilis, and the third from humans (c).

amino acids. This cornerstone of the basic protein-synthesizing apparatus appears difficult, if not impossible, to change, and has been evolutionarily conserved in all organisms. Thus the secondary structure of 5S RNA seems universally the same, a feature that enables all the various 5S RNAs to be aligned for every nucleotide position. When such alignments are effected, researchers can use differences between 5S RNAs to generate a phylogenetic tree in which nucleotide changes measure evolutionary distance.

As Figure 3.11 shows, this tree illustrates a divergence between primitive prokaryotes and primitive eukaryotes at a time perhaps 50 percent earlier than the divergence between fungi (e.g., yeast) and plants and animals. Since biologists estimate the latter divergence to have occurred about 1.2 billion years ago, the earlier prokaryote-eukaryote separation may well have occurred 1.8 billion years ago, a finding supported by the existence of fossil eukaryotic-type cells of Proterozoic age.

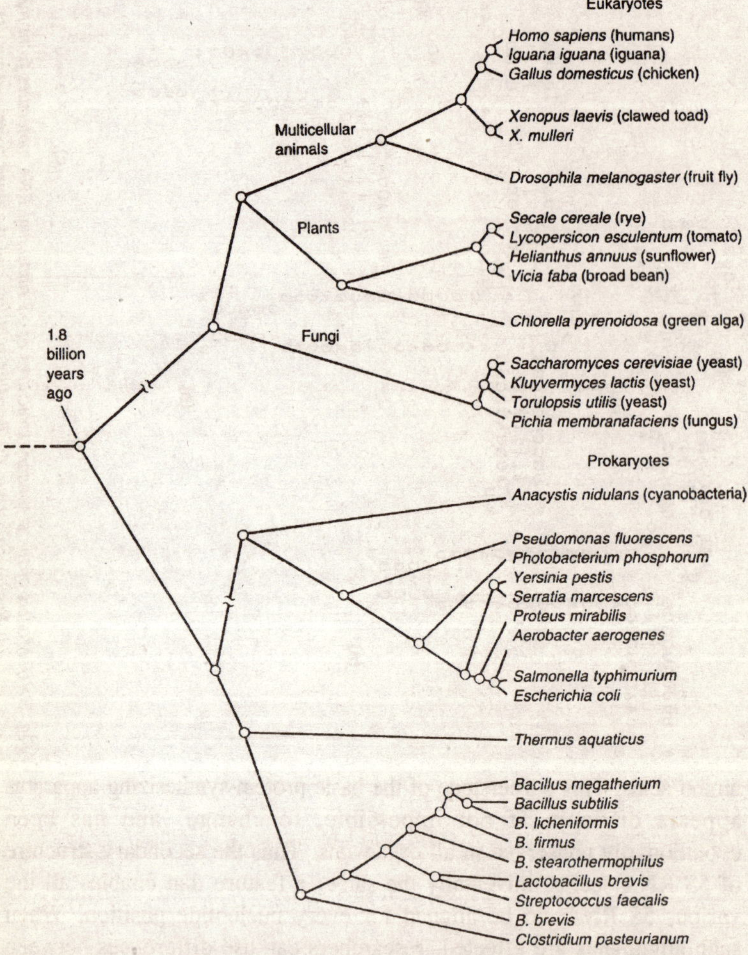

Fig. 3.11. A phylogenetic tree that best explains the data gathered from comparing 5S RNA nucleotide sequences between many different species.

In those instances where biologists are comparing a narrower spectrum of organisms, such as a group of vertebrates, mammals, or humans, they now commonly use mitochondrial DNA sequencing. Among its advantages, mitochondrial DNA is easily isolated and it also evolves at a sufficiently rapid rate to allow recognition of distinctions and similarities between organisms that have only recently diverged. Many findings have emerged that would otherwise have been difficult to establish.

In one example, mitochondrial DNA sequence analysis indicates that the many morphologically diverse species of cichlid fishes in southern Africa have a common monophyletic origin. Moreover, mitochondrial DNA analysis has clarified phylogenetic relationships among these species in different South African lakes. Thus, the similarities between cichlids in Lake Tanganyika and Lake Malawi have been shown to be caused by convergence rather than derived from common ancestry. The obvious lesson emerging from these and other molecular studies is that morphology alone is not always sufficient to establish phylogeny. For example, convergence can cause what seem to be morphological homologies between different aschelminth phyla, but molecular findings show distinctly different ancestries.

In addition to procedures using enzymatic alleles, researchers are now analyzing relationships even closer than between species, such as between races or individuals within a race, by nucleic acid techniques. For example, mitochondrial DNA studies are showing relationships among the various present human races.

Other types of DNA analysis have begun using *microsatellites*, tandem repeats of very short nucleotide sequences such as cytosine-adenine-adenine (CAACAACAA ...), which may vary between individuals in numbers of repeats at some particular chromosomal position. Through techniques involving the isolation and analysis of microsatellites, researchers can distinguish an individual with 10 CAA repeats at a particular locus, for example, from individuals with greater or lesser numbers of repeats. Because there are many microsatellite loci (humans have at least 50,000 such loci per haploid genome), and they mutate at a relatively high rate compared to protein-coding sequences, opportunities for tracking individual differences and relationships abound.

Combined Nucleic Acid-Amino Acid Phylogenies

Barnabas and co-workers have offered a comprehensive phylogeny that takes into account both nucleotide sequences from 5S RNA and

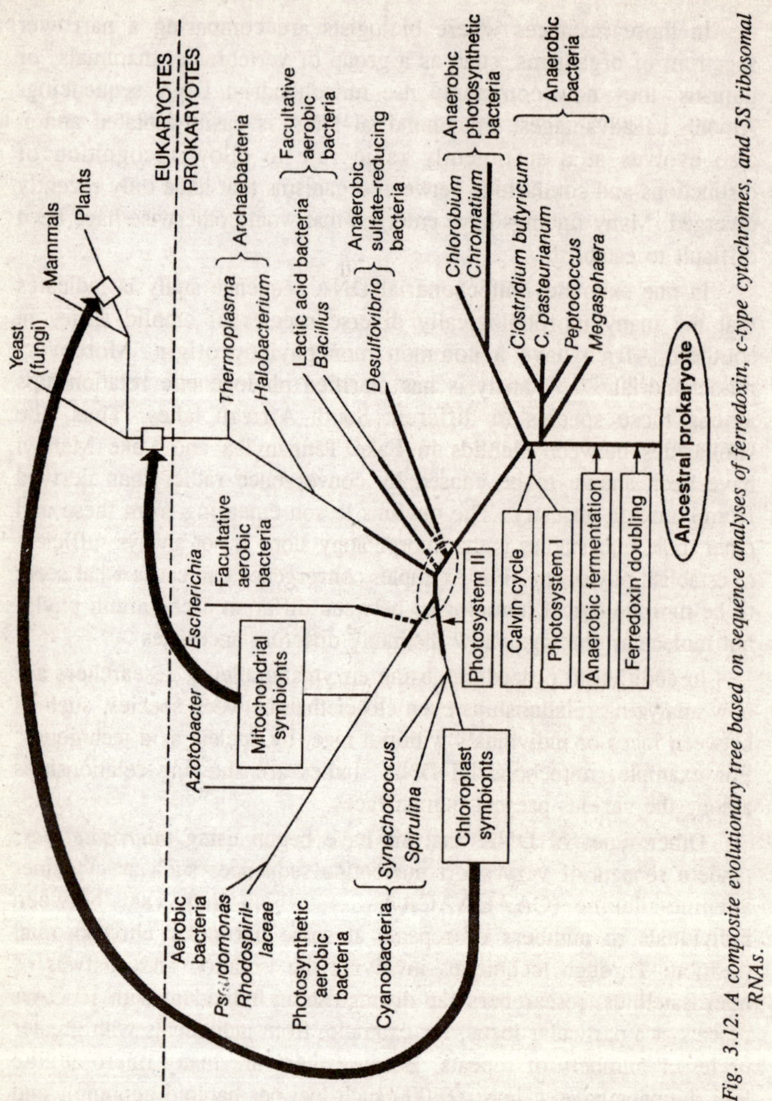

Fig. 3.12. A composite evolutionary tree based on sequence analyses of ferredoxin, c-type cytochromes, and 5S ribosomal RNAs.

amino acid sequences from ferredoxin and the c-type cytochromes. Since ferredoxin is a very primitive iron-containing protein used in a number of basic oxidative-eductive pathways, the doubling event that occurred early in its evolution provides a baseline for the phylogenetic tree. Organisms whose ferredoxins most resemble the inferred sequence of the primitive duplicated molecule are anaerobic and heterotrophic bacteria such as *Clostridium*. They later diverged, developing into anaerobic photosynthetic bacteria such as *Chromatium* and into the main line of aerobic respiratory organisms.

The cytochrome c analysis that Barnabas and co-workers used shows a phylogeny in which the eukaryotic sequences are most similar to the cytochrome c2 sequences of the nonsulfur purple photosynthetic bacteria (*Rhodospirillaceae*). Since cytochrome c functions exclusively in the eukaryotic mitochondrion (although coded by DNA in the nucleus), most likely it is the mitochondrion organelle itself (rather than the eukaryotic cell) that derives from this bacterial line. Studies on 5S RNA support this view, showing that eukaryotes actually diverged from an earlier prokaryotic form and not from the later *Rhodopseudomonas* line. Apparently the gene for cytochrome c incorporated into the eukaryotic nucleus after the inclusion of the mitochondrion organelle into the eukaryotic cell.

Similarly, the evidence of strong homologies between cyanobacteria and plant chloroplasts for their cytochrome and ferredoxin sequences indicates that eukaryotic plant cells must have incorporated a chloroplast organelle that may at one time have been a prokaryotic cyanobacterium. This phylogeny therefore offers very strong support to the symbiotic theory of organelle function.

Rates of Molecular Change: Evolutionary Clocks

Inherent in all the various phylogenies is the concept that evolutionary differences between organisms arise from mutational differences, and therefore, in general, the greater the number of mutational differences between organisms, the greater their evolutionary distance. In some of the phylogenies considered so far, evolutionists have used this notion to furnish evolutionary time scales; that is, they have assumed that mutations are incorporated (or *fixed*) at fairly regular rates over time, and the degree of mutational distance for a phylogenetic interval therefore correlates with the length of time in which such phylogenetic evolution took place. In other words, regarding changes in a specific gene, this assumption suggests that an *evolutionary clock* on the molecular level determines the rates at which many mutations become fixed. Since fixation of these mutations mostly depends on the clock rather than on their adaptive or selective value, Kimura and other geneticists have proposed that mutations are primarily "neutral" in their effect.

One prominent finding used to support the concept of an evolutionary clock is the constant number of differences in amino acid sequence for the same hemoglobin chain derived from different vertebrates. Specifically, if we look at comparisons with the shark sequence for α-hemoglobin, other vertebrates differ from it by similar numbers of amino acid changes: carp 85, salamander 84, chicken 83, mouse 79, and human 79. This indicates that although considerable morphological changes have occurred in these different lineages over a 400 million-year period, constant rates of mutation may have been occurring for at least some proteins. Therefore, if evolutionary clocks exist, two consequent's might be expected:

1. The lines of descent leading from a common ancestor to all contemporary descendants should have similar rates of fixed mutations, because they have experienced similar durations.
2. The proportional rate of fixation that occurs in one gene relative to the rates of fixation in other genes stays the same throughout any line of descent.

In a classical study, Fitch and Langley tested these attributes of the evolutionary clock hypothesis for seven proteins whose amino acid sequences they examined in 18 vertebrate taxa. Using commonly accepted dates of divergence for the various common ancestors of these taxa, the researchers obtained the temporal length of each separate line of descent, thus letting them compare the number of nucleotide substitutions that occurred within a given time period for all proteins together and for each protein individually. The results of the test showed that the rate at which all proteins have changed together varies significantly among the branches in the different lines of descent, indicating that molecular changes are not uniform for these geological periods.

Moreover, we cannot simply explain these differences in rates of protein change as arising from different generation times in different lines of descent, because the rate at which individual proteins changed relative to other proteins differs significantly within single branches. If changes in generation time caused molecular rate changes, we would expect all proteins to behave similarly within any particular branch, and their individual relative rates to remain unchanged. This analysis thus indicates that the ticking of the molecular evolutionary clock in each of these seven proteins is not constant in each branch of this phylogeny, whether scored in respect to time or generation.

Nevertheless, when we average the nucleotide substitutions over all seven proteins for each branching point in the phylogeny (rather than sum them or consider them individually), we do find a marked

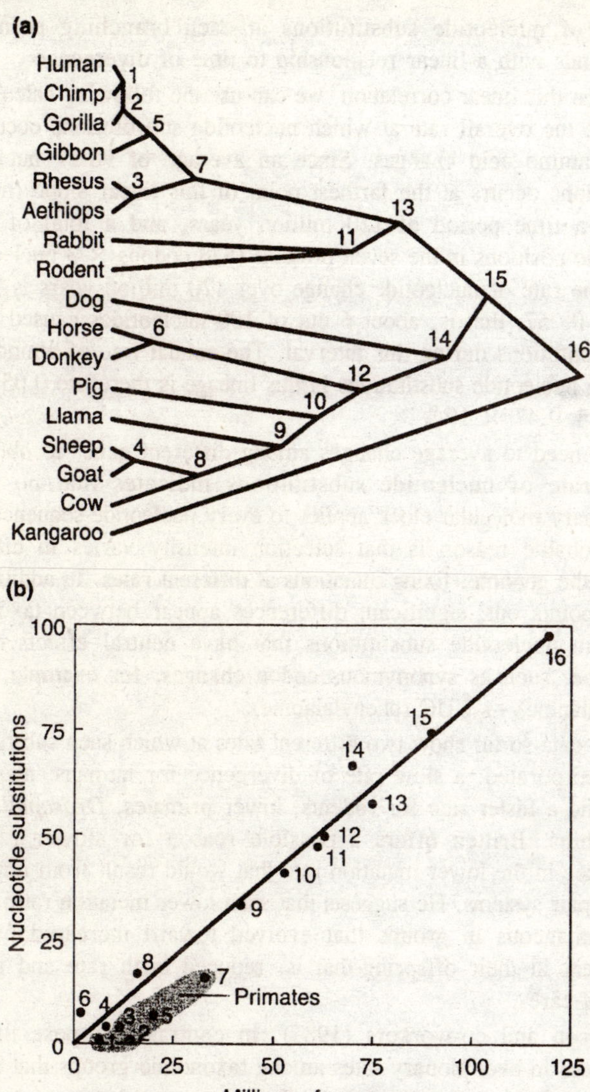

Fig. 3.13. (a) A phylogeny of 17 vertebrate species determined from amino acid sequences examined in seven different proteins. (b) The relationship between the average number of nucleotide substitutions at each nodal point to the estimated time at which its branches diverged.

uniformity in the rate of molecular change over time. A mammalian phylogeny derived from the mutational distance data provides an average number of nucleotide substitutions at each branching point that corresponds with a linear relationship to time of divergence.

Given this linear correlation, we can use the following calculations to derive the overall rate at which nucleotide substitutions occur that lead to amino acid changes. Since an average of 98.17 nucleotide substitutions occurs at the farthest point of this linear slope (number 16) for a time period of 120 million years, and a total of 1,734 nucleotide positions in the seven proteins (578 codons $\times$ 3 nucleotides) exists, the rate of nucleotide change over 120 million years is $98.17/1734 = 0.057$; that is, about 6 out of 100 nucleotides caused amino acid substitutions during this interval. The annual rate of amino-acid-changing nucleotide substitutions in this lineage is therefore $0.057/(120 \times 10^6) = 0.47 \times 10^{-9}$.

The need to average changes among different genes to obtain an annual rate of nucleotide substitutions indicates that no single evolutionary molecular clock applies to every nucleotide sequence. The most probable reason is that selection intensity varies in different parts of the genome, fixing mutations at different rates. In addition, as Britten points out, significant differences appear between taxonomic groups in nucleotide substitutions that have neutral effects on the phenotype, such as synonymous codon changes, for example, UUU (phenylalanine) $\rightarrow$ UUC (phenylalanine).

The data so far show two different rates at which such substitutions have incorporated: a slow rate of divergence for humans, apes, and birds, and a faster rate for rodents, lower primates, *Drosophila*, and sea urchins. Britten offers a possible reason for slower rates of divergence in the lower mutation rate that would result from improved DNA repair systems. He suggests that such lower mutation rates would be advantageous in groups that evolved toward increased parental investment in their offspring-that is, reduced birth rate and greater postnatal care.

Wilson and co-workers (1987), in contrast, propose that the differences in evolutionary rates among taxonomic groups that Britten cites are probably exceptional. Instead, they suggest that an examination of many different genes in both bacteria and mammals shows fairly similar rates of nucleotide substitutions at synonymous codon sites. Riley has disputed this view for such sites in *Drosophila* species, so all one can say at present is that although evidence may exist for an evolutionary clock in various lineages, it apparently does not tick at the same rate in all taxonomic groups.

Regulatory Genes and Some Evolutionary Consequences

One frequent observation that has emerged from comparing different organisms on the molecular level is that so many share the same kinds of proteins. For example, whether organisms arc prokaryotes or eukaryotes, they all share similar enzymes involved in basic biochemical processes such as glycolysis, amino acid synthesis, DNA replication, and protein synthesis. These dehydrogenases, kinases, polymerases, proteases, ferredoxins, cytochromes, nucleases, and numerous other gene products are common enzymes that are universally distributed. When we examine them closely, the distinctive structural features of different organisms within any group, such as vertebrates, seem less dependent on differences among the kinds of proteins organisms have than on how they organize and regulate various common proteins such as actin, myosin, collagen, and albumin.

Thus in comparing the anatomy of cat with dog or cat with mouse, differences seem to primarily depend on the extent and location of their common tissues (e.g., bone, muscle, nerve). This control over the quantity and placement of tissues arises from regulatory events during development that can easily be modified by mutational changes in the eukaryotic counterparts of prokaryotic regulator genes, simple gene rearrangements such as deficiencies, duplications, inversions, and transpositions (translocations) can markedly change regulatory control over gene function. Because regulation is so important, a number of evolutionists have therefore emphasized changes in regulation as being responsible for varicus major changes in evolution— "new bottles for old wine."

Wilson and co-workers (1974) based one such study on comparisons between frogs and placental mammals. Early frogs appear in Triassic deposits of about 200 million years ago, whereas placental mammals do not appear in the fossil record until some time during the Cretaceous, about 90 million years ago. Nevertheless, despite their more ancient fossil history and their numerous array of species, frogs have undergone few phenotypic changes in evolution compared to the enormous adaptive radiation of placental mammals. The more than 3,000 species of frogs look very much alike and taxonomists have consigned them to a single order (Anura), whereas the 3,800 species of placental mammals (of which about 1,700 are rodents) diverged widely and have usually been classified into 15 or 16 orders ranging from bats to primates to whales. Since the amino acid sequences in proteins of both these groups seem to have evolved at approximately

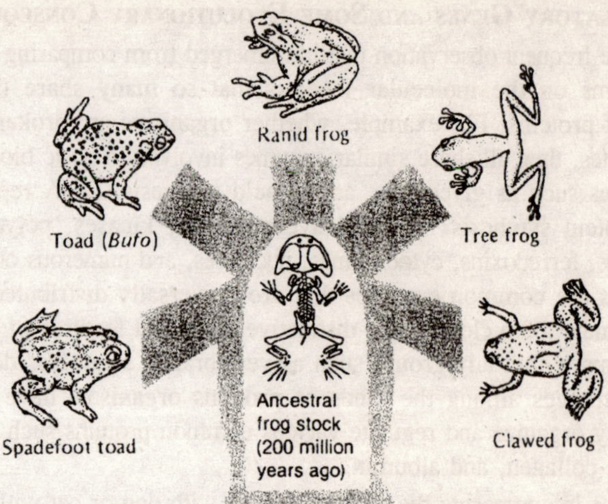

Fig. 3.14. Some of the major forms of present-day frogs along with a Triassic froglike fossil dating back to about 200 million years ago.

the same rate, Wilson suggests that the phenotypic similarity among frogs indicates that relatively few regulatory mutations have established themselves in frogs, compared to mammals.

The importance of *regulatory changes* seems obvious in the sharp phenotypic contrasts among related species such as humans and African apes. These two groups differ enough (brain size, facial structure, bipedal locomotion, etc.) for many anthropologists to place them in different taxonomic families (Hominidae, Pongidae), yet their composition of structural proteins is strikingly similar: they have almost identical α- and β-hemoglobin chains, cytochrome c proteins, and even fibrinopeptides. In fact, comparisons for any given protein between these two groups show an average of about 99 percent identity in amino acid sequence.

Clearly, these and other examples show that regulatory mutations can play a larger role in the morphological and functional differentiation of species than many structural gene changes. Some evolutionists have used this observation to support the notion that new species, new genera, or new orders (*macroevolutionary* events) can arise from regulatory changes that have a large phenotypic effect occurring over relatively short periods of time. That is, a population undergoing only small gradual changes (*microevolution*) may persist that way for long periods until pronounced regulatory changes are incorporated that allow one or more of its small isolated groups to evolve into a higher taxonomic category.

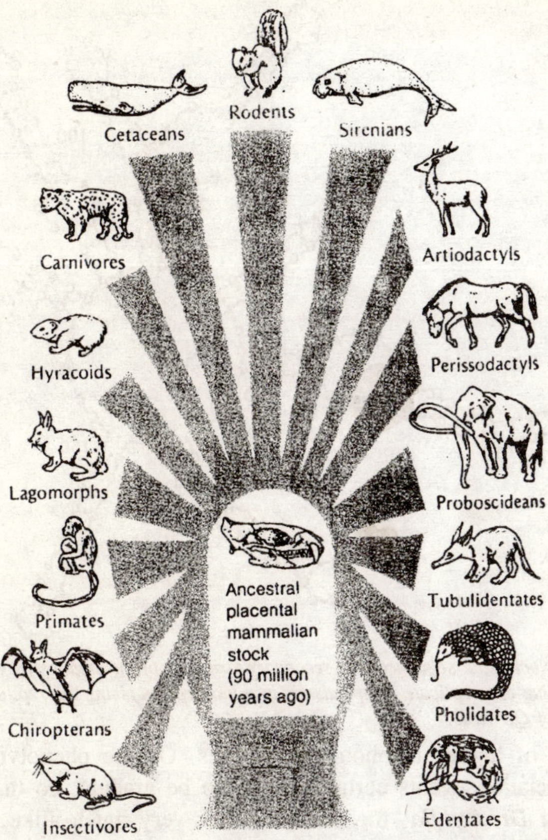

Fig. 3.15. Representatives of 15 present-day orders of placental mammals and the fossil skull of a Cretaceous placental-like mammal dating back to about 90 million years ago.

Since some of the fossii data for vertebrates, mollusks, and other animals show such periodic bursts of evolutionary activity, paleontologists such as Eldredge and Gould have given the name *punctuated equilibria* to what they see as long-term fossil uniformity of populations punctuated by geologically short periods of rapid speciation. Others argue that because the fossil record has gaps, punctuated equilibria are only apparent, not real: what seems rapid in geological time may actually involve many thousands of generations. However, although this dispute has generated many arguments and counterarguments, all evolutionists agree that both gradual and rapid changes occur during evolution. What we have not yet resolved is the relative importance of these changes in explaining speciation and the

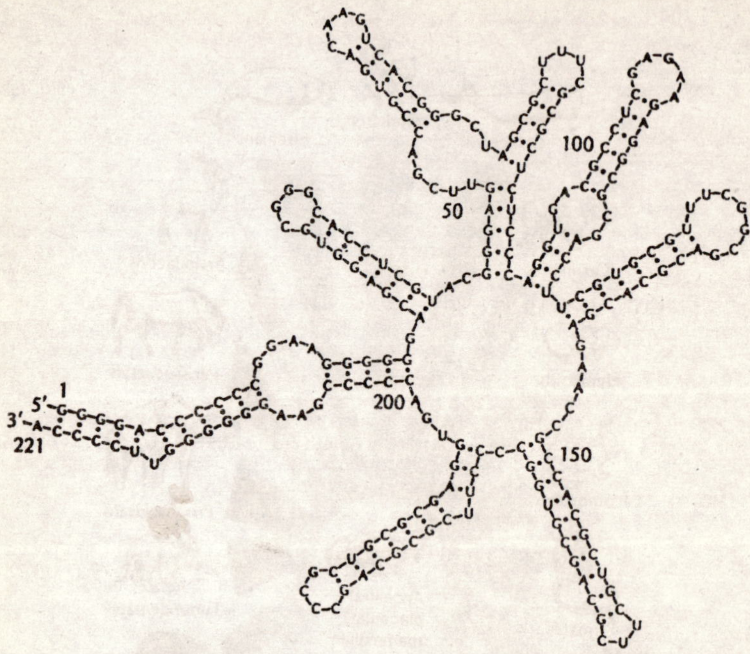

Fig. 3.16. Nucleotide sequence and secondary structure of the smallest RNA molecule that can replicate independently in a mixture containing the replicase enzyme of Qβ virus.

evolution of higher taxonomic categories. On the phenotypic level, some speciation events certainly seem to be gradual, so that sibling species of *Drosophila*, for example, look very much alike, whereas other related species that are probably no older than some *Drosophila* sibling species (e.g., chimpanzees and humans) look very different. It would obviously be helpful if the genetic and evolutionary events in each lineage could be analyzed.

Not surprisingly, prospects for obtaining detailed phylogenetic answers to these and other questions are becoming more favourable. Nucleotide-sequencing techniques are continually improving, and sequencing data are rapidly accumulating, enabling evolutionists to propose phylogenies for many groups whose relationships long seemed difficult to establish or inaccessible. Thus, ribosomal RNA analysis has defined a new major group among prokaryotes, the Archaebacteria, and is also being used to delineate relationships among eukaryotic phyla. Investigators have even extended the use of sequencing comparisons to studying molecular evolution on the laboratory level.

4

MOLECULAR PHYLOGENETICS

Molecular phylogenetics is the study of evolutionary relationships among organisms or genes by a combination of molecular biology and statistical techniques. It is also commonly called *molecular systematics*, if the relationships of organisms are the concern.

The molecular approach to systematics was initiated at the turn of the century by Nuttall (1904), who used serological cross-reactions to study the phylogenetics relationships among various groups of animals. However, extensive use of molecular data in phylogenetics studies did not begin until the early 1960s, i.e., until after the introduction of protein sequencing, protein electrophoresis, and other molecular techniques into the field. Protein sequence data allowed, for the first time, the investigation of long-term evolution such as the relationships among mammalian orders or even more distantly related taxa. On the other hand, less expensive and more expedient methods, such as protein electrophoresis, DNA-DNA hybridization, and immunological methods, though less accurate than protein sequencing, were extensively used to study the phylogenetic relationships among populations or closely related species. The advent of various recombinant DNA techniques since the 1970s has led to a rapid accumulation of DNA sequence data, thus stimulating even greater interest in molecular systematics. Yet the development of the polymerase chain reaction (PCR) method has made systematic studies even easier, resulting in an unprecedented high level of activities in phylogenetic reconstruction.

There are several reasons why molecular data, particularly DNA sequence data, are much more powerful for evolutionary studies than morphological and physiological data. First, DNA and protein sequences generally evolve in a much more regular manner than do morphological and physiological characters and therefore can provide a clearer picture of relationships of organisms. Second, molecular data are often much more amenable to quantitative treatments than are morphological data. In fact, sophisticated mathematical and statistical theories have already been developed for analyzing DNA sequence data. Third, molecular data are much more abundant. This abundancy is especially useful in studying microorganisms, in which only a limited number of morphological or physiological characters are available for phylogenetic studies. Indeed, molecular approaches have completely revolutionized the study of microbial taxonomy. Of course, we should not abandon traditional means of evolutionary inquiry such as morphology, anatomy, physiology, and paleontology. Rather, different approaches provide complementary data. Morphological and anatomical data are necessary for taxonomical studies, and paleontological information is required for constructing a time frame for evolutionary studies.

Molecular data have proved useful for studying not only the phylogenetic relationships among closely related populations or species such as the relationships among human populations and those between humans and apes but also very ancient evolutionary occurrences such as the origin of mitochondria and chloroplasts and the divergence of phyla and kingdoms. In the future, most phylogenetic issues are likely to be resolved by molecular data and we may eventually be able to fulfill Darwin's dream of having "a fairly true genealogical tree of each great kingdom of Nature."

To be able to take advantage of molecular data, however, one must understand the methodology of molecular phylogenetics. The aim of this chapter is to provide basic principles of tree reconstruction; a number of interesting examples of phylogenetic trees based on molecular data.

Phylogenetic Trees

All life forms are related by descent to one another. Closely related organisms are descended from more recent common ancestors than are distantly related ones. The purposes of phylogenetic studies are (1) to reconstruct the correct genealogical ties between organisms and (2) to estimate the time of divergence between organisms since they last shared a common ancestor.

In phylogenetic studies, the evolutionary relationships among a group of organisms are illustrated by means of a *phylogenetic tree*. A phylogenetic tree is a graph composed of nodes and branches, in which

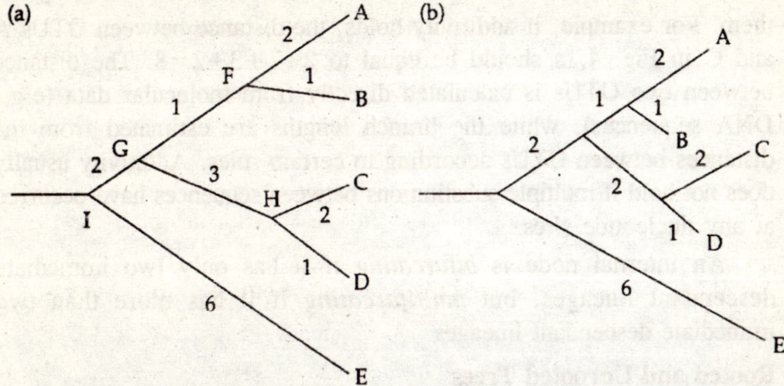

Fig. 4.1. Two alternative representations of a phylogenetic tree for five OTUs. (a) Unscaled branches. (b) Scaled branches.

only one branch connects any two adjacent nodes. The *nodes* represent the taxonomic units, and the *branches* define the relationships among the units in terms of descent and ancestry. The branching pattern of a tree is called the *topology*. The *branch length* usually represents the number of changes that have occurred in that branch. The taxonomic units represented by the nodes can be species, populations, individuals, or genes.

When referring to the nodes of a tree, we distinguish between *external* (or *terminal*) nodes and *internal* nodes. For example, in Fig 4.1 nodes A, B, C, D, and E are external, whereas all others are internal. External nodes represent the extant taxonomic units under comparison and are referred to as *operational taxonomic units* (OTUs). Internal nodes represent ancestral units. The branches of a tree are also known as the *edges* and can be classified as *internal* or *external*. For example, the branches leading to nodes A, B, C, D, and E are external, whereas all others are internal. External branches are also called *peripheral branches*.

Fig. 4.1 illustrates two common ways of drawing a phylogenetic tree. In Fig. 4.1a, the branches are *unscaled*; their lengths are not proportional to the number of changes, which are indicated on the branches. This presentation allows us to line up the extant OTUs and also to place the nodes representing divergence events on a *time scale* on which the times of divergence are known or have been estimated. In Fig. 4.1b, the branches are *scaled*, that is, their lengths are proportional to the numbers of changes.

A tree is said to be *additive* if the distance between any two OTUs is equal to the sum of the lengths of all the branches connecting them. For example, if additivity holds, the distance between OTUs A and C in Fig. 4.1a should be equal to 2+1+3+2=8. The distance between two OTUs is calculated directly from molecular data (e.g., DNA sequences), while the branch lengths are estimated from the distances between OTUs according to certain rules. Additivity usually does not hold if multiple substitutions between sequences have occurred at any nucleotide sites.

An internal node is *bifurcating* if it has only two immediate descendant lineages, but *multifurcating* if it has more than two immediate descendant lineages.

Rooted and Unrooted Trees

Phylogenetic tree can be either *rooted* or *unrooted*. In a rooted tree three exists a particular node, called the root, from which a unique path leads to any other node. The direction of each path corresponds to evolutionary time, and the root is the common ancestor of all the OTUs under study. An unrooted tree is a tree that only specifies the relationships among the OTUs but does not define the evolutionary path. Unrooted trees do not make assumptions or require knowledge about common ancestors.

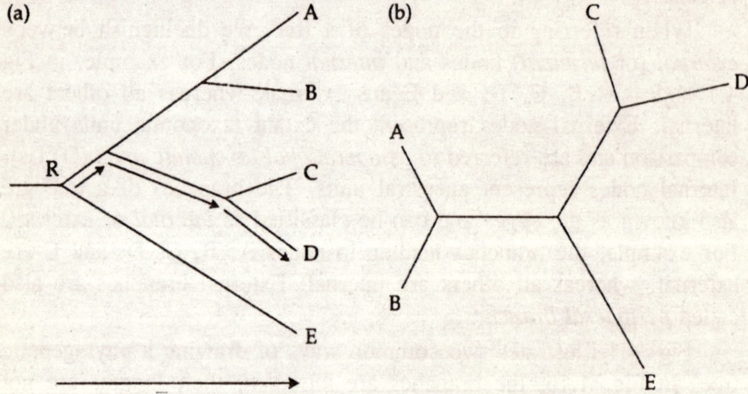

Fig. 4.2. (a) Rooted and (b) unrooted phylogenetics trees.

The number of bifurcating unrooted trees (N_U) for n OTUs ($n \geq 3$) is given by

$$N_U = (2n-5)!/\left[2^{n-3}(n-3)!\right] \quad ...(1)$$

while the number of bifurcating rooted trees (N_R) for $n \geq 2$ is

$$N_R = (2n-3)!/\left[2^{n-2}(n-2)!\right] \quad ...(2)$$

Note that the number of possible rooted trees for n OTUs is equal to the number of possible unrooted trees for n+1 OTUs. Table 4.1 shows that both N_U and N_R increase very rapidly with n—for 10 OTUs there are already more than 2 million bifurcating unrooted trees and close to 35 million rooted ones. Since only one of these trees represents correctly the true evolutionary relationships among the OTUs, it is usually very difficult to infer the true phylogenetic tree when n is large.

The majority of tree-making methods yield unrooted trees. To root an unrooted tree, we usually need an *outgroup* (i.e., an OTU for which external information, such as paleontological evidence, clearly indicates that it has branched off earlier than the taxa under study). The root is then placed between the outgroup and the node connecting it to the other OTUs.

Table 4.1. Possible numbers of rooted and unrooted trees up to 10 OTUs

Number of OTUs	*Number of rooted trees*	*Number of unrooted trees*
2	1	1
3	3	1
4	15	3
5	105	15
6	954	105
7	10,395	954
8	135,135	10,395
9	2,027.025	135,135
10	34,459,425	2,027,025

True and Inferred Trees

The sequence of speciation events that has led to the formation of any group of OTUs is historically unique. Thus, only one of all the possible trees that can be built with a given number of OTUs represents the true evolutionary history. A tree that is obtained by using a certain set of data and a certain method of tree reconstruction is called an *inferred tree*. An inferred tree may or may not be identical with the true tree.

Gene Trees and Species Trees

A phylogenetic tree that represents the evolutionary pathways of a group of species is called a *species tree*. When a phylogenetic tree is

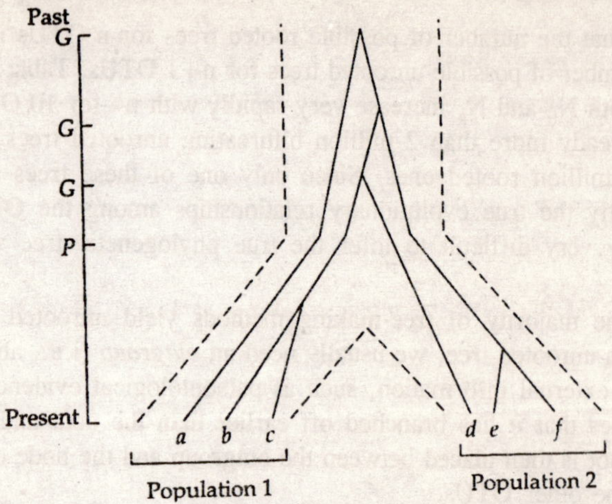

Fig. 4.3. Diagram showing that gene splitting (G) usually occurs earlier than population splitting (P) if the population is genetically polymorphic at time P.

constructed from one gene from each species, the inferred tree is a *gene tree*. It can differ from the species tree in two respects. First, the divergence of two genes sampled from two different species may predate the divergence of the two species. Such predating will result in an overestimate of the branch length and may present a problem if we are concerned with short-term evolution, in which the component of divergence due to genetic polymorphism within species cannot be ignored.

The second problem with gene trees is that the branching pattern of a gene tree (i.e., its topology) may be different from that of the species. Fig. 5.4 shows three different possible relationships between the two trees The topologies of gene trees in (a) and (b) are identical with those of the corresponding species tree (e.g., X and Y form a cluster). The gene tree in (c), however, is different from the true species tree, since now Y and Z are sister groups when considering this gene. The probability of obtaining the erroneous tree (c) is quite high when the time interval between the first and second species splitting (t_2–t_1) is short, as is probably true in the case of the phylogenetic relationships among humans, chimpanzees, and gorillas. To avoid this type of error, one needs to use many genes in the reconstruction of the phylogeny. A large amount of data is also required to avoid stochastic errors, which can occur because nucleotide substitutions occur randomly, so that, for instance, lineage Z may have by chance accumulated fewer substitutions than lineages X and Y, despite the fact that it has branched off earlier in time.

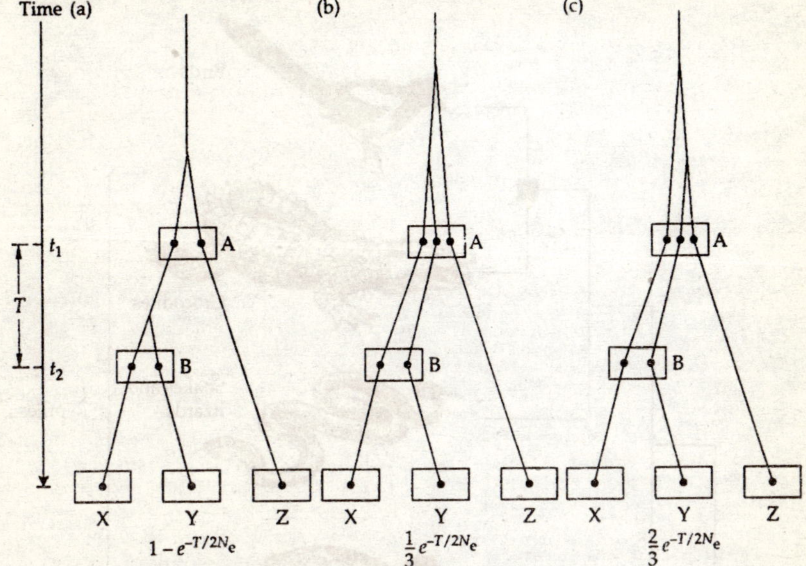

Fig. 4.4. Three possible relationships between a species tree (rectangles) and a gene tree (dots). In (a) and (b), the topologies of the species trees are identical to those of the gene trees. In (c), on the other hand, the time of divergence between X and Y greatly predates the time of divergence between the respective population.

Monophyletic Groups and Clades

A group of taxa is said to be *monophyletic* if they are derived from a single common ancestor but is said to be *polyphyletic* if they are derived from more than one common ancestor. For example, OTUs A and B form a monophyletic group because they are derived from the common ancestor F, whereas OTUs A and C are polyphyletic because they are derived from two different ancestors, F and H, respectively. However, the entire group of A, B, C, and D is monophyletic because of common ancestor G. A group of taxa is said to be *paraphyletic* if the taxa are derived from a common ancestor but the group does not include all decendent taxa of the same common ancestor. For example, the group of OTUs A, B, and E is paraphyletic because it does not include all the descendent OUTs of the common ancestor I, i.e., it does not include OTUs C and D.

In systematic studies one is often interested in the identification of clades. A *clade* is defined as a monophyletic group that includes all the descendent species of the common ancestor (in the literature, a clade and a monophyletic group are often used exchangeably). The

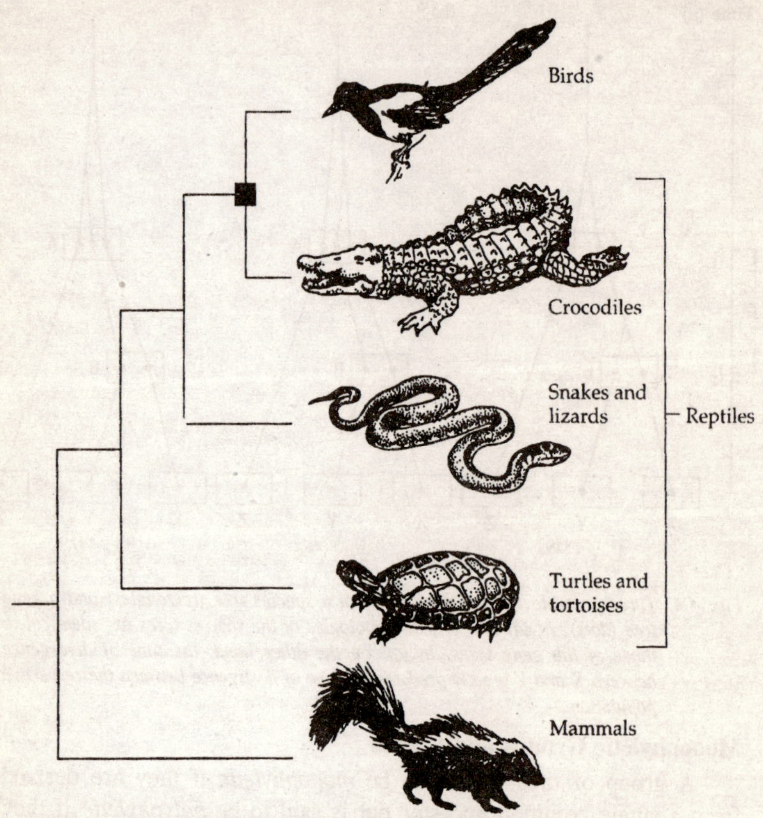

Fig. 4.5. Phylogenetic tree of birds, reptiles, and mammals.

classical taxonomic assignment of reptiles to a separate class does not fit the definition of a clade, for the three groups of reptiles share a common ancestor with another group, the class of birds, which is not included within the definition of the Reptilia. In other words, the class Reptilia is paraphyletic. Birds and crocodiles, on the other hand, do constitute a natural clade, the Archosauria, since they share a common ancestor not shared by any extant organism other than birds and crocodiles. Similarly birds and reptiles taken together constitute a natural clade, and of course, a monophyletic group.

Methods of Tree Reconstruction

Numerous tree-making methods have been proposed, because no single method performs well under all circumstances. A number of methods that have been frequently used or are convenient for illustrating basic principles will be discussed in this section. For simplicity, we consider nucleotide sequence data, though most of the methods described are also applicable to other types of molecular data such as amino acid sequences.

The methods described below can be classified into four types: distance matrix methods, maximum parsimony methods, maximum likelihood methods and methods of invariants. In *distance matrix methods*, evolutionary distances (usually numerous of nucleotide or amino acid substitutions between sequences) are computed for all pairs of taxa, and a phylogenetic tree is constructed by using an algorithm based on some functional relationships among the distance values. The first five methods described below are distance matrix method. In *maximum parsimony methods*, character states (e.g., the nucleotide or amino acid at a site) are used, and the shortest pathway leading to these character states is chosen as the best tree. In *maximum likelihood methods* one searches for the maximum likelihood (ML) value for the character state configurations among the sequences under study for each possible tree and chooses the one with the largest ML value as the preferred tree. (A nucleotide configuration at a site means the pattern of nucleotide differences at that site among the sequences involved.) The *methods of invariants* study some particular functions of the character states that have the expected value 0 under certain trees but have nonzero expectations under other trees. The invariant methods can be skipped if a reader is not deeply concerned with tree reconstruction or finds the material difficult.

Unweighted Pair-Group Method with Arithmetic Mean (UPGMA)

The *unweighted pair-group method with arithmetic mean* (UPGMA), the simplest method for tree reconstruction, was originally developed for constructing taxonomic phenograms, i.e., trees that reflect the phenotypic similarities between OTUs. However, it can also be used to construct phylogenetic trees if the rates of evolution are approximately constant among the different lineages so that an approximately linear relation exists between evolutionary distance and divergence time. For such a relation to hold, linear distance measures such as the number of nucleotide (or amino acid) substitutions should be used.

The UPGMA method employs a sequential clustering algorithm, in which local topological relationships are inferred in order of decreasing similarity and a phylogenetic tree is built in a stepwise manner. That is, we first identify the two OTUs that are most similar

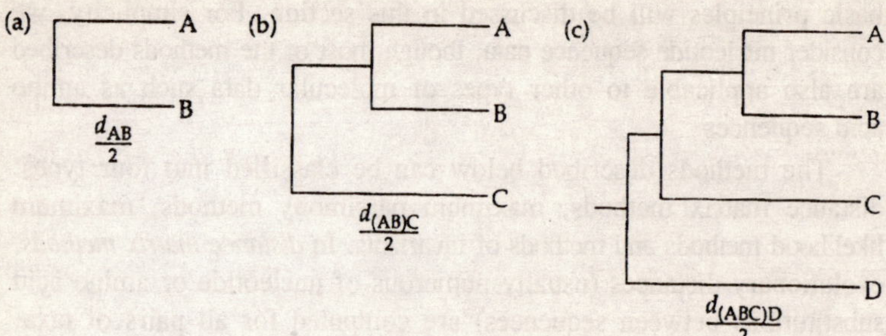

Fig. 4.6. Diagram illustrating the stepwise construction of a phylogenetic tree for four OTUs according to the UPGMA method.

to each other (i.e., have the shortest distance) and treat them as a new single OTU. Such an OTU is referred to as a *composite OTU*. Subsequently, from among the new group of OTUs, we identify the pair with the highest similarity, and so on, until only two OTUs are left.

To illustrate the method, let us consider a case of four OTUs. The pairwise evolutionary distances, such as Jukes and Cantor's (1969) estimates, are given by the following matrix:

	OTU		
OTU	A	B	C
B	d_{AB}		
C	d_{AC}	d_{BC}	
D	d_{AD}	d_{BD}	d_{CD}

In this matrix, d_{ij} stands for the distance between OTUs i and j. The first two OTUs to be clustered are the ones with the smallest distance. Let us assume that d_{AB} has the smallest value. Then, OTUs A and B are the first to be clustered, and the branching point is positioned at a distance of $d_{AB}/2$ substitutions.

After the first clustering, A and B are treated as a single composite OTU, and a new distance matrix is computed.

	OTU	
OTU	(AB)	C
C	$d_{(AB)C}$	
D	$d_{(AB)D}$	d_{CD}

In this matrix, $d_{(AB)C}=(d_{AC}+d_{BC})/2$, and $d_{(AB)D}=(d_{AD}+d_{BD})/2$. In other words, the distance between a simple OTU and a composite OTU is the average of the distances between the simple OTU and the constituent OTUs of the composite OTU. If $d_{(AB)C}$ turns out to be the smallest distance in the new matrix, then OTU C will be joined to the composite OTU (AB) with a branching node at $d_{(AB)C}/2$.

The final step consists of clustering the last OTU, D, with the composite OTU (ABC). The root of the entire tree is positioned at $d_{(ABC)D}/2=[(d_{AD}+d_{BD}+d_{CD})/3]/2$.

In the UPGMA method, the distance between two composite OTU is computed as the arithmetic mean of the pairwise distances between the constituent OTUs of the two composite OTUs. For example, the distance between the composite OTU (ij) and (mn) is

$$d_{(ij)(mn)} = \left(d_{im} + d_{in} + d_{jm} + d_{jn}\right) / 4 \quad ...(3)$$

In general, the distance between two clusters of OTUs X and Y is

$$d_{XY} = \sum_{i,j} d_{ij} / (n_X n_Y) \quad ...(4)$$

where the summation is over every i in cluster X and every j in cluster Y and n_X and n_Y are the numbers of OTUs in X and Y, respectively. For example, for the composite OTUs (ijk) and (mn),

$$d_{(ijk)(mn)} = \left(d_{im} + d_{in} + d_{jm} + d_{jn} + d_{km} + d_{kn}\right) / 6$$

Transformed Distance Method

If the assumption of rate constancy among lineages does not hold, UPGMA may give an erroneous topology. For example, suppose that the phylogenetic tree is the true tree. The pairwise evolutionary distance are given by the following matrix:

	OTU		
OTU	A	B	C
B	8		
C	7	9	
D	12	14	11

By using the UPGMA method, we obtain an inferred tree that differs from the true tree in its branching pattern. For example, OTUs A and C are grouped together, whereas in the true tree, A and B are sister OTUs. (Note that additivity does not hold in this case; the true distance between A and B is 8, whereas the sum of the lengths of the branches connecting A and B is 3.50+0.75+4.25=8.50.)

The topological errors might be remedied, however, by using a correction called the *transformed distance method*. Briefly, this method

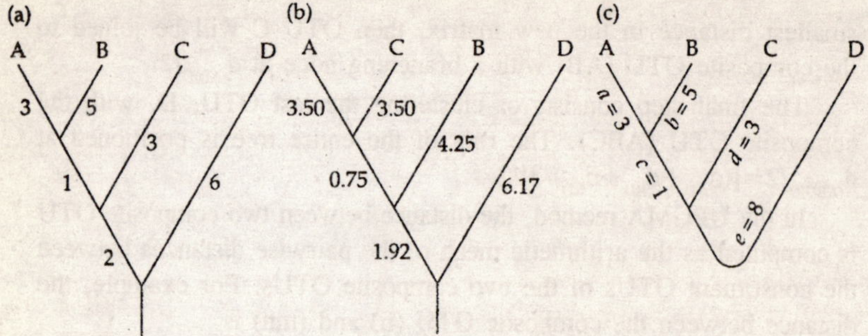

Fig. 4.7. (a) The true phylogenetic tree. (b) The erroneous phylogenetic tree. (c) The tree inferred by the transformed distance method.

uses an outgroup as a reference to make corrections for unequal rates of evolution among the lineage under study and then applies UPGMA to the new distance matrix to infer the topology of the tree.

In the present case, let us assume that taxon D is an outgroup to all other taxa. D can then be used as a reference to transform the distances by the following equation:

$$d'_{ij} = \left[\left(d_{ij} - d_{iD} - d_{jD}\right)/2\right] + \bar{d}_D \qquad ...(5)$$

where d'_{ij} is the transformed distance, i=A, B, C, and $\bar{d}_D = (d_{AD} + d_{BD} + d_{CD})/3$. The term $\bar{d}_D$ was introduced to make all d'_{ij} values positive, because in practice a distance can never be negative. For the general case of n OTUs (not including the outgroup), $\bar{d}_D = \sum d_{iD} / n$.

In our example, $d_D = 37/3$, and the new distance matrix for taxa A, B, and C is

	OTU	
OTU	A	B
B	10/3	
C	13/3	13/3

Since d'_{AB} has the smallest value, A and B are the first to be clustered together, and subsequently, C is added to the tree. By definition, the outgroup OUT, D, determines the root of the tree and is the last to be added. This gives the correct topology. In the above example, we considered only three taxa with one outgroup, but the method can be easily extended to more taxa and/or more outgroups. In many instances it is impossible to decide a *priori* which of the taxa under consideration is an outgroup. To overcome this difficulty, a two-stage approach has been proposed. In the first step, one infers the root of the tree by using the UPGMA method; of course, the inferred root may not be the true root. After that, the taxa on one side of the root are used as references (outgroups) for making corrections for the unequal rates of evolution among the lineages on the other side of the root, and vice versa. In our example, this approach also identifies the correct tree.

Neighbours-Relation Methods

In an unrooted bifurcating tree, two OTUs are said to be *neighbours* if they are connected through a single internal node. For example A and B are neighbours and so are C and D. In comparison, neither A and C nor B and C are neighbours. However, if we combine OTUs A and B into one composite OTU, then the combined OTU (AB) and OTU C become a new pair of neighbours.

Let us now assume that the tree shown is the true tree. The, if additivity holds, we should have

$$d_{AC} + d_{BD} = d_{AD} + d_{BC} = a + b + c + d + 2x = d_{AB} + d_{CD} + 2x \quad ...(7)$$

where x is the length of the internal branch. Therefore, the following two conditions hold:

$$d_{AB} + d_{CD} < d_{AC} + d_{BD} \quad ...(7a)$$

$$d_{AB} + d_{CD} < d_{AD} + d_{BC} \quad ...(7b)$$

These two conditions are collectively known as the *four-point condition*. They may hold even if additivity holds only approximately.

Conversely, for four OTUs with unknown phylogenetic relationships, the above two conditions can be used to identify the neighbours (A and B; C and D). Once the two pairs of neighbours are determined, so is the topology of the tree.

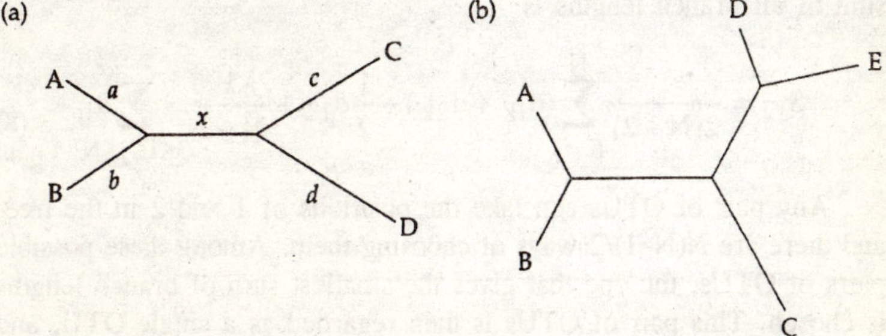

Fig. 4.8. Bifurcating unrooted trees with (a) four OTUs and (b) five OTUs.

Sattath and Tversky (1977) proposed the following method for dealing with more than four OTUs. First, compute a distance matrix as in the case of UPGMA. For every possible quadruple, say OTUs i, j, m, and n, compute $d_{ij}+d_{mn}$, $+d_{im}+d_{jn}$, and $d_{in}+d_{jm}$. Suppose that the first sum is the smallest. Then, assign a score of 1 to both the pair i and j and the pair m and n; pairs i and m, i and n, j and n, and j and m are assigned a score of 0. If, on the other hand, $d_{im}+d_{jn}$ has the smallest value, then we assign the pair i and m and pair j and n each a score of 1 and assign the other four possible pairs a score of 0. When all the possible quadruples are considered, the pair with the highest total score is selected as the first pair of neighbours and treated as a single OTU. Next, compute a new distance matrix as in the case of UPGMA and then repeat the same process to select the second pair of neighbours. This process is continued until the number of OTUs is reduced to three, which can form only one unrooted tree. A detailed illustration of this method will be given when we consider the phylogeny of apes and humans.

Another method (the neighbourliness method) based on the neighbours-relation concept has been proposed by Fitch (1981). A similar approach, called the neighbour-joining method is discussed in the following section.

Neighbour-Joining Method

The principle of the neighbour-joining method is to find neighbours sequentially that may minimize the total length of the tree. This method starts with a starlike tree, in which there is no clustering of OTUs. The first step is to separate a pair of OTUs from all the others. In this tree there is only one interior branch, that is, the branch connecting nodes X and Y, where X is the common node for OTUs 1 and 2 and Y is the common node for the others (3, 4, ..., N). For this tree the sum of all branch lengths is

$$S_{12} = \frac{1}{2(N-2)}\sum_{K=3}^{N}(d_{1k}+d_{2k})+\frac{1}{2}d_{12}+\frac{1}{N-2}\sum_{3\le i\le j\le N}d_{ij} \quad ...(8)$$

Any pair of OTUs can take the positions of 1 and 2 in the tree, and there are N(N–1)/2 ways of choosing them. Among these possible pairs of OTUs, the one that gives the smallest sum of branch lengths is chosen. This pair of OTUs is then regarded as a single OTU, and the arithmetic mean distances between OTUs are computed to form a new distance matrix. The next pair of OTUs that gives the smallest

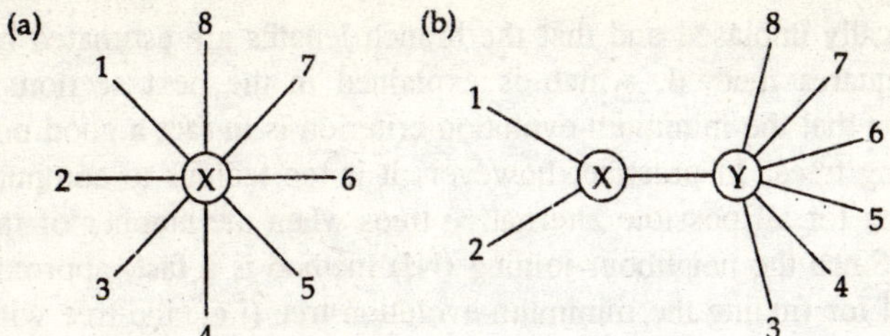

Fig. 4.9. (a) A starlike tree with no hierachical structure. (b) A tree in which OTUs 1 and 2 are clustered.

sum of branch lengths is then chosen. This procedure is continued until all N–3 interior branches are found. Saitou and Nei (1987) showed that in the case of four OTUs the necessary condition for this method to obtain the correct tree topology is also given by the four-point condition.

Instead of the S criterion defined in Equation (8), Studier and Keppler (1988) proposed the Q criterion, defined as follows:

$$Q_{12} = (N-2)d_{12} - \sum_{i=1}^{N} d_{1i} - \sum_{i=1}^{N} d_{2i} \qquad ...(9)$$

Gascuel (1994) showed that S and Q are related by the following equation.

$$S_{12} = \frac{1}{2(N-2)} Q_{12} + \frac{1}{(N-2)} \sum_{1 \le i \le j \le N} d_{ij} \qquad ...(10)$$

Since the last term in the above equation is a constant for all possible trees, minimizing S is equivalent to minimizing Q. However, the Studier-Keppler algorithm is computationally faster than the Saitou-Nei algorithm. Gascuel (1994) further provided a theoretical explanation for why the neighbour-joining method and the neighbours-relation method of Sattath and Tversky (1977) give identical or similar tree topologies.

Minimum-Evolution Method

For each possible alternative tree one can estimate the length of each branch from the estimated pairwise distances (d_{ij}) between taxa and then compute the sum (S) of all branch-length estimates. The minimum-evolution criterion is to choose the tree with the smallest S value as the best tree. Rzhetsky and Nei (1993) showed that the expectation of S for the true tree is indeed the smallest among all possible trees, provided that the evolutionary distances (d_{ij}) used are statistically unbiased and that the branch lengths are estimated by the least-squares method, which is explained in the next section. This suggests that the minimum-evolution criterion is in fact a good one for choosing trees. In practice, however, it is too tedious to compute the S values for all possible alternative trees when the number of taxa is large. Since the neighbour-joining (NJ) method is a fast, approximate method for finding the minimum-evolution tree (i.e., the tree with the smallest S), the true tree is likely to be the NJ tree or a tree that differs from the NJ tree by only a few topological differences. Thus, Rzhetsky and Nei (1992) proposed to compare the S values for the NJ tree and all the trees that are very similar to the NJ tree in topology and then to choose the tree with the minimum S value as the best tree. Note that under the minimum-evolution criterion a tree is worse than another tee only if its S value is significantly larger than that of the other tree. Thus, all trees whose S values are not significantly different from the minimum S value should be regarded as candidates for the true tree.

Maximum Parsimony Methods

The principle of maximum parsimony searches for a tree that requires the smallest number of evolutionary changes to explain the differences observed among the OTUs under study. Such a tree is called a *maximum parsimony tree*. Often more than one tree with the same minimum number of changes are found, so that no unique tree can be inferred. The method discussed below was first developed for amino acid sequence data and was later modified for use on nucleotide sequences.

We start with the definition of *informative sites*. A nucleotide site is phylogenetically informative only if it favours some trees over the others. To illustrate the distinction between informative and noninformative sites, consider the following four hypothetical sequences:

	Site								
Sequence	1	2	3	4	5	6	7	8	9
1	A	A	G	A	G	T	G	C	A
2	A	G	C	C	G	T	G	C	G
3	A	G	A	T	A	T	C	C	A
4	A	G	A	G	A	T	C	C	G

There are three possible unrooted trees for four OTUs. Site 1 is not informative because all sequences at this site have A, so that no change is required in any of the three possible trees. At site 2, sequence

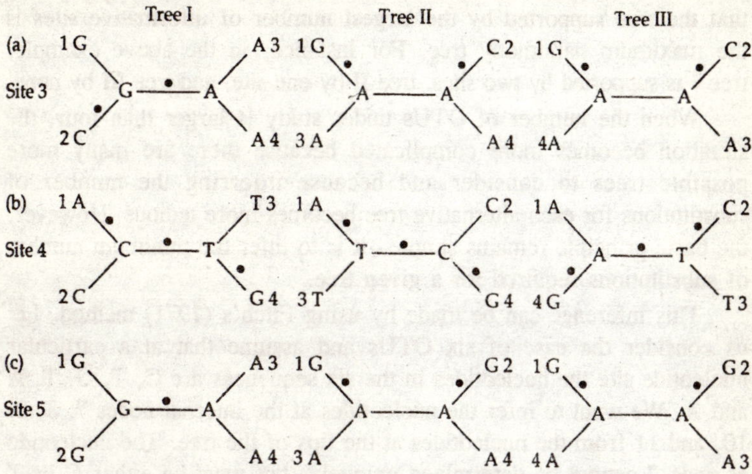

Fig. 4.10. Three possible unrooted trees for four DNA sequences that have been used to choose the most parsimonious tree.

1 has A while all other sequences have G, and so a simple assumption is that the nucleotide has changed from G to A in the lineage leading to sequence 1. Thus, this site is also not informative, because each of the three possible trees requires 1 change. As for site 3 each of the three possible trees requires 2 changes and so it is also not informative. Note that if we assume that the nucleotide at the node connecting OTUs 1 and 2 in tree I is C in Fig. 4.10a is C instead of G, the number of changes required for the tree remains 2. Fig. 4.10b shows that for site 4, each of the three trees requires 3 changes and thus site 4 is also noninformative. For site 5, tree I requires only 1 change, whereas tree II and III require 2 changes each. Therefore, this site is informative.

From these examples, we see that a site is informative only when there are at least two different kinds of nucleotides at the site, each of which is represented in at least two of the sequences under study. In the above, the informative sites are indicated by asterisks. For these sites, tree I requires 1, 1, and 2 changes, respectively tree II requires 2, 2 and 1 changes; and tree III requires 2, 2 and 2 changes. Thus, tree I is chosen because it requires the smallest number of changes (4) at the informative sites.

In the case of four OTUs, an informative site favours only one of the three possible alternative trees. For example, site 5 favours tree I over trees II and III, and is said to support tree I. It is easy to see that the tree supported by the largest number of informative sites is the maximum parsimony tree. For instance, in the above example, tree I is supported by two sites, tree II by one site, and tree III by none.

When the number of OTUs under study is larger than four, the situation becomes more complicated because there are many more possible trees to consider and because inferring the number of substitutions for each alternative tree becomes more tedious. However, the basic principle remains simple—it is to infer the minimum number of substitutions required for a given tree.

This inference can be made by using Fitch's (1971) method. Let us consider the case of six OTUs and assume that at a particular nucleotide site the nucleotides in the six sequences are C, T, G, T, A and A. We want to infer the nucleotides at the internal nodes 7, 8, 9, 10, and 11 from the nucleotides at the tips of the tree. The nucleotide at node 7 cannot be determined uniquely, but must be either C or T under the parsimony rule. Therefore, the set of candidate nucleotides at node 7 consists of C and T. Similarly, the set of candidate nucleotides at node 8 consists of G and T and that at node 9 consists of A, G, and T. However, at node 10, T is chosen because it is shared by the sets at the two descendant nodes, 7 and 9. Finally, the nucleotide at node 11 cannot be determined uniquely but parsimony requires it be either A or T.

In mathematical terms, the rule used is as follows: the set at an interior node is the intersection of its two immediately descendant sets if the intersection is not empty (e.g., the nucleotide at node 10 is the intersection of the sets at nodes 7 and 9); otherwise, it is the union of the descendant sets (e.g., the set at node 9 is the union of the sets at nodes 8 and 5). For every occasion that a union is required to form the nodal set, a nucleotide substitution at this position must have occurred at some point during the evolution of this position. Thus, counting the number of unions gives the minimum number of substitutions required to account for descendant nucleotides from a common ancestor, given the phylogeny assumed at the outset. In the present example, this number is 4.

The alternative trees in Fig. 4.11c and d each require 3 unions for forming the nodal sets and thus 3 nucleotide substitutions. This is the minimum number of substitutions required to explain the differences among the nucleotides at this position because all of the four types of nucleotides are present at this position. Therefore, there must be at least 3 mutational changes. There are many other alternative trees,

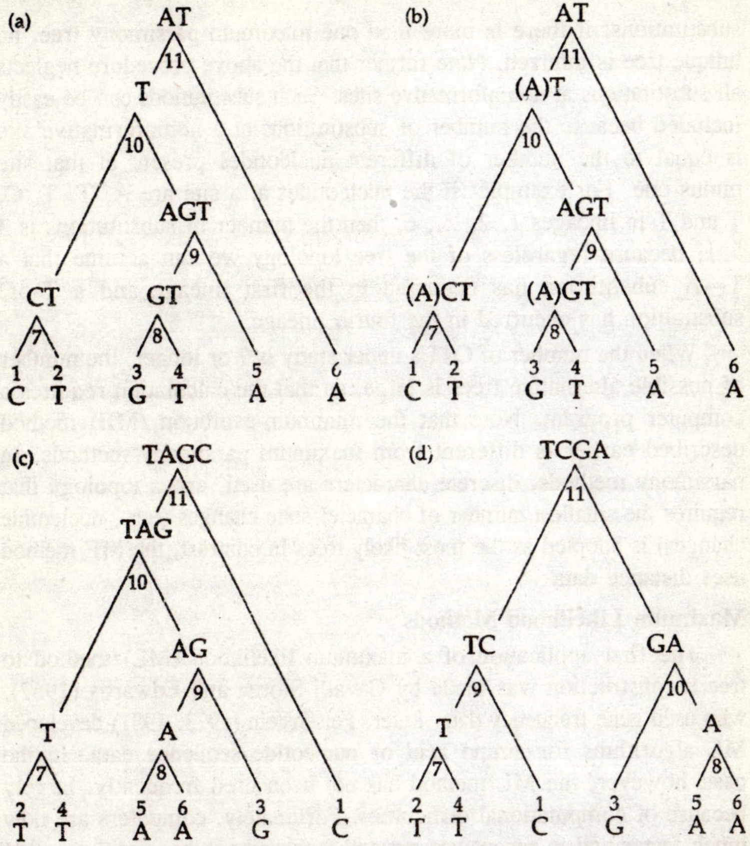

Fig. 4.11. Nucleotides in six extant species and the possible nucleotides in five ancestral species.

each of which requires 3 substitutions. Thus, unlike the case of four OTUs, when more than four OTUs are involved, an informative site may favour many alternative trees.

Although inferring the minimum number of substitutions is straightforward, inferring the evolutionary path (i.e., the true sequence in which the nucleotide had changed in each step) is often difficult for the procedure.

The procedure for inferring a maximum parsimony tree can be summarized as follows: First, identify all the informative sites. Next, for each possible tree, infer the minimum number of substitutions at each informative site and sum up the numbers over all informative sites. Finally, choose the tree or trees with the smallest number of substitutions; if there is more than one maximum parsimony tree, no unique tree is inferred. Note further that the above procedure neglects all substitutions at noninformative sites. Such substitutions can be easily included because the number of substitutions at a noninformative site is equal to the number of different nucleotides present at that site minus one. For example. if the nucleotides at a site are A, T, T, C, T and T in lineages 1, 2, ..., 6, then the number of substitutions is 3 – 1, because regardless of the tree topology we can assume that a T→A substitution has occurred in the first lineage and a T→C substitution has occurred in the fourth lineage.

When the number of OTUs under study is 7 or longer, the number of possible alternative trees is large, so that the calculation requires a computer program. Note that the minimum-evolution (ME) method described earlier is different from maximum parsimony methods. In parsimony methods, discrete characters are used, and a topology that requires the smallest number of character-state changes (e.g., nucleotide changes) is adopted as the most likely tree. In contrast, the ME method uses distance data.

Maximum Likelihood Methods

The first application of a maximum likelihood (ML) method to tree reconstruction was made by Cavalli-Sforza and Edwards (1967), who used gene frequency data. Later, Felsenstein (1973, 1981) developed ML algorithms for amino acid or nucleotide sequence data. In the past, however, the ML method has not been used frequently, largely because of computational difficulties. Fortunately, computers are now much faster and in recent years many authors have become interested in this method.

The ML method requires a probabilistic model for the process of nucleotide substitution. That is, we must specify the transition probability from one nucleotide state to another in a time interval in each branch. For example, consider the one-parameter model with the rate of substitution λ per site per unit time. Assume that the nucleotide at a given site is i at time 0. Then, the probability that the nucleotide at time t is i is given by

$$P_{ii}(t) = \frac{1}{4} + \frac{3}{4}e^{-4\lambda t/3} \quad ...(11)$$

and the probability that the nucleotide at time t is j, j≠i, is given by

$$P_{ij}(t) = \frac{1}{4} - \frac{1}{4}e^{-4\lambda t/3} \quad ...(12)$$

The next step is to set up the likelihood function. Let us use the case of four sequences (taxa) as an example. First, we assume a constant rate of substitution and consider the hypothetical tree. The likelihood function for a nucleotide site with nucleotides i, j, k, and l in sequence 1, 2, 3 and 4, respectively, can be computed as follows. If the nucleotide at the ancestral node (the root) was x, the probability of having nucleotide l in sequence 4 is $P_{xl}(t_1+t_2+t_3)$ because $t_1+t_2+t_3$ is the total amount of time between the two nodes, the probability of having nucleotide y at the common ancestral node of sequences 1, 2 and 3 is $P_{xy}(t_1)$, and so on. Therefore, given x, y, and z at the ancestral node and the two other internal nodes, the probability of observing i, j, k, and l at the tips of the tree is equal to

$$P_{xl}(t_1+t_2+t_3)P_{xy}(t_1)P_{yk}(t_2+t_3)P_{yz}(t_2)P_{zi}(t_3)P_{zj}(t_3)$$

All the above transition probabilities can be computed by using Equation (11) and (12). Since in practice we do not know the ancestral nucleotide, we can only assign a probability g_x, which is usually assumed to be the frequency of nucleotide x in the sequence. Noting that x, y, and z can be any of the four nucleotides, we sum over all possibilities and obtain the following likelihood function

$$h(i,j,k,l)-\sum_x g_x P_{xl}(t_1+t_2+t_3)$$

$$\times\sum_y P_{xy}(t_1)P_{yk}(t_2+t_3)\sum_z P_{yz}(t_2)P_{zi}(t_3)P_{zj}(t_3) \quad \text{...(13)}$$

Note that this likelihood function represents the probability of observing the configuration of nucleotides i, j, k, and l in sequences 1, 2, 3, and 4 for the given hypothetical tree. If the hypothetical tree is different from that in Fig. 4.12a, then the likelihood function is different from that of Equation (13). That is, the likelihood function depends on the hypothetical tree.

A formulation without the assumption of rate constancy can be done in a similar manner. In this case, it is usually more convenient to consider the transition probability in terms of the branch length; for example, we consider $P_{ij}(\upsilon_\alpha)$ instead of $P_{ij}(t_\alpha)$, where $\upsilon_\alpha=\lambda_\alpha t_\alpha$ and λ_α and t_α are have values of λ and t for the αth branch. For the unrooted tree given in Fig. 4.12b, we obtain the following likelihood function

$$h(i,j,k,l)=\sum_x g_x P_{xl}(\upsilon_4)P_{xk}(\upsilon_3)\sum_y P_{xy}(\upsilon_5)P_{yi}(\upsilon_1)P_{yj}(\upsilon_2) \quad \text{...14)}$$

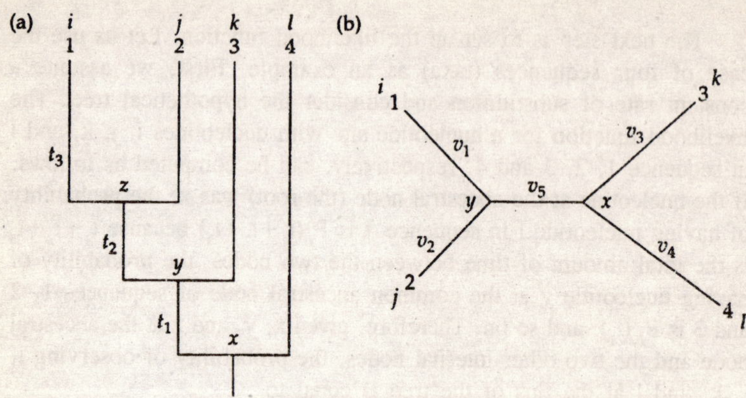

Fig. 4.12. Model trees for the derivation of the likelihood function.

if we assume that the internal node connecting taxa 3 and 4 is the ancestral node. If the process is time reversible, which is the case for the one-parameter model and many other current models of nucleotide substitution, then any node or point in the tree can be taken as the "ancestral" node. This is called the pulley principle. However, if the process is not time reversible, then we must assume or infer the root of the tree.

In the above we considered a single site. The likelihood for all sites is the product of the likelihoods for individual sites if all the nucleotide sites evolve independently. Suppose that there are s homologous sequences each with N nucleotides. Let $X_k=(x_{1k}, \ldots, x_{sk})$ be the vector representing the nucleotide configuration at the kth site, i.e., $x_{\zeta k}$ is the nucleotide at the kth site in the ζth sequence. For a given phylogenetic tree T, let $f(X_k|\theta_1, \ldots, \theta_\eta, T)$ be the likelihood of tree T for the kth site, where $\theta_1, \ldots, \theta_\eta$ are the unknown parameters such as the branching dates and the rates of nucleotide substitution. For example, for the configuration in the tree in Fig. 4.12b, we have $X_k=(i, j, k, l)$, $\theta_i=\upsilon_i$, and $f(X_k|\theta_1, \ldots, \theta_\eta, T) = h(i, j, k, l)$, which is given by Equation (14). For simplicity, let us assume that the sequences are homogeneous so that all sites on the sequences are equivalent (i.e., they evolve at the same rates). Then the likelihood function for the entire sequence for tree T is

$$L(\theta_1,\ldots,\theta_\eta|X_1,\ldots,X_N,T) = \prod_{k=1}^{N} f(X_k|\theta,T) \quad \ldots(15)$$

where $L(\theta_1, \ldots, \theta_\eta|X_1, \ldots, X_N,T)$ signifies the fact that given the data (observation) X_1, ..., X_N and tree T, the likelihood is a function of θ_1, ..., θ_η.

After the likelihood function L is derived under a hypothetical tree, one can use a computer to search for the θ_1, ..., θ_η values that maximize L. For example, for the tree in Fig. 4.12a, one searches for the t_α values that maximize L. (The maximum likelihood value is usually denoted ML). Finally, one chooses from among all possible trees the one with the largest ML value as the tree see. Note that since the likelihood functions such as Equation (13) and (14) depend on the model of nucleotide substitution, so do the ML values for different trees. Thus, a tree with the largest ML value under one substitution model may not have the largest ML value under another model. So, one should have an explicit model when deriving the likelihood functions.

Table 4.2. Nucleotide configuration for four sequences

	Configuration			
No.	*A*	*B*	*C*	*D*
1.	i	i	i	i
2.	i	i	i	j
3.	i	i	j	i
4.	i	j	i	i
5.	j	i	i	i
6.	i	i	j	j
7.	i	j	i	j
8.	i	j	j	i
9.	i	i	j	k
10.	i	j	i	k
11.	j	i	i	k
12.	j	k	i	i
13.	j	i	k	i
14.	i	j	k	i
15.	i	j	k	l

Under the assumption that all sites are independent and equivalent, the likelihood function can also be derived by considering the nucleotide configurations among the sequences. Let us consider the case of 4 sequences. As shown in Table 4.2 there are 15 configuration patterns. The first pattern (i, i, i, i) means that the nucleotide is the same in all four sequences, the second pattern (i, i, i, j) means that the nucleotide is the same for the first three sequences, but is different for the fourth sequence, and so on. For the unrooted tree given in Fig. 4.12b and under the one-parameter substitution model, we can show that the probability of obtaining the first pattern is equal to

$$U_1 = 4h(i,i,i,i)$$

where h(i, i, i, i) is given by Equation (14), and the factor 4 arises because there are four possible nucleotides. The probability of obtaining the second pattern is equal to

$$U_2 = 12h(i,i,i,j)$$

in which the factor 12 arises because there are four possibilities for i and for each i there are three possibilities for j. In the same manner, we can obtain the probabilities for the other pattern in Table 4.2. The likelihood function, L_j, for the jth topology is

$$L_j = C\prod_{i=1}^{15} U_{ji}^{Ni} \quad \text{...(16)}$$

where U_{ji} is the probability of obtaining the ith pattern for the jth topology, N_i is the observed number of the ith pattern, $C=N!/(N_2!\ldots N_{15}!)$, and N is the total number of nucleotide sites examined and is equal to the sum of N_i´s. This approach reduces the amount of computation considerably.

Methods of Invariants

Let us start with the definition of invariants. First, consider the evolution of a nucleotide site in two descendant lineages A and B. Assume that the probability of changing from nucleotide i to nucleotide j per unit time is a_{ij} in lineage A and b_{ij} in lineage B. Let P(xy) be the probability that at time t the nucleotides in lineages A and B are x and y, respectively. Consider, for example, the function

$$f = P(AT) - P(TA) \quad \text{...(17)}$$

If the nucleotide at the root of the two lineages, i.e., at t=0, is G, then at t=1, $P(AT)=a_{GA}b_{GT}$ because the probability of changing from G to A in lineage A is a_{GA} and the probability of changing from G to T in lineage B is b_{GT}. Similarly, $P(TA)=a_{GT}b_{GA}$. Therefore, at t=1,

$$f = a_{GA}b_{GT} - a_{GT}b_{GA}$$

If we use the one-parameter model of nucleotide substitution such that $a_{ij}=\alpha_1$ for i≠j and $b_{ij}=\alpha_2$ for i≠j, then

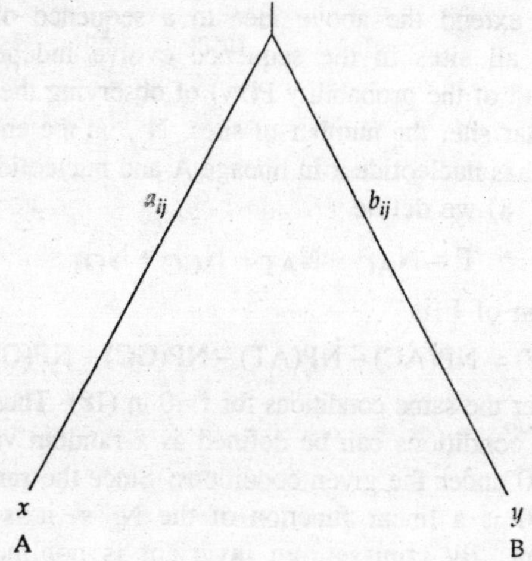

Fig. 4.13. Evolution of a nucleotide site in two lineages.

$$f = \alpha_1\alpha_2 - \alpha_1\alpha_2 = 0$$

for any values of α_1 and α_2. Actually, f=0 for any time t; this can be shown by using formula (12). Therefore, under the above conditions the function defined by (15) does not change with time, i.e., f=0 for all t, and it is said to be an invariant. (Note that if $\alpha_1 \neq \alpha_2$, the rates in two lineages are unequal.) However, if the nucleotide at the root is A instead of G, then at t=1

$$f = a_{AA}b_{AT} - a_{AT}b_{AA}$$
$$= (1-3\alpha_1)\alpha_2 - \alpha_1(1-3\alpha_2)$$

because $a_{AA}=1-3\alpha_1$ and $b_{AA}=1-3\alpha_2$. So, if $\alpha_1 \neq \alpha_2$, then f is not 0 and thus not an invariant. Therefore, whether the function defined by (17) is an invariant depends on the initial nucleotide at the root and on the assumption of equal rates in the two lineages.

In general, it is more useful to have invariants that are independent of the initial nucleotide, which is usually unknown. Further, it is better to remove the assumption of rate constancy, which often does not hold. The following function is such an invariant under the one-parameter model

$$f = P(AC) - P(AT) - P(GC) + P(GT) \qquad ...(18)$$

because it can be shown that f=0 for any initial nucleotide at the root and for any values of α_1 and α_2.

Next, we extend the above idea to a sequence of any length, assuming that all sites in the sequence evolve independently. We consider, instead of the probability P(xy) of observing the configuration xy at a particular site, the number of sites, N_{xy}, in the entire sequence, each of which has nucleotide x in lineage A and nucleotide y in lineage B. Instead of (18) we define

$$F = N_{AC} - N_{AT} - N_{GC} + N_{GT} \quad ...(19)$$

The expectation of F is

$$E(F) = NP(AC) - NP(AT) - NP(GC) + NP(GT)$$

which is 0 under the same conditions for f=0 in (18). Thus, an *invariant* under a set of conditions can be defined as a random variable whose expectation is 0 under the given conditions. Since the random variable defined in (19) is a linear function of the N_{in}'s, it is said to be a linear invariant. By contrast, an invariant is nonlinear if it is a nonlinear function of the N_{ij}'s. Here, we consider only linear invariants, an approach initiated by Lake (1987); Lake used the term *evolutionary parsimony*, but it is not based on any parsimony principle. For the theory of nonlinear invariants, which was initiated by Cavender and Felsenstein (1987).

We now use the case of four DNA sequences A, B, C and D to illustrate how the theory of linear invariants can be used to infer a phylogeny for four taxa. Since there are four possible types of nucleotides, there are 4^4=256 different configurations (combinations) of nucleotides at each site for four sequences. Lake (1987), however, pools these configurations into 36 configuration groups by the following rule. At a given site, any nucleotide of sequence A is denoted "1" and so are the nucleotides in the other sequences if they are the same as the one in sequence A; those related to the latter by a transitional change are represented by "2", and the two possible transversions are represented by "3" and "4", in the order we come across them, if at all. [Recall that exchange between A and G (purines) and those between T and C (pyrimidines) are transitions, whereas all other types are transversions.] With this notation, any configuration of four nucleotides can be represented by one of 36 four-dimensional vectors. For example, the four configurations AGCT, GCCA, ACTA, and GCTA are recorded as vectors 1234, 1332, 1341, and 1342, respectively.

Denote the three possible unrooted trees for four species by X, Y, and Z. For a set of four nucleotide sequences, let N_{ijmn} be the total score of the configuration ijmn's over the length of the sequences. Lake (1987) defines the following three quantities:

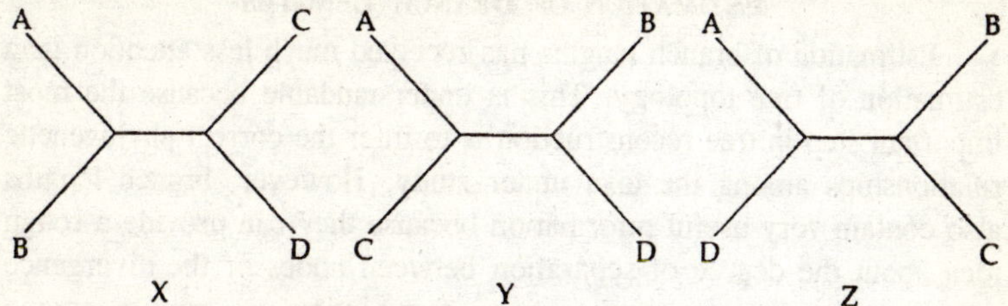

Fig. 4.14. Three alternative unrooted trees for four taxa.

$$X = N_{1133} + N_{1234} - N_{1233} - N_{1134}$$
$$Y = N_{1313} + N_{1324} - N_{1323} - N_{1314}$$
$$Z = N_{1331} + N_{1342} - N_{1332} - N_{1341}$$

Under a particular six-parameter model of nucleotide substitution, we have the following results: If tree X is the true tree, then Y and Z are invariants whereas X is not. If tree Y is the true tree, then X and Z are invariants whereas Y is not. If tree Z is the true tree, then X and Y are invariants whereas Z is not. Therefore, if one of X, Y and Z, say X, is significantly greater than 0 whereas the other two (i.e., Y and Z) are not, then tree X may be taken as the true tree. (However, if more than one of X, Y, and Z are significantly greater than 0, the problem is unresolved.)

Lake (1987) proposed to test the significance of X, say, as follows. Let $A=N_{1133}+N_{1234}$ and $B=N_{1233}+N_{1134}$. Then, $(A-B)^2/(A+B)$ is taken as a χ^2 with one degree of freedom; if A and B are small, one should test the equality of A and B by the binomial test. A tee is inferred if its associated χ^2 is significant whereas the other two χ^2's are not. Note, however, since three independent χ^2's instead of one are tested, the probability for one of the c^2's to be significant at the level of α can be up to $1-(1-\alpha)^3 \approx 3\alpha$; however, under the assumption that one of the three trees is the true tree, the probability of misidentifying the true tree is about 2α.

Computer Programs

A computer program can usually be obtained from the original authors. In addition, there are software packages available for phylogenetic analysis. PHYLIP contains computer programs for most of the methods discussed above.

PAUP contains an excellent program for the maximum parsimony method and programs for maximum likelihood and distance methods. MEGA contains programs for distance methods.

Estimation of Branch Lengths

Estimation of branch lengths has received much less attention than estimation of tree topology. This is understandable because the most important step in tree reconstruction is to infer the correct phylogenetic relationships among the taxa under study. However, branch lengths also contain very useful information because they can provide a rough idea about the degree of separation between nodes or the divergence time between lineages if the rates of evolution are similar among lineages. We shall discuss how to estimate branch lengths, assuming that the tree topology has already been inferred.

For the UPGMA method, the procedure for estimating branch lengths has already been presented together with the procedure for inferring the branching order. For the maximum parsimony method, we have also described the procedure for inferring the minimum number of substitutions in each branch. This procedure tends to underestimate the branch lengths, for it is intended to minimize the number of substitutions required. The degree of underestimation may not be severe if the tree contains only short branches; otherwise, the underestimation may become serious. For the maximum likelihood method, branch lengths and branching order are usually estimated together; since the computation is tedious, one usually uses computer programs. For distance matrix methods, branch lengths are usually estimated by Fitch and Margoliash's (1967) method or the (ordinary) least-squares method, which will be described below. It should be emphasized that the distances between sequences should be the distances corrected for multiple hits but not the observed distances, for the latter are likely to give underestimates of branch lengths.

We start with Fitch and Margoliash's (1967) method. First, let us consider the simplest case, i.e., an unrooted tree with three OTUs (A, B, and C) and a single node. Let x, y, and z be the lengths of the branches leading to A, B, and C, respectively. The following equations hold:

$$d_{AB} = x + y \quad ...(20a)$$

$$d_{AC} = x + z \quad ...(20b)$$

$$d_{BC} = y + z \quad ...(20c)$$

From these equations, we obtained the following solutions:

$$x = (d_{AB} + d_{AC} - d_{BC}) / 2 \quad ...(21a)$$

$$y = (d_{AB} + d_{BC} - d_{AC}) / 2 \quad ...(21b)$$

$$z = (d_{AC} + d_{BC} - d_{AB}) / 2 \quad ...(21c)$$

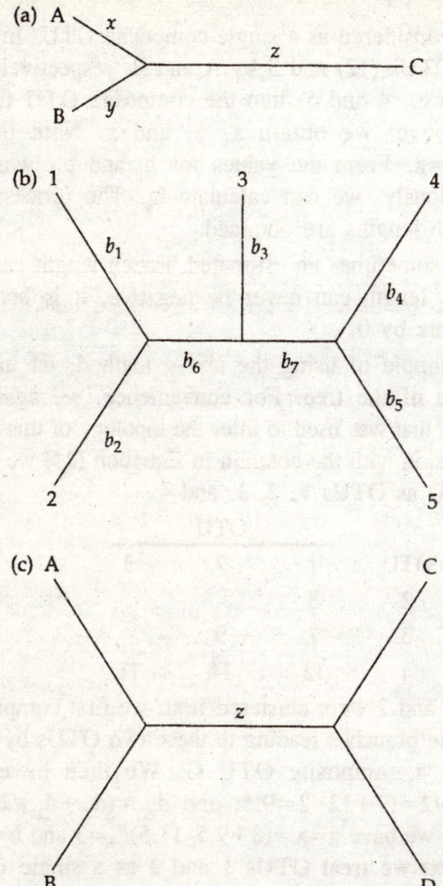

Fig. 4.15. Unrooted phylogenetic trees used to illustrat the computation of branch length by Fitch and Margoliash's and the least-squares methods: (a) a tree with three OTUs, (b) a tree with five OTUs, (c) a tree with four composite OTUs.

Let us now deal with the case of more than three OTUs. For simplicity, let us assume that there are five OTUs (1, 2, 3, 4, and 5) and that the topology and the branch lengths are as shown in Fig. 4.15b. Suppose that OTUs 1 and 2 are the first OTUs to be clustered together in the tree reconstruction process. We then use A and B to denote OTUs 1 and 2, respectively, and put all the other OTUs into a composition OTU denoted as C. By this arrangement, we can apply Equation (20a-c) to estimate the lengths of the branches leading to A, B, and C, except that now $d_{AC}=d_{1(345)}=(d_{13}+d_{14}+d_{15})/3$, and $d_{BC}=d_{2(345)}=(d_{23}+d_{24}+d_{25})/3$. Then we have $b_1=x$ and $b_2=y$. OTUs 1 and 2 are subsequently considered as a single composite OTU. In the next step, suppose that OTUs (12) and 3 by A and B, respectively, and put the other OTUs (i.e., 4 and 5) into the composite OTU C. In the same manner as above, we obtain x, y, and z. Note that $b_3=y$ and $b_6+(b_1+b_2)/2=x$. From the values for b_1 and b_2, which have been obtained previously, we can calculate b_6. The process is continued until all branch lengths are obtained.

Note that sometimes an estimated branch length can be negative. Since the true length can never be negative, it is better to replace such an estimate by 0.

As an example of using the above method, let us compute the branch lengths of the tree. For convenience, we again present the distance matrix that was used to infer the topology of this tree; however, to avoid confusion with the notation in Equation (21) we rename OTUs A, B, C and D as OTUs 1, 2, 3, and 4.

	OTU		
OTU	1	2	3
2	8		
3	7	9	
4	12	14	11

Since OTUs 1 and 2 were clustered first, we first compute the lengths (a and b) of the branches leading to these two OTUs by putting OTUs 3 and 4 into a composite OTU C. We then have $d_{AB}=d_{12}=8$, $d_{AC}=(d_{13}+d_{14})/2=(7+12)/2=9.5$, and $d_{BC}=(d_{23}+d_{24})/2=11.5$. From Equation (21), we have $a=x=(8+9.5–11.5)/2=3$ and $b=y=(8+11.5–9.5)/2=5$. Next we treat OTUs 1 and 2 as a single OTU (12) and denote it by A. Denote OTU 3 by B and OTU 4 by C. We then have $d_{AB}=d_{(12)3}=(d_{13}+d_{23})/2=(7+9)/2=8$, $d_{AC}=d_{(12)4}=(d_{14}+d_{24})/2=(12+14)/2=13$, and $d_{BC}=d_{34}=11$. Therefore, from Equation 21, we have $x=(8+13–11)/2=5$, and $d=y=(8+11–13)/2=3$, and $e=z=(13+11–8)/2=8$. We note that $(a+b)/2+c=x$, and so $c=1$. This completes the computation. Note, however, that since we do not know the exact location of the root, we cannot estimate the length of the branch connecting the root and OTU D but can only estimate the length from the common ancestral node of OTUs A, B, and C through the root to OTU D, i.e., $e=8$.

We now discuss the least-squares method. Let us use the tree as an example. For the least-squares method we set up the following equations.

$$
\begin{aligned}
d_{12} &= b_1 + b_2 + && && && e_{12} \\
d_{13} &= b_1 + && b_3 + && && b_6 + e_{13} \\
d_{14} &= b_1 + && && b_4 + && b_6 + b_7 + e_{14} \\
d_{15} &= b_1 + && && && b_5 + b_6 + b_7 + e_{15} \\
d_{23} &= && b_2 + b_3 + && && b_6 + e_{23} \\
d_{24} &= && b_2 + && b_4 + && b_6 + b_7 + e_{24} \\
d_{25} &= && b_2 + && && b_5 + b_6 + b_7 + e_{25} \\
d_{34} &= && && b_3 + b_4 + && b_7 + e_{34} \\
d_{35} &= && && b_3 + && b_5 + b_7 + e_{35} \\
d_{45} &= && && b_4 + b_5 + && + e_{45} \qquad \text{...(22)}
\end{aligned}
$$

where e_{ij}'s are sampling errors; e_{ij} has mean 0 and variance $V(d_{ij})$. The least-squares estimates of the b_i's are estimates $\hat{b}_i$ such that the following sum of squares is the smallest:

$$\sum_{i,j}\left(d_{ij} - \hat{d}_{ij}\right)^2$$

where $\hat{d}_{12} = \hat{b}_1 + \hat{b}_2, \hat{d}_{13} = \hat{b}_1 + \hat{b}_3 + \hat{b}_6$, and so on.

Rzhetsky and Nei (1993) have developed a simple computational procedure for implementing the least-squares estimators. Consider Fig. 4.15b as an example. If we choose one particular interior branch of this tree, the tree can be drawn in the form of Fig. 4.15c, where A, B, C, and D each represent a cluster of OTUs. For example, for branch b_6 in Fig. 4.15b, A=(1), B=(2), C=(3), and D=(4,5). In general, let d_{AB} be the distance between clusters A and B as defined by Equation (4); d_{AC}, d_{AD}, d_{BC}, d_{CD} are similarly defined. Then the least-squares estimator of the branch length z in Fig. 4.15c is given by

$$\hat{z} = \frac{1}{2}\left[\gamma(d_{AC} + d_{BD}) + (1-\gamma)(d_{BC} + d_{AD}) + d_{AB} - d_{CD}\right] \qquad \text{...(23)}$$

where $\gamma = (n_B n_C + n_A n_D)/[(n_A + n_B)(n_C + n_D)]$ and n_A, n_B, n_C and n_D are the number of OTUs in clusters A, B, C, and D, respectively. By contrast, the least-squares estimator of the length z of the exterior branch of the tree in Fig. 4.15a is given by Equation (21c), except that A, B, and C now may be composite OTUs and d_{AB}, d_{AC}, and d_{BC} are computed by using Equation (4).

Note that the Fitch-Margoliash method gives the least-squares estimates when the number of OTUs is five or smaller, but this may not be the case when the number of OTUs is six or larger.

Phenetics versus Cladistics

A long-standing controversy in taxonomy has been the often acrimonious dispute between *cladists* and *pheneticists*. The term *cladistics* can be defined as the study of the pathways of evolution. In other words, cladists are interested in such questions as: how many branches are there among a group of organisms; which branch connects to which other branch; and what is the branching sequence. A treelike network that expresses such ancestor-descendant relationships is called a *cladogram*. To put it another way, a cladogram refers to the topology of a rooted phylogenetic tree.

On the other hand, *phenetics* is the study of relationships among a group of organisms on the basis of the degree of similarity between them, be that similarity molecular, phenotypic, or anatomical. A treelike network expressing phenetic relationships is called a *phenogram*. While a phenogram may serve as an indicator of cladistic relationships, it is not necessarily identical with the cladogram. If there is a linear relationships, it is not necessarily identical with the cladogram. If there is a linear relationship between the time of divergence and the degree of genetic (or morphological) divergence, the two types of trees may become identical to each other.

Among the methods discussed above, the maximum parsimony method is a typical representative of the cladistic approach, whereas the UPGMA method is a typical phenetic method. The other methods, however, cannot be classified easily according to the above criteria. For example, the transformed-distance method and the neighbour-joining method have often been said to be phenetic methods, but this is not an accurate description. Although these methods use similarity (or dissimilarity, i.e., distance) measures, they do not assume a direct connection between similarity and evolutionary relationship, nor are they intended to infer phenetic relationships.

In molecular phylogenetics, a better classification of methods would be to distinguish between *distance* and *character-state approaches*. Method belonging to the former approach are based on distance measures, such as the number of nucleotide or amino acid substitutions, while methods belonging to the latter approach rely on the state of the character, such as the nucleotide or amino acid at a particular site, or the presence or absence of a deletion or an insertion at a certain DNA location. According to this classification, the first five methods described in this chapter are distance methods, while the maximum parsimony method is a character-state method. The method of linear invariants also uses character-state information and thus is also a character-state method. Likelihood methods use character-state information to form the likelihood function but search for branch lengths (or times) that maximize the likelihood for a given tree. Therefore, they cannot be easily classified into either of the two categories.

It has often been argued that character-state methods are more powerful than distance methods, because the raw data are a string of character states (e.g., the nucleotide sequence) and in transforming character-state data into distance matrices some information is lost. Note, however, that while the maximum parsimony method indeed uses the raw data, it usually uses only a small fraction of the available data. For instance, in the example in the section on maximum parsimony methods, only three of the nine sites are used. For this reason, this method is often less efficient than some distance matrix methods. Of course, if the number of informative sites is large, the maximum parsimony method is generally very effective. The maximum likelihood method uses the character-state information at all sites and thus can be said to use the "full" information; however, it requires assumptions on the model of nucleotide substitution. The next section gives a more detailed comparison of methods.

Comparison of Methods

Phylogenetic reconstruction has always been controversial and is likely to continue to be so far a long time, because the issue is complex and because there are differences in personal preference or philosophy. Since currently no method works well under all conditions or can be claimed to be better than all others, it is useful to know the strengths and weaknesses of different methods. We therefore consider the assumptions explicitly or implicitly involved in a method, the computational time, the statistical consistency of a method, and the relative performances of different methods under various conditions.

The performance of a method is usually studied by using either computer simulation or empirical tests. Empirical tests use cases where the evolutionary history is known. Fitch and Atchley (1985) and Atchley and Fitch (1991) used genetic data to test whether current methods or phylogenetic reconstruction can indeed recover the known relations among inbred strains of laboratory mice. Gouy and Li (1989) tested Lake's method of linear invariants, the maximum parsimony method, and the neighbour-joining method by using the small subunit rRNA sequences from human, *Drosophila*, rice, and *Physarum*. Hiliis et al. (1992, 1994) developed an ingenious experimental approach by taking advantage of the facts that viruses (e.g., bacteriophage T7) can be manipulated in the laboratory through thousands of generations per year and that mutation rates of viruses can be easily elevated through the use of mutagens. However, the number of empirical studies is still too limited to draw any general conclusion about the relative performances of different methods; as molecular evolution is stochastic, a large number of replications is needed for testing methods.

Assumptions and Computational Time

Few methods are completely explicit in assumptions about the pattern of and rate of nucleotide substitution; indeed, some methods (e.g., the maximum parsimony method) make no explicit assumption at all. Knowing the assumptions (explicitly or implicitly) stipulated in a method may help us understand its range of applicability and become cautious about an inferred tree when any of the assumptions appear to be violated; for testing the assumptions of a model or method. It should be emphasized that when a method makes no explicit assumption, it does not mean that the method works under all conditions. It should also be emphasized that a method requiring fewer assumptions may not necessarily perform better than one requiring more assumptions. The performance of a method depends not only on assumptions but also on the computational procedure or the optimality criterion used.

The UPGMA method implicitly assumes a constant rate for all branches. There is now strong evidence that this assumption is often violated. Moreover, if the sequences used are not long, some branches may by chance happen to have evolved faster than others, thus in effect violating the rate-constancy assumption. For this reason, an inferred UPGMA tree often involves errors in branching order (i.e., topological errors), even if the assumption of equal rates holds.

The transformed distance method, the neighbours-relation method, the neighbour-joining method and many other distance matrix methods do not assume rate-constancy. However, since all these methods assume that the effect of unequal rates among branches can be "corrected" from the distances in the distance matrix, their performance is affected by the accuracy of the distances estimated. When the distances are small, fairly accurate estimates of distances can be obtained, and these methods may perform well even under nonconstant rates of evolution, if the sequences used are long. Indeed, when the distances are small, one may even use the observed (i.e., uncorrected) distances. Note, however, if the sequences are short, then estimates of distances are subject to large statistical errors. Moreover, if some distances are large or if the rate varies greatly among sites, then accurate estimation of distances is difficult. Under any of these situations, the performance of a distance method can be compromised.

The maximum parsimony (MP) method makes no explicit assumptions, except that the parsimony criterion assumes that a tree that requires fewer substitutions is better than one that requires more. Note that a tree that minimizes the number of substitutions also minimizes the number of "homoplastic events," i.e., parallel, convergent, and back (reversal) substitutions. When the degree of divergence between sequences is small so that homoplasies are rare, the parsimony criterion usually work well. However, when the degree of divergence is large so that homoplasies are common, it may not work well. In particular, if some sequences have evolved much faster than others, then homoplastic events are likely to have occurred more often among these sequences than among others and the parsimony criterion may become misleading. Note also that the chance of homoplasy depends on the substitution pattern. For example, if transitional substitutions occur more often than transversional substitutions, then the chance of homoplasy will be higher than the case of random substitution among the four nucleotides. Therefore, the performance of the MP method is affected by unequal rates of evolution and by the pattern of nucleotide substitution, through it makes no explicit assumptions about either rates of evolution or the substitution model. Of course, some of these effects can be taken into account in a computer algorithm; e.g., transitional bias can be taken into account by assigning different weights to transitional and transversional changes, but this requires an assumption of the transition/transversion ratio.

The maximum likelihood (ML) method makes explicit assumptions about the rate of evolution and the pattern of nucleotide substitution. It has been commonly believed that this method is insensitive to violations of assumptions, but the simulation stud of Tateno et al. (1994) suggested that the ML method may not be very robust. One advantage of this method is that one may develop a computer program for it with options for choosing rates and patterns of substitution, so that it may be able to handle various situations. However, the computation may become very tedious and current computer packages may not have many user options for choosing substitution models and rates.

The method of linear invariants has become explicit in assumptions, owing to works by Cavender (1989, 1991), Jin and Nei (1990), Fu and Li (1992a, b), and others. For example, these authors showed that the most general conditions for the existence of linear invariants is a six-parameter model of nucleotide substitution. One advantage of linear invariants is that such invariants exist without the assumption of equal rates among lineages.

It has often been said that distance matrix methods such as the neighbour-joining (NJ) method use no optimality criterion. That is not quite true—although these pairwise clustering methods use no global criterion, they do use local criteria. For example, the neighbours-relation method uses the score of neighbours and the NJ method uses the S_{ij} value defined by Equation 8 as a criterion for clustering OTUs. Notice also that the minimum-evolution method does use a global criterion, though, like the maximum parsimony method, it may be too tedious to do an exhaustive search for a tree with the minimum total tree length.

We now consider the computational time. In the first four distance matrix methods described in this chapter, the OTUs are sequentially clustered to derive a final tree. For this reason, the computer algorithms for these methods are fast. In comparison, the minimum-evolution (ME) and maximum parsimony (MP) methods compare all possible alternative trees. This comparison is feasible only when the number of OTUs is small or moderate and the sequences under study are not long. For example, for 10 OTUs there are more than 2 million possible unrooted trees to be considered and so the computer time required becomes very large if the sequences are long. Thus, when the number of OTUs is large, say > 11, one does not do an exhaustive search. We mentioned earlier how this difficulty might be overcome in the case of the ME method. The MP method uses the branch-and-bound technique to reduce the number of potentially optimal trees to be considered or even uses a heuristic approach such as "stepwise addition" or "branch swapping," which does not guarantee obtaining the optimal tree. The maximum likelihood (ML) method is even more time-consuming because it considers all alternative trees and for each tree it searches for the maximum likelihood value. Thus, when the number of OTUs is large, it uses heuristic approaches to reduce the number of trees to be considered. The method of linear invariants is computationally much less tedious than the MP and ML methods but more tedious than distance matrix methods.

Consistency

In statistic an estimation method (estimator) is said to be consistent if it approaches the true value of he quantity to be estimated as more

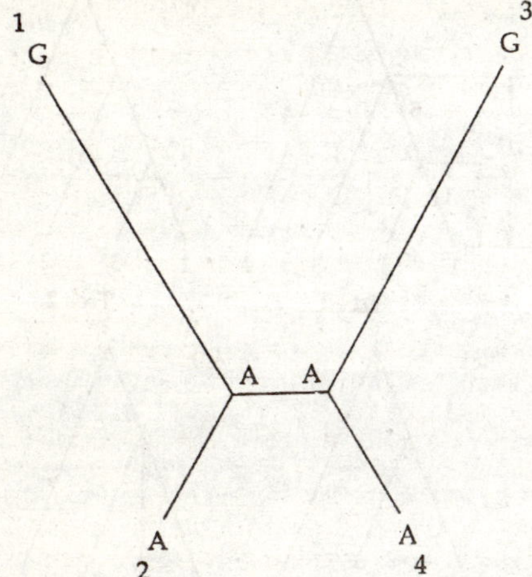

Fig. 4.16. A schematic illustration for parallel changes in the two long lineages 1 and 3.

than more data are accumulated. Phylogenetic reconstruction can be regarded as a type of statistical estimation—the "quantity" to be estimated is the true phylogeny. Obviously, a desired property of any tree reconstruction method is that it be consistent.

The fact that the maximum parsimony method can become inconsistent under unequal rates of evolution was first discovered by Felsenstein (1978). Suppose that branches 1 and 3 are long, whereas the other branches are short. Then, the nucleotides at the tips of lineages 2 and 4 are likely to remain the same as the nucleotide, say A, at their common ancestral node. On the other hand, the nucleotides at the tip of lineages 1 and 3 may be different from A and by chance both may become G (C or T). Under the parsimony criterion, such a site supports the erroneous grouping of lineages 1 and 3 in a cluster.

The above example provides an intuitive argument for how the MP method can become inconsistent. To have a better understanding of the conditions for inconsistency to occur, let us consider a more rigorous analysis using the model tree, in which OTU 4 is known to be an outgroup. Let p_1, p_2 and p_3 be the probabilities that a randomly chosen site will support trees 1, 2, 3, respectively. The MP method is consistent as long as p_1 is larger than both p_2 and p_3. This condition is violated only when substantial differences occur among the rates in lineages 1, 2 and 3. Let T_1, T_2 and T_3 be the divergence times. Assume that the substitution rate is u in all branches except that $u_1=wu$ in branch 1. Further, assume that $T_1=100$ million years (Myr), $T_2=50$ Myr, and $T_3=45$ Myr. For $u=10^{-8}$ substitutions per site per

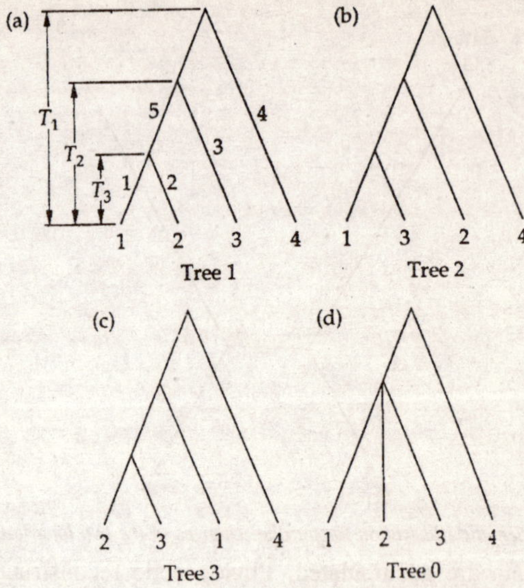

Fig. 4.17. Tree 1 is the model tree for the evolution of DNA sequences in four lineages. Trees 2 and 3 are the erroneous alternative trees and tree 0 is a trifurcating tree.

year, if $u_1 > 1.3u$ (i.e., $w > 1.3$), then $p_3 > p_1 > p_2$ and the MP method becomes inconsistent. In contrast, for $u = 10^{-9}$ the MP method becomes inconsistent only when $w > 4$. This example shows that the conditions for inconsistency to occur depend not only on rate differences but also on the degree of sequence divergence and that one way to avoid the inconsistency of the MP method is to use more slowly evolving sequences. Note that when w exceeds a critical value so that $p_3 > p_1 > p_2$, the data will tend to support tree 3 or, in other words, lineages 1 and 4, which are the longer edges, will tend to be clustered together in an unrooted tree. This tendency has been dubbed "long edges attract".

In Fig. 4.17a if the rate is the same in all branches except that $u_2 = wu$ for lineage 2, then the effect is the same as that in the above case except now that tree 2 instead of tree 3 is favoured when inconsistency occurs. Inconsistency of the MP method can also occur when the rate in lineage 3 evolves more slowly than lineages 1 and 2. However, it cannot occur if lineage 3 evolves faster than lineages 1 and 2, because lineages 3 and 4 will then be the longer edges.

Surprisingly, in contrast to the case of four taxa, in the case of five or more taxa the MP method can become inconsistent even under equal rates of evolution. Under some models of nucleotide substitution, the inconsistency problem of the MP method may be avoided by making certain nonlinear transformations of the nucleotide configuration patterns among sites, but it may be difficult to find such transformations in general.

Distance matrix methods are consistent under equal rates of evolution. When unequal rates occur, the UPGMA method often becomes inconsistent, whereas the transformed distance method, the neighbours-relation method, and the neighbour-joining method remain consistent if the distances are estimated accurately. The latter methods, however, may become inconsistent if distances are not estimated accurately. For example, in the case of four taxa if the observed distances instead of the distances corrected for multiple hits are used, then the conditions for the transformed distance method, the neighbours-relation method and the neighbour-joining method to become inconsistent are similar to those for the MP method. As discussed, accurate estimation of distances is difficult when the rate varies among sites or when the substitution pattern is complicated.

Maximum likelihood methods are commonly believed to be consistent. However, if the assumptions used to construct the likelihood function are unrealistic. ML methods may become inconsistent. For example, inconsistency may occur if the substitution model used is not realistic for sequences under study or if the rate of evolution is assumed to be uniform (either among sites or among lineages) but this is in fact not the case.

The method of linear invariants can become inconsistent if some of the necessary conditions for the existence of linear invariants are violated. For example, Lake's method can become inconsistent if the condition of balanced transversion is violated; "balanced transversion" means that the rates for the two types of transversion for a nucleotide are equal (e.g., $f_{AT}=f_{AC}$).

Simulation Studies

Computer simulation provides a convenient means for comparing the performances of different methods because many different conditions can be investigated and because the true tree (i.e., the model tree) is known, so that the probability for a method to recover the true tree can be evaluated. In contrast, in empirical studies the true tree is often unknown and there is usually little information about the conditions under which the molecules used have evolved. Moreover, in computer simulation a large number of replications can be conducted, a situation that is usually not feasible in empirical or experimental studies. Of course, one should be cautious about the direct applicability of simulation results to real-world situations because simulation studies usually make simplifying assumptions. However, if a method does not perform well under simple conditions, it is unlikely to do well in complex situations. Thus, simulation studies can help us understand the limitations of a method.

There is a wealth of literature on the simulation study of tree-making methods. Here we review some simulation studies that show the relative performances of different methods under certain conditions. The criterion for the performance of a method is the probability of recovering the correct tree. Since the simulation study of Saitou and Nei (1987) showed that the UPGMA method is inferior to the transformed distance method, the neighbours-relation method, and the neighbour-joining method even under the assumption of rate-constancy and that the neighbours-relation method and the neighbour-joining (NJ) method perform approximately equally well, the UPGMA and the neighbours-relation methods will not be considered in subsequent comparisons. For Lake's (1987) method of linear invariants, we note that this method was intended to deal with unequal rates of evolution. Simulation studies showed that when rates are highly unequal among lineages, the method of linear invariants is indeed superior to the MP method, provided that the necessary conditions for the existence of linear invariants hold. However, because it relies mainly on transversional substitutions and thus uses only a small fraction of the sequence data, it is inferior to the NJ method, if a proper distance measure is used for the latter. Indeed, it is even inferior to the MP method when the rates of substitution are nearly equal among lineages.

Let us first consider the case of a constant rate of evolution. Using tree, as a model tree, Sourdis and Nei (1988) simulated the evolution of a nucleotide sequence along each branch starting from a random sequence of nucleotides at the root of the tree. They used the one-parameter or Kimura's two-parameter model of nucleotide substitution to simulate the change of nucleotide at each site. For each simulation replicate they used the simulated nucleotide sequences at the tips of the model tree to reconstruct a tree by the maximum parsimony (MP) method, the neighbour-joining (NJ) method and the transformed distance (TD) method. For model tree a (i.e., a constant rate), the MP method is as good as or slightly better than the TD and NJ methods only when the number of nucleotide substitutions is small

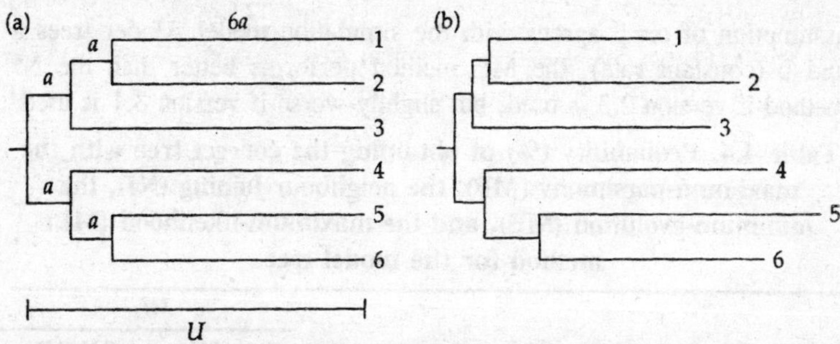

Fig. 4.18. Model trees. (a) Constant rate of nucleotide substitution. (b) Varying rates among lineages.

(U=0.05) and when the number of nucleotides in a sequence is L ≥1200. In all other cases, the TD and NJ methods perform better than the MP method.

Table 4.3. Probabilities (%) of obtaining the correct tree for three tree-making methods for model trees a and b.

	Constant rate (Tree a)			*Varying rate (Tree b)*		
Method	*300*	*600*	*1,200*	*300*	*600*	*1,200*
		U=0.05				
MP	46.3	79.0	97.0	16.7	39.3	58.0
NJ	58.7	82.0	95.3	30.0	46.7	63.3
TD	54.7	82.0	95.0	25.3	42.3	62.3
		U=0.5				
MP	44.7	69.3	84.0	16.7	22.3	31.0
NJ	62.7	84.3	97.0	25.0	34.7	49.3
TD	64.3	84.3	96.0	25.0	33.3	48.0

Saitou and Imanishi (1989) and Hasegawa et al. (1991) compared the MP method, the NJ method, the minimum-evolution (ME) method and the maximum likelihood (ML) method, using the model trees. Again, the MP method is less efficient than the NJ method. The performances of the ME and NJ methods are about the same. The performance of the ML method depends to some extent on the version of the DNAML algorithm of Felsenstein's PHYLIP package; version 2.3 uses the one-parameter model of nucleotide substitution [i.e., equal transition (α) and transversion (β) rates], which was the model used in the computer simulation, while version 3.1 assumes α=2β. In the present case, version 2.3 performs better than version 3.1 because the assumption of α=β agrees with the simulation model. Under trees a and b (constant rate), the ML method performs better than the NJ method if version 2.3 is used, but slightly worse if version 3.1 is used.

Table 4.4. Probability (%) of obtaining the correct tree with the maximum-parsimony (MP), the neighbour-joining (NJ), the minimum-evolution (ME), and the maximum-likelihood (ML) method for the model tree.

				ML	
Model tree	*NJ* *MP*	*ME* (α=β)	(α=β)	*DNAML version 2.3* (α=β)	*DNAML version 3.1* (α=2β)
Tree a (constant rate)					
8a=0.05					
300 bp	34	40	40	52	38
600 bp	76	82	80	88	80
8a=0.50					
300 bp	60	46	42	48	48
600 bp	84	82	82	78	70
Tree b (constant rate)					
8a=0.50					
300 bp	54	70	68	80	62
600 bp	84	86	80	90	88
8a=0.50					
300 bp	48	60	60	66	56
600 bp	58	70	70	80	76
Tree c (nonconstant rate)					
a=0.01					
300 bp	64	72	72	84	78
600 bp	90	92	92	96	98
a=0.05					
300 bp	24	68	68	92	92
600 bp	20	96	96	100	100
Tree d (nonconstant rate)					
a=0.01					
300 bp	68	74	74	86	80
600 bp	94	92	92	96	96
a=0.05					
300 bp	26	78	78	96	96
600 bp	46	100	100	100	100

In an extensive simulation study with Kimura's (1980) two-parameter model and with 10 OTUs, Kuhner and Felsenstein (1994) found that under a constant rate of evolution the MP method was slightly superior to the NJ method for the cases with short sequences and low rates or with long sequences and high rates, but inferior to the NJ method for the intermediate cases. The ML method was superior to both the NJ and MP methods in all cases considered.

We now consider nonconstant rates of substitution. For the model tree, the MP method is less efficient than the TD and NJ methods. Moreover, Table 4.4 shows that for model trees, the MP method is also less efficient than the NJ method, which is in turn less efficient than the ML method. Using a four-taxa tree, Hasegawa et al. (1991) showed that the ML method performed well under unequal rates. Therefore, the ML method seems to be quite robust against the effects of unequal rates. The same conclusions were reached in the above-mentioned study by Kuhner and Felsenstein (1994). However, if transitional substitution occurs several times more frequently than transversional substitution, then the NJ method performs somewhat better than the ML method.

Only a few studies of the effect of rate variation among sites on the performance of a method have been conducted, though this is an important situation. Kuhner and Felsenstein's (1994) study showed that the MP method performed less well than the NJ method, which was

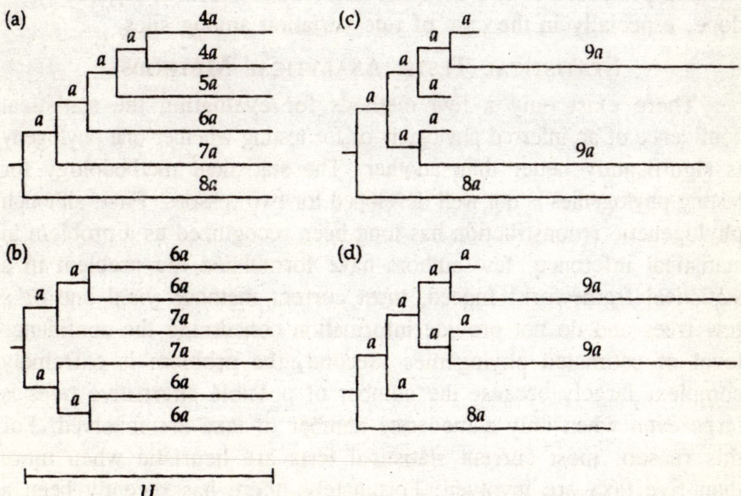

Fig. 4.19. Model trees used by Saitou and Imanishi's computer simulation.

in turn inferior to the ML method. In this study, however, unweighted parsimony (all changes were given equal weights) was used. Tateno et al. (1994) showed that weighting (i.e., transversions were given more weight than transitions and more conservative sites were given more weight than less conservative sites) improved the performance of the MP method, but it remained inferior to the NJ and ML methods. In Kuhner and Felsenstein (1994), the distances used in the NJ method were estimated without correction of rate heterogeneity. Jin and Nei (1990) and Tateno et al. (1994) showed that corrections for rate heterogeneity with the assumption of a gamma distribution of substitution rates can improve considerably the performance of the NJ method, even when the underlying distribution of substitution rates is not close to a gamma distribution; indeed, under certain conditions the performance of the NJ method may become better than the ML method. To date, no simulation study seems to have been conducted on the performance of the ML method with rate variation among sites; Yang (1993) has developed such a method for four taxa. Such a study would require a large amount of computer time, but should be of great interest.

In summary, simulation studies have shed much light on the relative performances of tree reconstruction methods. In general, under nonuniform rates either among lineages or among sites, the MP method tends to perform less well than the NJ method, while the ML method seems to be more robust than the MP and NJ methods. However, because performance is a complex issue, more research remains to be done, especially in the case of rate variation among sites.

Statistical Tests: Analytical Methods

There exist only a few methods for evaluating the statistical confidence of an inferred phylogeny or for testing whether one phylogeny is significantly better than another. The statistical methodology for testing phylogenies is not well developed for two reasons. First, although phylogenetic reconstruction has long been recognized as a problem in statistical inference, few authors have formulated the problem in a statistical framework. Indeed, most current methods yield one or a few trees and do not provide information concerning the confidence level of estimated phylogenies. Second, the problem is extremely complex, largely because the number of possible alternative trees is large even when only a moderate number of taxa are involved. For this reason, most current statistical tests are heuristic when more than five taxa are involved. Fortunately, there has recently been a strong trend to make phylogenetic reconstruction more statistical.

Statistical tests can be classified as analytical or resampling. Resampling methods (e.g., the bootstrap and the jackknife) resample the data to infer empirically the variability to the estimate obtained by a tree-making method. In phylogenetic study, bootstrap is the most popular resampling method. Analytical tests can be based on parsimony methods, distance methods, likelihood methods, or invariant methods. The likelihood approach is more involved and readers are referred to the review by Li and Gouy (1991) and the studies by Kishino et al. (1980) and Teteno et al. (1994). Invariant methods have been discussed earlier. We discuss below the other approaches.

Parsimony Tests

The analytic approach using parsimony tests was initiated by Cavender (1978). He studied the confidence limits that can be placed on a phylogeny inferred from a four-species data set. The results were extended by Felsenstein (1985b), who obtained narrower confidence limits by adding the assumption of a constant rate of evolution, i.e., an evolutionary clock. Felsenstein's results will be discussed below. We shall also discuss the case of more than four species.

Four species with a molecular clock

We assume that species 4 is a known outgroup, so that it can be used as a reference to infer the branching order of the other three species. Tree 1, 2, and 3 are the three possible rooted bifurcating trees. We assume that each randomly chosen informative site supports tree i with probability p_i, i=1, 2, or 3. Then the probability that among n random informative sites there are i sites supporting tree 1, j sites supporting tree 2 and k sites supporting tree 3 is trinomial, that is,

$$\text{Prob}\,(i, j, k) = \frac{n}{i!j!k!} p_1^i p_2^i p_3^k \qquad \text{...(24)}$$

For the moment let us assume that tree 1 is the true tree. Under the assumption of rate constancy, an informative site has a higher probability of supporting the true tree than supporting an incorrect tree, i.e., $p_1 \geq p_2 = p_3$. The probability that among n random informative sites the number (C) of sites supporting a particular incorrect tree is m or larger is given by

$$\text{Prob}\,(C \geq m) = \sum_{i=m}^{n} \frac{n!}{i!(n-i)} p_2^i (p_1 + p_2)^{n-i} \qquad \text{...(25)}$$

which is obtained by expanding $[p_2+(p_1+p_3)]^n$ with the condition $p_3=p_2$. Since $p_1+2p_2=1$ and since $p_2 \le p_1$, Formula 25 assumes the maximum value when $p_1=p_2=1/3$, i.e., if the three species represent a true trichotomy.

Now suppose that we do not know which tree is the true tree. Suppose further that in a sequence data set with n informative sites the best supported tree is favoured by m sites. Is this tree the true tree? Let us assume that this tree is incorrect and has by chance been supported by so many sites. The probability for a particular incorrect tree to be supported by m or more sites is given by Equation 25. Therefore, the probability for one of the two incorrect trees to be supported by m or more sites is

$$P = 2\,\text{Prob}\,(C \ge m) \qquad ...(26)$$

If P is smaller than α, then the support for the best tree is significant at the level of α. This is Felsenstein's (1985b) test.

In practice, p_1 is unknown and it is necessary to use the least favourable value, 1/3. A table for the test based on Equation 25 has been given by Li and Gouy (1991). If n is greater than 50, Equation 25 can be computed by assuming that m/n follows a normal distribution with mean equal to 1/3 and variance equal to $(1/3 \times 2/3)/n$.

In addition to the above test, Felsenstein (1985b) has also developed a statistic for testing whether the best supported tree is significantly better than the next best tree. Suppose that these two trees are supported by n_1 and n_2 sites, respectively. Let $s=n_1-n_2$. Then the probability that the best supported tree is at least s steps better is given by

$$\text{Prob}\,(S \ge s) = \sum_{(i,j,k)} \text{Prob}\,(i, j, k) \qquad ...(27)$$

where

$$j - \max\,(i, k) \ge s$$
$$\text{or } k - \max\,(i, j) \ge s$$

If Prob $(S \ge s)$ is smaller than α, then the best supported tree is significantly better than the next best tree at the α level. Again we do not know p_1 and so we do the worst-case analysis assuming $p_1=p_2=p_3=1/3$. Felsenstein (1988) gave a table of results computed from this analysis.

The test given by Formula 27 is computationally somewhat more complicated than that given by Formula 26, but the powers of the two tests are about the same. This result is not surprising because if the best supported tree is significant, it is likely to be significantly better than the next best tree (and also the worst tree).

Equation 26 is based on the assumption that the chance to have a truly trifurcating tree is negligible, so that one of the three bifurcating trees is the true tree. It is interesting to see if one can indeed reject the trifurcating tree. Under this null model we have $p_1=p_2=p_3=1/3$. So, testing the null hypothesis is equivalent to testing whether the trinomial distribution with $p_1=1/3$ can fit the observed data. If the expected number of sites supporting a bifurcating tree, which is equal to n/3 under the null model, is not too small, say ≥ 5, the following statistic

$$\chi^2 = \sum_{i=1}^{3} \frac{(n_i - n/3)^2}{n/3} \qquad \text{...(28)}$$

has a χ^2 distribution with 2 degree of freedom, where n_i is the observed number of sites supporting tree i, i=1, 2, 3. (The degree of freedom of the test is 2 because of the constraint $n_1+n_2+n_3=n$.) It should be noted that rejection of the null hypothesis does not necessarily imply significant support for one of the bifurcating trees; to draw this conclusion, one needs to conduct test (26) or (27). For example, if $n_1=12$, $n_2=10$, and $n_3=2$, then $\chi^2=15$ so that the null model can be rejected, but tree 1 is obviously not significantly better than tree 2.

The above formulation did not take into account the effect of unequal base compositions among sequences, but parsimony tends to group sequences together according to their base compositions. This problem has recently been treated by Steel et al. (1993), who proposed a frequency-dependent significance test (the FD test).

Although the above results were derived under the assumption that one of the four species is known to be an outgroup, they apply equally well in the absence of this knowledge if we consider unrooted trees and if the rate-constancy assumption holds.

More than four species

In theory the above analysis can be extended to the case of more than four species. However, the situation now becomes very complicated because the number of possible alternative trees increases rapidly with the number of species; for example, for five species there are 15 possible unrooted trees. Moreover, the above formulation is based on the worst-case analysis, so when the number of possible alternative trees is large, the test developed is likely to have a very low power.

In view of the complexity of the problem Kishino and Hasegawa (1989) developed a heuristic test in a manner similar to that of Templeton (1983). They considered whether two trees X and Y differ significantly in terms of the number of substitutions. Assume that all nucleotide sites are independent and equivalent. Let the minimum number of base substitutions at the ith information site for tree X be X_i and let

$$X = X_1 + X_2 + \ldots + X_n \qquad \ldots(29)$$

We define Y in the same manner. Let Z=Y–X. The mean of Z can be estimated by the inferred values of X and Y. The variance of Z can be estimated by the sample variance over sites, that is,

$$V(Z) = \frac{n}{n-1}\sum_{i=1}^{n}\left(Z_i - \frac{1}{n}\sum_{k=1}^{n} Z_k\right)^2 \qquad \ldots(30)$$

where $Z_i = X_i - Y_i$. When n is large, $[Z-E(Z)]/V(Z)^{1/2}$ approaches the standard normal distribution, and the standard normal test can be used to test whether Z is significantly different from 0. Note that a two-sided test should be used because by chance Y can be larger or smaller than X. If $|Z|$ is significantly greater than 0, then $Z > 0$ implies that tree X is better than tree Y where $Z < 0$ implies that tree Y is better than tree X. To take account of variation in base composition among sequences, Steel et al. (1993) proposed a modified test.

It should be noted that when one tree is significantly more parsimonious than another tree, it does not mean that it is the most parsimonious tree. To draw this conclusion one must show that the tree is more parsimonious than all of the other possible trees. In practice, however, when the number of possible trees is large, we cannot compare all of them and so instead we usually compare only those that seem to be most plausible. In this respect, one should note that two trees usually differ only in a small part of the whole topology, and in comparing two trees we implicitly assume that the other part of the tree topology is correct. Since this assumption may not be true, it can complicate the test.

Distance Tests

Two types of test are available for testing the significance of a phylogeny inferred from a distance matrix method. One approach is to test the significance of estimated internodal distances; if all internodal distances are significantly larger than 0, the inferred tree may be considered significant. Li (1989) and Tajima (1992) have developed methods for this test. The other approach is based on the minimum-evolution criterion: if the total length of a tree is significantly shorter than every other alternative tree, it is considered significant. This approach appears to be superior to the first test and will be discussed below.

Under the minimum-evolution (ME) criterion, tree A is significantly better than tree B if the sum (S_A) of branch-length estimates for tree A is significantly smaller than that (S_B) for tree B. The test proposed by Rzhetsky and Nei (1992) is as follows. The sum (S) of branch lengths for a tree is estimated from the pairwise distances between taxa. Let us use tree A in Fig. 4.20 as an example. Let b_i be the least-squares of b_i. Then, by definition

$$S_A = \hat{b}_1 + \cdots + \hat{b}_7$$

In terms of d_{ij}, we have

$$S_A = d_{12}/2 + d_{13}/4 + d_{14}/8 + d_{15}/8 + d_{23}/4 + d_{24}/8 + d_{25}/8 + d_{34}/4 + d_{35}/4 + d_{45}/2 \quad ...(31)$$

Similarly, for tree B we have

$$S_B = d_{12}/4 + d_{13}/2 + d_{14}/8 + d_{15}/8 + d_{23}/4 + d_{24}/4 + d_{25}/4 + d_{34}/8 + d_{35}/8 + d_{45}/2 \quad ...(32)$$

Therefore,

$$D = S_B - S_A = -d_{12}/4 + d_{13}/4 + d_{24}/8 + d_{25}/8 - d_{34}/8 - d_{35}/8 \quad ...(33)$$

The null hypothesis is D=0 and a two-sided test should be used because D can be greater or smaller than 0. If $|D|$ is significantly greater than 0 and if S_A is smaller than S_B, then tree A is better than tree B. The variance of D is given in Rzhetsky and Nei (1992) and can be used to test the significance of D.

D can also be written as

$$D = \frac{1}{2}b_6 - \frac{1}{8}(2e_{12} - 2e_{13} - e_{24} - e_{25} + e_{34} + e_{35}) \quad ...(34)$$

where e_{ij} is the sampling error of d_{ij} as defined in Equation 5.22. Therefore, testing D=0 is equivalent to testing b_6=0 and if the branch length b_6 is significantly greater than 0, then tree A is better than tree B. Similarly, for the comparison of tree A and C in Fig. 4.20, we have

$$S_C - S_A = \frac{3}{4}(b_6 + b_7) - \frac{3}{8}(e_{12} - e_{14} - e_{25} + e_{45}) \quad ...(35)$$

Therefore, we are testing the null hypothesis that both b_6 and b_7 in tree A are equal to 0.

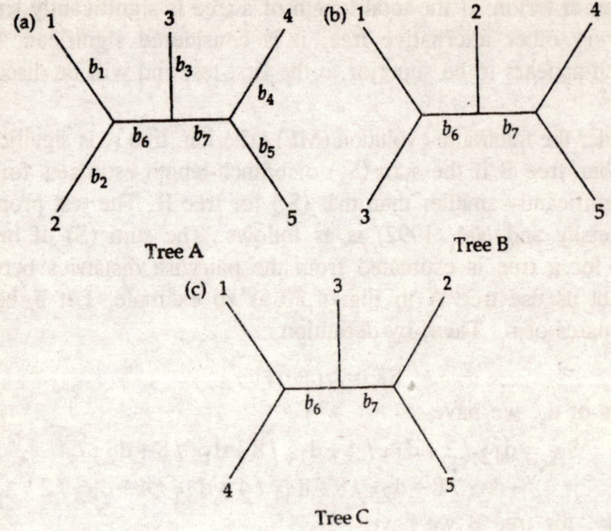

Fig. 4.20. Model trees illustrating computation of branch lengths and S values.

The above principle applies to any pair of bifurcating trees, and if a tree is found to have an S value significantly smaller than that for any other tree, it can be taken as the best tree. However, when the number of taxa is not small, it is impractical to compare all the trees. As mentioned earlier, Rzhetsky and Nei (1992) proposed a fast approximate search for the ME tree. Their search scheme is based on the observation that the neighbour-joining (NJ) tree is almost always identical with the ME tree. Briefly their procedure is as follows: (1) Construct an NJ tree by using Saitou and Nei's (1987) method. (2) Obtain all trees that differ from the NJ tree by only one or two topological differences. (3) Estimate S for each of these topologies and compute $D=S-S_{NJ}$ and the standard error of D for each tree, where S_{NJ} is the S value for the NJ tree. (4) If D is significantly greater than 0 for each tree, adopt the NJ tee as the most probable tree. However, if there are alternative trees whose S values are not significantly different from S_{NJ}, they should be regarded as good as the NJ tree. Furthermore, if there is any tree whose S is significantly smaller than S_{NJ}, it should be considered better than the NJ tree.

STATISTICAL TESTS: THE BOOTSTRAP

In statistics, the bootstrap is a computational technique for estimating statistics or parameters when the distribution is difficult to derive analytically. Since its introduction into phylogenetic study by

Felsenstein (1985a), the bootstrap technique has been frequently used as a means to estimate the confidence level of phylogenetic hypotheses. The statistical properties of this technique in the context of phylogenetic study appear to be complex, but recent studies have led to a better understanding of the technique. For an explanation of the technique under simple models.

Procedure

To explain the technique, let us use the simple example shown in Fig. 4.21; OTU 5 is assumed to be a known outgroup. In principle, one should specify the null hypothesis before taking or analyzing data, and we assume this has been done. Note that the null hypothesis as depicted by the tree consists of two subhypotheses: (1) OTUs 1 and 2 belong to one clade and (2) OTUs 3 and 4 belong to another clade. Now suppose that one DNA sequence of length N is obtained from each OTU and that the sequences are as follows:

Taxa	Sequence
1	GCAGTACT...
2	GTAGTACT...
3	ACAATACC...
4	ACAACACT...
5	GCGGCATT...

To estimate the confidence levels of subhypotheses 1 and 2, we do bootstrapping on the sample sequences to generate N_b pseudosamples. Note that sampling is with replacement, i.e., a site sampled can be sampled again with the same probability as can each of all other sites. Each pseudosample is generated as follows. First, a site is

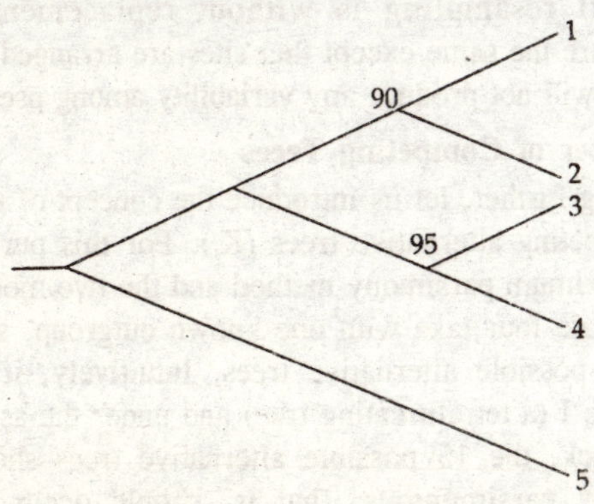

Fig. 4.21. Hypothetical tree for illustrating the bootstrap technique.

randomly chosen from the original sample; say it is the sixth site, and this site is placed as the first site in the following figure.

A pseudosample:

1	AGATACTC...
2	AGATGCTT...
3	AAACACTC...
4	AAACACCC...
5	AGACATCC..

Next, another site is randomly chosen from the *original* sample; say, it is the first site, and this site is placed in the second position in the pseudosample. (Note that the sixth site has now been chosen twice.) This sampling procedure is continued until N sites are chosen. This pseudosample is then used to construct a tree and the subset (1,2) is given a score of 1 if OTUs 1 and 2 appear in one group, but a score of 0 otherwise; the score for the subset (3,4) is similarly decided. This completes one bootstrap replicate. The above procedure is repeated N_b times, and the proportion of bootstrap replicates in which the subset (1,2) appeared is taken as the bootstrap value supporting this subset; the bootstrap supporting the subset (3,4) is similarly obtained. These bootstrap proportions are indicated at the nodes of the tree and are usually taken as the confidence levels for the subsets.

In essence, the bootstrap technique assumes that the original sample can represent the original sample space, so that each pseudosample can be treated as a real sample, i.e., a sample from the original sample space. Obviously, for this assumption to hold well, the original sample has to be large enough. Note also that resampling is with replacement; if resampling is without replacement, then all pseudosamples are the same except that sites are arranged in different orders, and this will not produce any variability among pseudosamples.

Effective Number of Competing Trees

Before going further, let us introduce the concept of the effective number of competing alternative trees (K_e). For this purpose, let us consider the maximum parsimony method and the two model trees. In each tree there are four taxa with one known outgroup, so that there are in total 15 possible alternative trees. Intuitively, it seems that under model tree I (a tetrafurcating tree) and under the assumption of a molecular clock, the 15 possible alternative trees should be, on average, equally parsimonious, that is, should occur with equal probability, but this is not the case. In fact, each of the 12

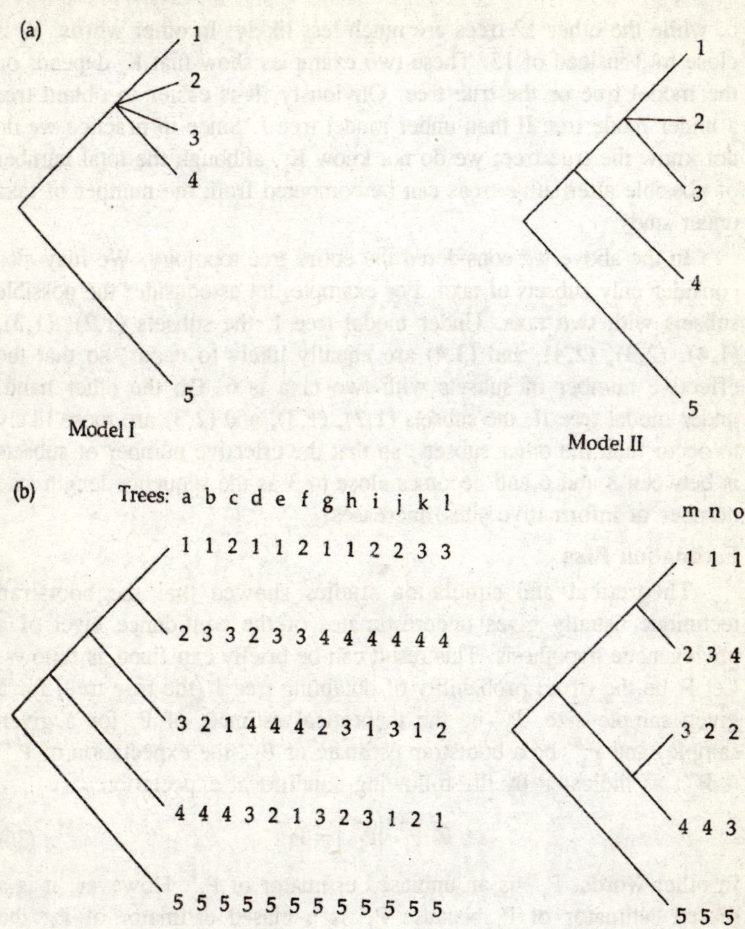

Fig. 4.22. (a) Two model trees for four taxa with one known outgroup. (b) Fifteen possible alternative trees.

asymmetrical bifurcating trees (trees a to l) is more likely to be inferred (i.e., more parsimonious) than each of the three symmetrical bifurcating trees (trees m to o). Indeed, when the sequence length, L, is large, the probability of inferring a symmetrical tree becomes very small, and hence the probability of each of the 12 asymmetrical trees becomes close to 1/12 instead of 1/15. In other words, the effective number of competing alternative topologies (K_e) is close to 12 rather than 15 (it may not be an integer). Under model tree II, taxa 4 branched off much earlier than the other three taxa, which form a trichotomy. In this case the major competing trees are trees a, b, and c, while the other 12 trees are much less likely. In other words, K_e is close to 3 instead of 15. These two examples show that K_e depends on the model tree or the true tree. Obviously, it is easier to obtain tree a under mode tree II than under model tree I. Since in practice we do not know the true tree; we do not know K_e, although the total number of possible alternative trees can be computed from the number of taxa under study.

In the above we considered the entire tree topology. We may also consider only subsets of taxa. For example, let us consider the possible subsets with two taxa. Under model tree I, the subsets (1,2), (1,3), (1,4), (2,3), (2,4), and (3,4) are equally likely to occur, so that the effective number of subsets with two taxa is 6. On the other hand, under model tree II, the subsets (1,2), (1,3), and (2,3) are more likely to occur than the other subsets, so that the effective number of subsets is between 3 and 6 and becomes close to 3 as the sequence length (the number of informative sites) increases.

Estimation Bias

Theoretical and simulation studies showed that the bootstrap technique usually gives underestimates of the confidence level of a phylogenetic hypothesis. This result can be briefly explained as follows. Let P_1 be the (true) probability of obtaining tree 1 (the true tree) for a given sample size, P_1^* be the theoretical estimate of P_1 for a given sample, and P_1^{**} be a bootstrap estimate of P_1^*, the expectation of P_1^{**} is P_1^*, as indicated by the following conditional expectation:

$$E\left(P_1^{**} \middle| P_1^*\right) = P_1^* \quad ...(36)$$

In other words, P_1^{**} is an unbiased estimator of P_1^*. However, it is a biased estimator of P_1 because P_1^* is a biased estimator of P_1; the bias arises because P_1^* is a nonlinear function of the proportions of informative sites. Note also that bootstrapping introduces an additional variance, i.e.,

$$V\left(P_1^{**} \middle| P_1^*\right) = P_1^*\left(1 - P_1^*\right) / N_b \quad ...(37)$$

The error in the estimation of P_1^{**} by bootstrapping can be reduced by increasing the number of bootstrap replications N_b. For example, if $P_1^*=0.95$ and $N_b=400$, then the standard error introduced by bootstrapping is 0.02, or the 95% confidence interval of P_1^* is approximately (0.93, 0.97). Thus, $N_b \geq 400$ is recommended in general for bootstrap estimation. Note, however, that the total variance of the estimate of P_1 by P_1^{**} contains an additional term $V(P_1^*)(1-1N_b)$, which is almost independent of N_b and cannot be reduced to 0 by increasing N_b.

The degree of underestimation of P_1 by P_1^{**} increases with the effective number of competing trees K_e. To see this, let us consider a simple artificial model, called the multinomial model of K alternatives. It assumes that there are K types of informative site, each of which supports one and only one topology. Let p_i be the probability of occurrence of the ith type site ($i=1,2, \ldots, K$). Then the distribution of the K types of informative site in a sample of N sites follows the multinomial distribution. For simplicity, let us assume that N is large, so that the chance of having more than one parsimonious tree is negligible. We assume further that $p_1 \geq p_2 = \ldots = p_K$. Then, P_1 increases from 1/K to 1 as the difference $p_1 - p_2$ increases from 0 to 1. Table 4.5 shows that the expected bootstrap value P_1^{**} is smaller than P_1 and the difference increases with K. This means that the underestimation of P_1 by bootstrapping tends to be more serious as K increases. Thus, if we are considering the probability (P_1) of obtaining the subset (1,2), then bootstrapping tends to underestimate P_1 more seriously under model tree I than under model tree II, because K_e is 6 for the former model but only between 3 and 6 for the latter.

Table 4.5. Relationship between P_1 and P_1^* (P_1^{**}) as a function of the number of alternatives (K) in the multinomial model

P_1	P_1^*			
	K=2	*K=3*	*K=5*	*K=10*
0.50	0.50	0.45	0.40	0.35
0.60	0.57	0.48	0.47	0.42
0.70	0.65	0.60	0.55	0.50
0.80	0.72	0.68	0.64	0.59
0.90	0.82	0.78	0.75	0.70
0.95	0.88	0.85	0.82	0.79
0.99	0.95	0.94	0.92	0.90

Complete-and-Partial Bootstrap Technique

To correct the bias in bootstrap estimation, Zharkikh and Li (1995) developed the *complete-and-partial (CP) bootstrap technique*. For simplicity, let us consider the multinomial model described above. Suppose that we want to estimate P_1. Then, we first obtain an estimate (P_1^{**}) by the standard bootstrap technique, which is called complete bootstrapping because the size of each pseudosample is the same as the original sample size N. Next, we obtain a second estimate (P_{1r}^{**}) by taking pseudosamples of smaller size, say $N_r=N/r$, where $r>1$; this second procedure is called partial bootstrapping. From these two estimates, an unbiased estimate of P_1 can be obtained. The reason for requiring two bootstrap estimates is because P_1 depends on two unknowns. K and the difference p_1 - max (p_2, ..., p_k), and a single bootstrap value is not sufficient for solving the two unknowns. Bias corrections can also be made by the iterated bootstrap technique, which is, however, more tedious than the CP bootstrap technique.

In the above we assumed that the null hypothesis is specified in advance. In phylogenetic study, however, the null hypothesis is usually not prespecified. The lack of prespecification can lead to overestimation of the confidence level. This effect can also be taken into account by the CP bootstrap technique. A computer program is available for the CP bootstrap technique with either the maximum parsimony method or the neighbour-joining method.

It should be emphasized that the bootstrap technique assumes that the sample is representative of the sample space. Thus, the technique is not recommended when the sample size is small. Note also that the current CP bootstrap technique has been developed under the assumption that the method is statistically consistent and that the sites on the DNA sequences are independently and identically distributed, i.e., all sites are independent of one another and evolve at the same rate. Since violations of these conditions are likely to occur, the results obtained by the current CP bootstrap technique should be taken with caution. How to remove the assumptions of independence and equivalence of sites is a subject of great interest for future research.

5

EXAMPLES OF MOLECULAR PHYLOGENY

Application of molecular biology techniques and advances in tree reconstruction methodology have led to tremendous progress in phylogenetic studies in the last two decades, resulting in a better understanding of the evolutionary history of almost every taxonomic group. This chapter presents several examples where molecular studies have resolved a long-standing issue, led to a drastic revision of a traditional view, or pointed to a new direction of research. As the field of molecular phylogenetics is progressing rapidly, some of the views presented here may soon be revised.

MAN'S CLOSEST RELATIVES

The question "Who are man's closest evolutionary relatives?" has always intrigued biologists. Darwin (1871), for instance, claimed that the African apes, the chimpanzee (*Pan*) and the gorilla (*Gorilla*), are man's closest relatives, and hence, he suggested that man's evolutionary origins were to be found in Africa. Darwin's view fell into disfavour for various reasons, and for a long time taxonomists believed that the genus *Homo* was only distantly related to the apes, and thus *Homo* was given a family of its own, Hominidae. Chimpanzees, gorillas, and orangutans (*Pongo*) were usually placed in a separate family, the Pongidae. The gibbons (*Hylobates*) were either classified separately or with the Pongidae. Goodman (1963) correctly recognized that this systematic arrangement is anthropocentric, because humans represent "a new grade of phylogenetic development, one which is 'higher' than the pongids and all other preceding grades." Indeed, placing the various

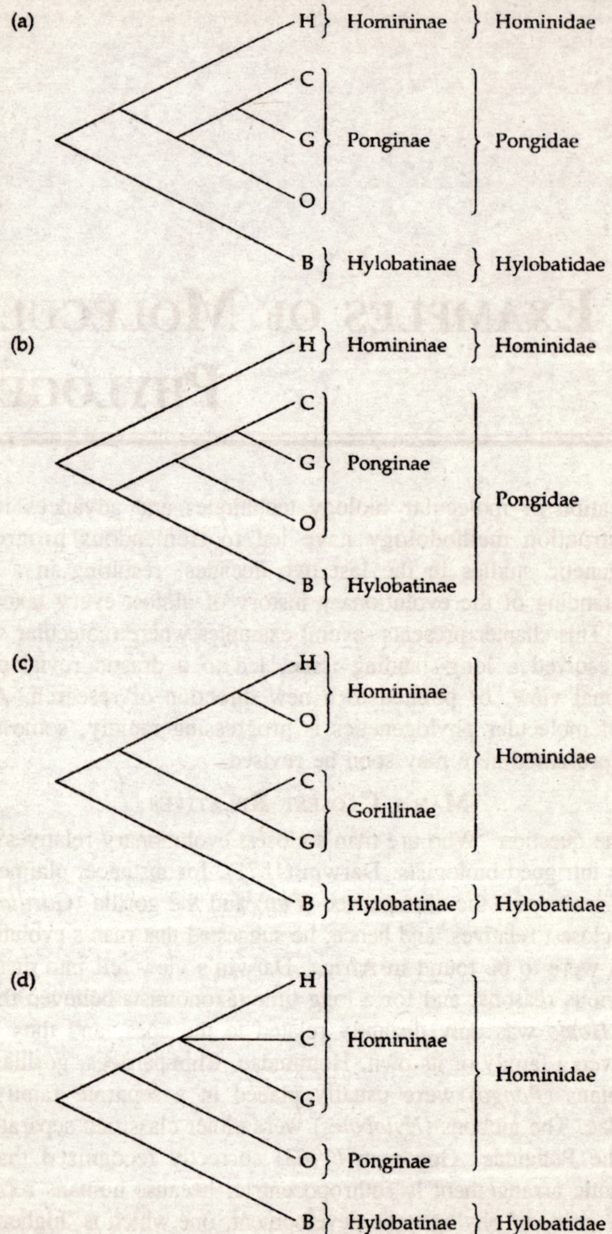

Fig. 5.1. Four alternative phylogenies and classifications of modern apes and humans (Hominoidea)

apes in one family and humans in another implies that the apes share a more recent common ancestry with each other than with humans. When *Homo* was put in the same clade with a living ape, it was usually with the Asian ape, the orangutan.

By using a serological precipitation method, Goodman (1962) was able to demonstrate that humans, chimpanzees, and gorillas constitute a natural clade, with orangutans and gibbons having diverged from the other apes at much earlier dates. From microcomplement fixation data, Sarich and Wilson (1967) estimated the divergence time between humans and chimpanzees or gorillas to be as recent as 5 million years ago, rather thana minimum date of 15 million years ago, as was commonly accepted by paleontologists at that time.

However, serological studies, electrophoretic studies, and amino acid sequences could not resolve the evolutionary relationships among humans and the African apes, and the so-called human-gorilla-chimpanzee trichotomy remained unsolved and continued to be an extremely controversial issue. Early studies of mitochondrial DNA by either restriction or sequence analysis slightly favoured a closer relation of chimpanzee and gorilla, whereas the DNA-DNA hybridization data of Sibley and Ahlquist (1984) supported the human-chimpanzee clade. The first large set of DNA sequence data for resolving the controversy was obtained by M. Goodman and colleagues. In the following discussion, this data set and that of Maeda et al. (1988) will be used to illustrate some tree-making methods. Further data concerning the issue will be discussed later.

Table 5.1 shows the number of nucleotide substitutions per 100 sites between each pair of the following OTUs: humans (H), chimpanzees (C), gorillas (G), orangutans (O), and rhesus monkeys (R). Let us first apply the UPGMA method to these distances. The distance between

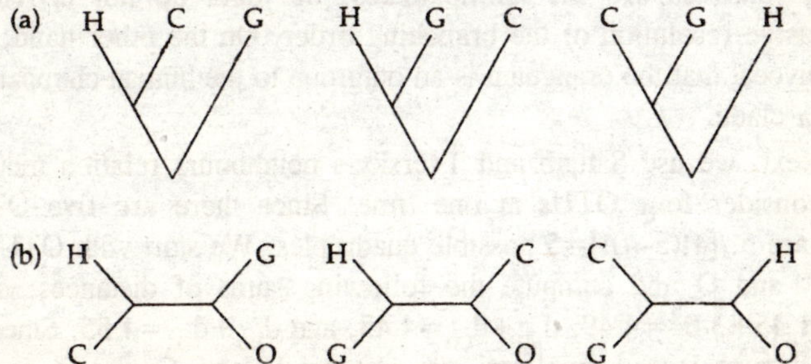

Fig. 5.2. (a) Three possible rooted tree for gorillas, chimpanzees, and humans. (b) Comparable unrooted trees with the orangutan as an outgroup.

humans and chimpanzees is the shortest (d_{HC}=1.45). Therefore, we join these two OTUs first and place the node at 1.45/2=0.73. Following the procedure (Methods of Tree Reconstruction), we compute the distances between the composite OTU (HC) and each of the other species, and obtain a new distance matrix"

	OTU		
OTU	(HC)	G	O
G	1.54		
O	2.96	3.04	
R	7.53	7.39	7.10

Table 5.1. Mean (below diagonal) and standard error (above diagonal) of the number of nucleotide substitutions per 100 sites between OTUs

	OTU				
OTU	*Human*	*Chimpanzee*	*Gorilla*	*Orangutan*	*Rhesus monkey*
Human		0.17	0.18	0.25	0.41
Chimpanzee	1.45		0.18	0.25	0.42
Gorilla	1.51	1.57		0.26	0.41
Orangutan	2.98	2.94	3.04		0.40
Rhesus monkey	7.51	7.55	7.39	7.10	

Since (HC) and G are separated by the shortest distance, they are the next to be joined together, and the connecting node is placed at 1.54/2=0.77. Continuing the process, we obtain the tree. We note that the estimated branching node for H and C is very close to that for (HC) and G. In act, the distance between the two nodes is smaller than all the standard errors for the estimates of the pairwise distances among H, C, and G. Thus, although the data suggest that man's closet living relatives are the chimpanzees, the data do not provide a conclusive resolution of the branching order. On the other hand, it is unequivocal that the orangutan is an outgroup to the human-chimpanzee-gorilla clade.

Next, we use Sattath and Tversky's neighbours relation method. We consider four OTUs at one time. Since there are five OTUs, there are 5!/[4!(5–4)!]=5 possible quadruples. We start with OTUs H, C, G, and O and compute the following sums of distances: $d_{HC}+d_{GO}$=1.45+3.04=4.49, $d_{HG}+d_{CO}$=4.45, and $d_{HO}+d_{CG}$=4.55. Since the second sum is the smallest, we choose H and G a one pair of neighbours and C and O as the other. Similarly, we consider the four

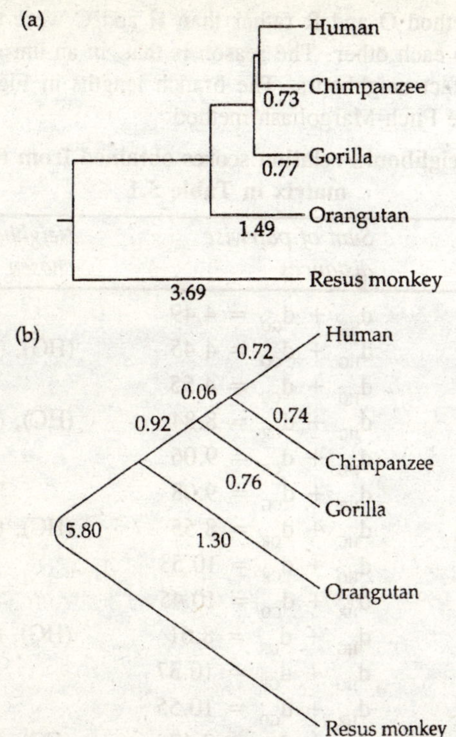

Fig. 5.3. Phylogenetic tree for humans, chimpanzees, gorillas, orangutans, and rhesus monkeys inferred (a) from the UPGMA method and (b) from Sattath and Tversky's method.

other possible quadruples. The results are shown in Table 5.2. Noting from the bottom of that table that (OR) as the first pair of neighbours. Treating this pair as a single OTU, we obtain the following new distance matrix:

	OTU		
OTU	H	C	G
C	1.45		
G	1.51	1.57	
(OR)	5.25	5.25	5.22

As only four OTUs are left, it is easy to see that $d_{HC}+d_{G(OR)} = 6.67 < d_{HG}+d_{C(OR)} = 6.76 < d_{H(OR)}+d_{CG} = 6.82$. Therefore, we choose H and C as one pair of neighbours and choose G and (OR) as the other. The final tree obtained by this method is shown in Fig.5.3b. The topology of this tree is identical to that in Fig. 5.3a Note, however, that in this method O and R rather than H and C were the first pair to be joined to each other. The reason is that, in an unrooted tree, O and R are in fact neighbours. The branch lengths in Fig. 5.3b were obtained by the Fitch-Margoliash method.

Table 5.2. Neighbours-relation scores obtained from the distance matrix in Table 5.1

OTUs compared	*Sum of pairwise distances*	*Neighbour pairs chosen*
H, C, G, O	$d_{HC} + d_{GO} = 4.49$	
	$d_{HG} + d_{CO} = 4.45$	(HG), (CO)
	$d_{HO} + d_{CG} = 4.55$	
H, C, G, R	$d_{HC} + d_{GR} = 8.84$	(HC), (GR)
	$d_{HG} + d_{CR} = 9.06$	
	$d_{HR} + d_{CG} = 9.08$	
H, C, O, R	$d_{HC} + d_{OR} = 8.55$	(HC), (OR)
	$d_{HO} + d_{CR} = 10.53$	
	$d_{HR} + d_{CO} = 10.45$	
H, G, O, R	$d_{HG} + d_{OR} = 8.61$	(HG), (OR)
	$d_{HO} + d_{GR} = 10.37$	
	$d_{HR} + d_{GO} = 10.55$	
C, G, O, R	$d_{CG} + d_{OR} = 8.67$	(CG), (OR)
	$d_{CO} + d_{GR} = 10.33$	
	$d_{CR} + d_{GO} = 11.59$	

Total scores: (HC)=2, (HG)=2, (HO)=0, (HR)=0, (CG)=1, (CO)=1, (CR)=0, (GO)=0, (GR)=1, (OR)=3

Third, we use the neighbour-joining method. It is simpler to use Studier and Keppler's (1988) algorithm than Saitou and Nei's (1987) algorithm. From equation, the Q value for the pair of OTUs i and j is

$$Q_{ij} = (N-2)d_{ij} - \sum_{k=1}^{N} d_{1k} - \sum_{k=1}^{N} d_{2k} \quad \text{...(1)}$$

In the present case, N=5, and for OTUs 1 and 2 we have

$$Q_{12} = 3d_{12} - (d_{12} + d_{13} + d_{14} + d_{15}) - (d_{12} + d_{23} + d_{24} + d_{25})$$

With the d_{ij} values in Table 5.1, we have Q_{12}=–2261. In a similar manner, we obtain the other Q value under step 1 in Table 5.3. Since Q_{45} is the smallest, we take OTUs 4 and 5 (i.e., orangutan and rhesus monkey) as the first pair of neighbours. This is the same as in the case of the neighbour-relation method, and when OTUs 4 and 5 are regarded as a composite OTU, the new distance matrix is the same as the preceding matrix. Using the new distance matrix and the same procedure as above, we obtain the Q values under step 2 in Table 5.3. Since Q_{12} and $Q_{3(45)}$ are the smallest, we choose OTUs 1 and 2 as one pair of neighbours. Thus, the final tree obtained by this method is exactly the same as that by the neighbour-relation method.

Table 5.3. The Q values for the distance matrix in Table 5.1.

Step 1	Step 2
Q_{12}=-2261	Q_{12}=-1357
Q_{13}=-2243	Q_{13}=-1348
Q_{14}=-2057	$Q_{1(45)}$=-342
Q_{15}=-2047	Q_{23}=-1342
Q_{23}=-2231	$Q_{2(45)}$=-1348
Q_{24}=-2075	$Q_{3(45)}$=-1357
Q_{25}=-2041	
Q_{34}=-2045	
Q_{35}=-2089	
Q_{45}=-2431	

Finally, let us consider the maximum parsimony method. For simplicity, let us consider only humans, chimpanzees, gorillas, and orangutans. Table 5.4 shows the informative sites for the 10.2-kb region including the η-globin pseudogene and its surrounding regions. For each site, the hypothesis supported is given in the last column. If we consider base changes only, then there are 15 informative sites, of which eight support the human-chimpanzee clade (hypothesis I), four support the chimpanzee-gorilla clade (hypothesis II), and three support the human-gorilla clade (hypothesis III). In addition, all four informative sites involving a gap support the human-chimpanzee clade. Therefore, the human-chimpanzee clade is chosen as the best representation of the true phylogeny. In another analysis with more sequence data, Williams and Goodman (1989) showed that support for the human-chimpanzee clade is statistically significant at the 1% level.

The clustering of humans and chimpanzees in one clade is, however, not supported by the repeat pattern in the involucrin gene, which favours instead the chimpanzee-gorilla clade. Nevertheless, the overall molecular evidence is now strongly in favour of the human-chimpanzee

Table 5.4. Informative sites among human, chimpanzee, gorilla, and orangutan sequences

Site	*Human*	*Chimpanzee*	*Gorilla*	*Orangutan*	*Hypothesis supported*
Data from Miyamoto et al. (1987)					
34	A	G	A	G	III
560	C	C	A	A	I
1287	*	*	T	T	I
1338	G	G	A	A	I
3057-3060	*****	*****	TAAT	TAAT	I
3272	T	T	*	*	I
4473	C	C	T	T	I
5153	A	C	C	A	II
5156	A	G	G	A	II
5480	G	G	T	T	I
6368	C	T	C	T	III
6808	C	T	T	C	II
6971	G	G	T	T	I
Data from Maeda et al. (1988)					
127-132	*****	*****	AATATA	AATATA	I
1472	G	G	A	A	I
Δ2131	A	A	G	G	I
Δ2224	A	G	A	G	III
2341	G	C	G	C	III
2635	G	G	A	A	I

clade. In addition to the 10.2-kb sequence data discussed above, this clade is supported by extensive DNA-DNA hybridization data, by two-dimensional protein electrophoresis data, and especially by extensive mitochondrial DNA sequence data, including the complete mtDNA sequences from three humans, one common and one pigmy chimpanzee, one gorilla and one orangutan. Thus, the closest extant relatives of humans are the two chimpanzee species, followed by gorillas, orangutans, and the nine gibbon species. This is contrary to the conclusion of a closer relatedness of chimpanzee and gorilla derived from morphological and physiological comparisons.

Whale Stories

Whales, dolphins, and porpoises, which form the order Cetacea, are among the most fascinating and spectacular of all vertebrates. They are intelligent and can be trained to play various tricks, they have an elaborate communication system (particularly, the sonar system in toothed whales), and large whales far exceed the bulk of the largest dinosaurs.

The origin of Cetacea has been an enduring evolutionary mystery since Aristotle, for the transition from terrestrial life to an aquatic way of life that is rivaled by no other vertebrates required many drastic yet coordinated changes in biological systems. For example, living cetaceans are unique among mammals in that they lack external hind limbs and that all swim by dorsoventral oscillation of a heavily muscled tail.

A link between cetaceans and ungulates was suggested more than a century ago by Flower (1883) on the basis of comparative anatomical information. This view was accepted by Gregory (1910), but Simpson (1945) and Romer (1966) suggested that the cetacean lineage goes back to the very root of the eutherian radiation. However, Flower's view was later endorsed by Van Valen (1966) and Szalay (1969), who argued on the basis of dental and other grounds for a connection between cetaceans and mesonychid condylarths, and early Tertiary assemblage of ungulates. The first paleontological evidence for a connection between cetaceans and artiodactyls was provided by remains of a middle-Eocene (~45 Myr old) whale exhibiting an artiodactyl-like paraxonic arrangement of the digits on its vestigial hind limbs. Recent fossil findings by Gingerich et al. (1994) and Thewissen et al. (1994) have provided some insight into the terrestrial-aquatic transition.

The molecular evidence for a close relationship between Cetacea and Artiodactyla has been increasing since the 1980s. Goodman et al. (1982) analyzed seven protein sequences and concluded that Cetacea belongs with Artiodactyla. This conclusion received further support from studies on mtDNA sequences. A recent phylogenetic analysis of mtDNA and protein sequence data suggested that the cetaceans are in fact not just related to but within the artiodactyls.

The order Artiodactyla is traditionally divided into three suborders: Suiformes (e.g., pigs and hippopotamuses), Tylopoda (camels and llamas), and Ruminantia (cows, goats, giraffes, deer, and elk). Graur and Higgins (1994) inferred the phylogenetic position of Cetacea in relation to the three artiodactyl suborders by using protein and DNA sequence data from pig, cow, camel, several cetacean species, and an outgroup (e.g., mouse). In Fig. 5.4a, tree I represents the traditional view that the cetacean lineage branched off prior to the divergence between the two artiodactyls, cow and pig (or camel). Trees II and III

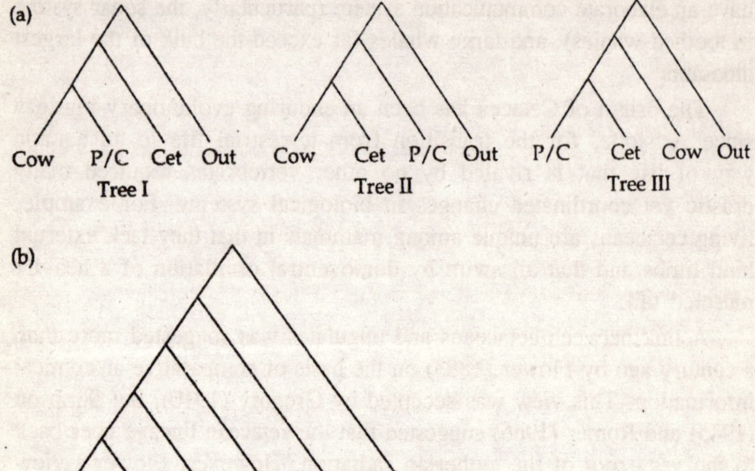

Fig. 5.4. (a) Three possible phylogenetic trees for cetacean (Cet), cow (Cow), pig or camel (P/C), and an outgroup (Out). (b) Schematic molecular phylogenetic tree for the taxa under study.

represent the alternatives, i.e., cetaceans are clustered with one of the two artiodactyl species.

Table 5.5 shows the results of the cow-pig-cetacean analysis of mtDNA data with mouse as the outgroup. Among the 131 informative transversions, tree II was supported by 55 informative sites, whereas trees I and III were each supported by only 38 sites, so that tree II was significantly more parsimonious than trees I and III (P < 5%). When Kishino and Hasegawa's (1989) test of difference between DNA maximum-likelihood trees was used, tree II was significantly better than tree I (P < 5%), though tree III could not be rejected. Moreover, none of the 100 maximum-likelihood bootstrap replicates supported tree I and only 45 of the 100 neighbour-joining bootstrap replicates supported tree I. By using Li's (1989) test, the length of the internal branch in tree II was found to be significantly greater than 0 (P < 1%). Thus, the mtDNA sequence data strongly supported tree II. Graur and Higgins also used the sequences of 11 proteins, and their analyses using the parsimony and the neighbour-joining methods again supported tree II, though not with statistical significance. Five of the 11 proteins used were also available for camel, and a parsimony analysis suggested that cetaceans are more closely related to pig than to camel.

Table 5.5. Analysis of mitochondrial genes for three alternative phylogenies among pigs, cows, and cetaceans, with mouse as outgroup

Methods	*Cow-Pig*	*Cow-Cetacean*	*Pig-Cetacean*
Parsimony (131 informative sites)	38 (641)	55 (624)	38 (641)
ML bootstrap (100 replicates)	0	89	11
NJ bootstrap (100 replicates)	45	908	47
ML test	–62.1 ±24.4	ML	–39.8±26.8
Internal branch test		0.0058±0.0021	

Based on the above analyses, Graur and Higgins proposed the phylogeny shown in Fig. 5.4b, in which Ruminantia and Cetacea are more closely related to each other, followed by the other two suborders Suiformes and Tylopoda. Pecora (e.g., cow) and Tragulidae (chevrotain) represent two of the most distantly related infraorders of Ruminantia and the placement of Cetacea outside the Pecora-Tragulidae clade was based on a parsimony analysis of mitochondrial 12S and 16S rRNA genes. On morphological and paleontological grounds, pig and cow are estimated to have diverged ~55-60 Myr ago while the divergence between Tragulus and cow is put at ~45 Myr ago. Using these dates and the transversion divergences of mtDNA, Graur and Higgins estimated that the cetacean and ruminant lineages diverged 45-49 Myr ago. This estimate is compatible with the early whale fossils with dated ages of 52 and 46-47 Myr.

Living cetaceans are subdivided into two highly distinct suborders, Odontoceti (the echolocating toothed whales, about 67 species) and Mysticeti (the filter-feeding baleen whales, 11 species) The two suborders are thought to have had long independent histories, having separated from the extinct suborder Archeoceti more than 35-40 Myr ago. This view was challenged by Milinkovitch et al. (1993), who proposed that the sperm whale (a toothed whale) is more closely related to the baleen whales (represented by the fin whale and the humpback whale) than to other toothed whales. This proposal was based on the parsimony, neighbour-joining, and maximum likelihood analyses of

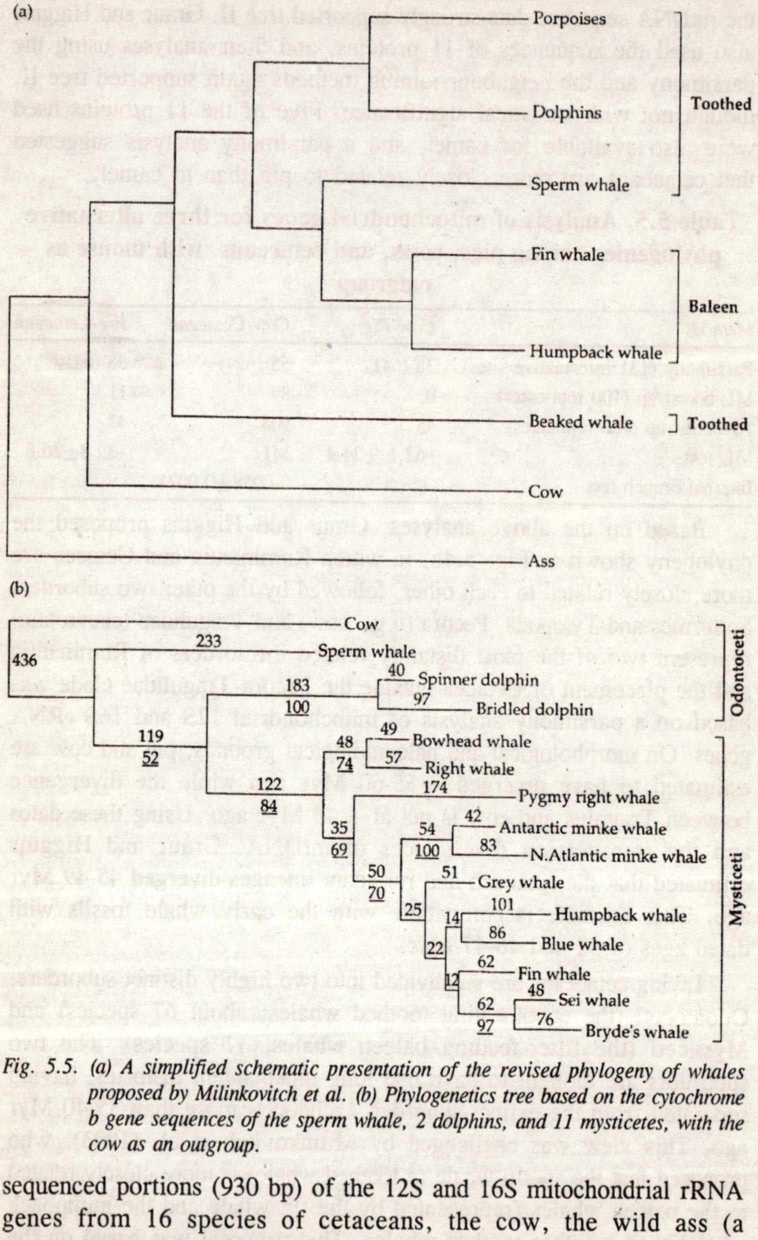

(a)

Porpoises
Dolphins
Sperm whale
Toothed
Fin whale
Humpback whale
Baleen
Beaked whale
Toothed
Cow
Ass

(b)

Cow
436
233
Sperm whale
Odontoceti
40
Spinner dolphin
183
100
97
Bridled dolphin
119
52
48
74
49
Bowhead whale
57
Right whale
122
84
174
Pygmy right whale
42
Antarctic minke whale
35
69
54
100
83
N.Atlantic minke whale
50
70
51
Grey whale
25
101
Humpback whale
14
86
Blue whale
22
62
Fin whale
12
48
Sei whale
62
97
76
Bryde's whale
Mysticeti

Fig. 5.5. (a) A simplified schematic presentation of the revised phylogeny of whales proposed by Milinkovitch et al. (b) Phylogenetics tree based on the cytochrome b gene sequences of the sperm whale, 2 dolphins, and 11 mysticetes, with the cow as an outgroup.

sequenced portions (930 bp) of the 12S and 16S mitochondrial rRNA genes from 16 species of cetaceans, the cow, the wild ass (a perissodactyl), a human, and a sloth. This revised phylogeny, which was also supported by the available myoglobin sequence data, suggests that baleen whales lost the capacity for echolocation rather than never evolved it; the other alternative would be less likely because it requires that the elaborate echolocating system evolved at least twice, once in the sperm whale and once in other toothed whales. The proposed revision, however, was claimed to be at variance with the cytochrome b gene sequence data, which suggested that the baleen whales are separated from the sperm whale and dolphins (toothed whales). However, Arnason and Gullenberg's study used only one outgroup (cow), but the inference was shown to be sensitive to the choice of an outgroup. Moreover, their analysis used equal weights for transitional and transversional changes, but when transitional changes were excluded or given less weight, the cytochrome b gene sequence data actually supported the sperm whale-baleen whale clade. The sister group relationship between sperm whales and baleen whales was also supported by an extended analysis that incorporated two mitochondrial rRNA segments and the cytochrome b gene from 21 cetacean species with three artiodactyl outgroups. Therefore, there is considerable support for the proposed major revision of the whale phylogeny.

An interesting point is that the grey whale is positioned between the minke whales and the remaining rorquals; in an equally parsimonious tree (not shown) the grey whale is placed between the blue/humpback whales and the fin whale. In either tree, the placement of the grey whale among the rorquals is of great interest and is supported by other molecular data: sequence analyses of the mtDNA control region, the common cetacean satellite DNA and the heavy mysticete satellite, and the taxonomic distribution of the light mysticete satellite. Traditionally, the grey whale has been placed in a distinct, monotypic family, the Eschrichtiidae, on the basis of its similarity to skeletal features of the extinct family Cetotheriidae and its unique adaptation to a benthic feeding ecology. Whereas the right whales and bowhead whales use their long baleen to skim the water surface for small planktonic prey and the rorquals use their vast pleated throats to engulf dense concentrations of plankton or school fish, the grey whale uses its lower jaw like a shovel to scoop up benthic amphipods and mysids, forcing the mud and water back through its coarse baleen plates. The traditional placement of the grey whale in a separate family would imply an ancestral origin of the grey whale's unique benthic ecology, whereas the new finding of the close relationship of the grey whale to the rorquals implies a strikingly rapid morphological adaptation of the grey whale to a new ecological niche.

ARE THE YEWS CONIFERS OR NOT?

The yews, which form the genus *Taxus*, are the best known members of the family Taxaceae, which is therefore commonly known as the yew family. (The anti-cancer drug Taxol is obtained from the bark of Pacific yew trees). This family includes about 20 species and has been suggested to consist of either five genera, *Taxus*, *Austrotaxus*, *Pseudotaxus*, *Torreya*, and *Amentotaxus*, or only the first four of the five genera. The phylogenetic position of Taxaceae is one of the oldest unsolved problems in gymnosperm systematics. The main problem is that the Taxaceae bear uniovulate seeds with the ovule at the terminal position, whereas most other conifers bear multiovulate cones with ovules in the axillary position. For this reason there has long been a strong dispute on whether the Taxaceae are conifers.

In the basic taxonomic scheme of extant conifers proposed by Pilger (1926), the order Coniferales included Taxaceae and six other

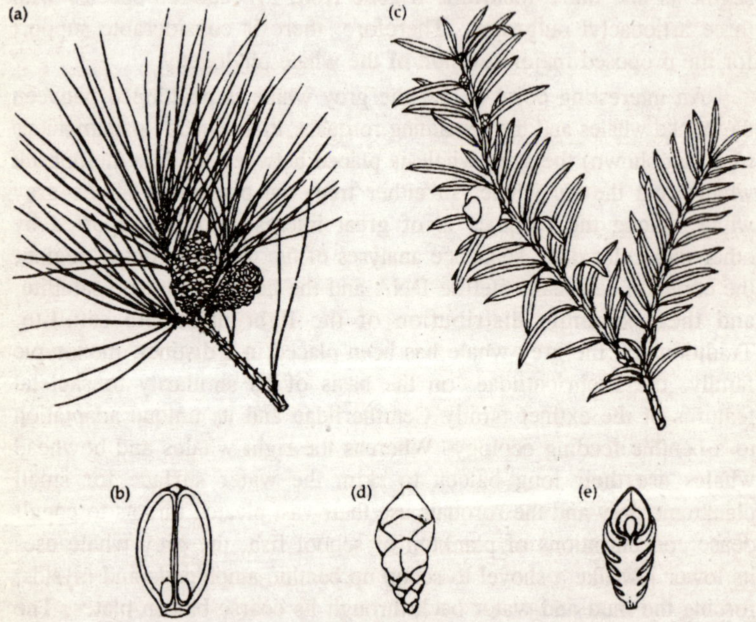

Fig. 5.6. (a) and (b) The pine Pinus luchuensis: (a) branch with mature female cones and (b) the adaxial view of ovuliferous scale dissected from cone, showing two winged seeds at each side. (c), (d), and (e) The yew Taxus mairei: (c) branch with mature fruit and ovuliferous branchlet, (d) the lateral view of ovuliferous branchlet, and (e) the longitudinal section of (d), showing the erect, single ovule.

families: Podocarpaceae, Araucariaceae, Cephalotaxaceae, Pinaceae, Taxodiaceae, and Cupressaceae. This scheme is different from that of Sahni (1920). Arguing that the seed structures of *Taxus*, *Torreya*, and *Cephalotaxus* were derived from the Paleozoic seeds of the extinct Cordaitales and that these three genera, like *Ginkgo*, are direct descendants of the Cordaitales, Sahni proposed to place them in a distinct phylum, Taxales, equivalent in systematic rank with the Coniferae. Florin (1948, 1954), largely following Sahni, strongly advocated that the terminal uniovulate seed of Taxaceae is a primitive character, its origin being entirely different from that of the conifers. He suggested that Taxaceae represents a distinct class, Taxales, that has evolved in parallel with Coniferales and Ginkgoales, probably since the upper Devonian. Florin's view has been well accepted. In contrast, Pulle (1937), Takhtajan (1953), and others thought that the Taxaceae are not fundamentally different from the rest of conifers but possess the most reduced solitary-ovule cones. Takhtajan (1953) suggested that the Taxaceae are descendants of the Podocarpaceae, as evidenced by their pseudoterminal development of the arillate ovules and the structure of the wood. Harris (1976) proposed that the single, terminal ovule of Taxaceae was derived from the fertile, lateral short shoot of Paleozoic conifers by reduction and followed by a shift of the single ovule onto the apex.

To resolve the above controversy, Raubeson and Jansen (1992) studied the phylogenetic distribution of a rare chloroplast-DNA structural mutation among conifers, nonconiferous seed plants, and pteridophytes. The basic structure of the chloroplast genome, a circular DNA with a large (10-25 kb) inverted repeat, was shown to be conserved across all major plant lineages. However, Strauss et al. (1988) found that one copy of the two repeats was lost in the genomes of Monterey pine and Douglas fir, two members of the family Pinaceae. Since this loss appeared to be a very rare event, Raubeson and Jansen (1992) used it as a cladistic character to examine the question of monophyly of conifers. They examined a broad taxonomic representation of conifers, including at least one member of each of the seven conifer families, plus representatives of each of the other major extant vascular plant lineages. Their results indicated that only one copy of the repeat is present in all conifers, whereas two copies were present in those pteridophytes and nonconiferous seed plants surveyed. The finding that the loss was shared by all conifers but by one of the outgroups studied supported conifer monophyly.

Table 5.6. Presence of absence of the inverted repeat in the chloroplast genome.

Present	*Absent*
Lycopsida	Taxaceae
Lycopodium obscurum	*Taxus X media*
Selaginella sp.	*Torreya nucifera*
Isoetes melanopoda	Araucariaceae
Psilotales	*Araucaria spp.*
Psilotum nudum	*Agathis australis*
Ophioglossales	Cephalotaxaceae
Botrychium virginianum	*Cephalotaxus fortunii*
Marattiales	Cupressaceae
Marattia sp.	*Calocedrus decurrens*
Filicales	*Juniperus communis*
Osmunda cinnamomea	*Cupressus macrocarpa*
Lygodium palmatum	*Callitris presii*
Polypodium vulgare	Pinaceae
Gingkoales	*Larix laricina*
Gingko biloba	*Pinus strobus*
Cycadales	*Tsuga canadensis*
Cycas revoluta	Podocarpaceae
Encephalartos sp.	*Podocarpus spp.*
Gnetales	*Dacrydium cupressinum*
Ephedra tweediana	*Phyllocladus trichomanoides*
Gnetum gnemon	Taxodiaceae
Welwitschia mirabilis	*Metasequoia glyptostroboides*
	Taxodium distichum
	Sequoia sempervirens
	Cryptomeria japonica

Although the above study provided evidence that Taxaceae and the other six conifer families form a monophyletic group, it provided no information on whether Taxaceae has branched off earlier than the other six families, that is, it remained possible that Taxaceae is actually an outgroup to the other conifers. The studies of Chaw et al. (1993, 1995) rejected this possibility and provided additional support for the monophyly of conifers. They obtained the 18S rRNA sequences from a ginkgo and two members of Gnetales (*Ephedra* and *Genetum*) and from representatives of the following conifer families: Taxaceae, Cephalotaxaceae, Cupressaceae, Taxodiaceae, Araucariaceae, Phyllocladaceae, Podocarpaceae, and Pinaceae. An analysis by the neighbour-joining method with a chlamydomonas as an outgroup suggested that Taxaceae, Cephalotaxaceae, Cupressaceae, Taxodiaceae, and Araucariaceae have very short distances to one another (at most only a few percent divergences) and form a monophyletic group supported by 98% of the bootstrap replicates; however, whether Cephalotaxaceae is the closest relative of Taxaceae as suggested by Hart (1987) remains to be seen. The analysis also suggested that all the conifers form a monophyletic group supported by 89% of the bootstrap replicates and that Pinaceae is an outgroup to the other conifers.

Fig. 5.7 also sheds some light on the evolutionary position of the genus *Phylocladus*, which was so named because its members possess rhomboidal, flattened, celery-leaf-like photosynthetic organs, i.e., the phylloclades. Robertson (1906) considered the phyloclades to be the expansion of certain stem branches into flattened leaflike structures, implying the phylloclade as a specialized and advanced feature. However, interpreting the phylloclade as a very ancient structure, Keng (1974, 1978) claimed that the genus is the most primitive member of the conifers and proposed to remove the genus from Podocarpaceae to form a monogeneric family, Phyllocladaceae. This proposition was supported by Clifford and Constantine (1980) and others, but opposed by Quinn (1987). From a claddistic analysis of a large set of morphological data, hart (1987) concluded that the genus is a terminal taxon within the Podocarpaceae. This conclusion is supported by the tree—*Phyllocladus* is an outgroup to the other genera of Podocarpaceae, which form a monophyletic group. Since *Phyllocladus* is not a natural clade, it should be included in the family Podocarpaceae. At any rate, it is clear that *Phyllocladus* is not a primitive member of the conifers and that the phylloclade is probably a derived feature.

The above inferences were supported by some insertion and deletion events and by a parsimony analysis. Further support for the monophyly of conifers was provided by the sequence data of the chloroplast *rbcL* gene. The *rbcL* data also placed Pinaceae as an outgroup to the other conifers.

In conclusion, all three sets of molecular data are not compatible with the proposal that the Taxaceae form a clade distinct from the other conifers; rather, they provide strong evidence for the monophyly

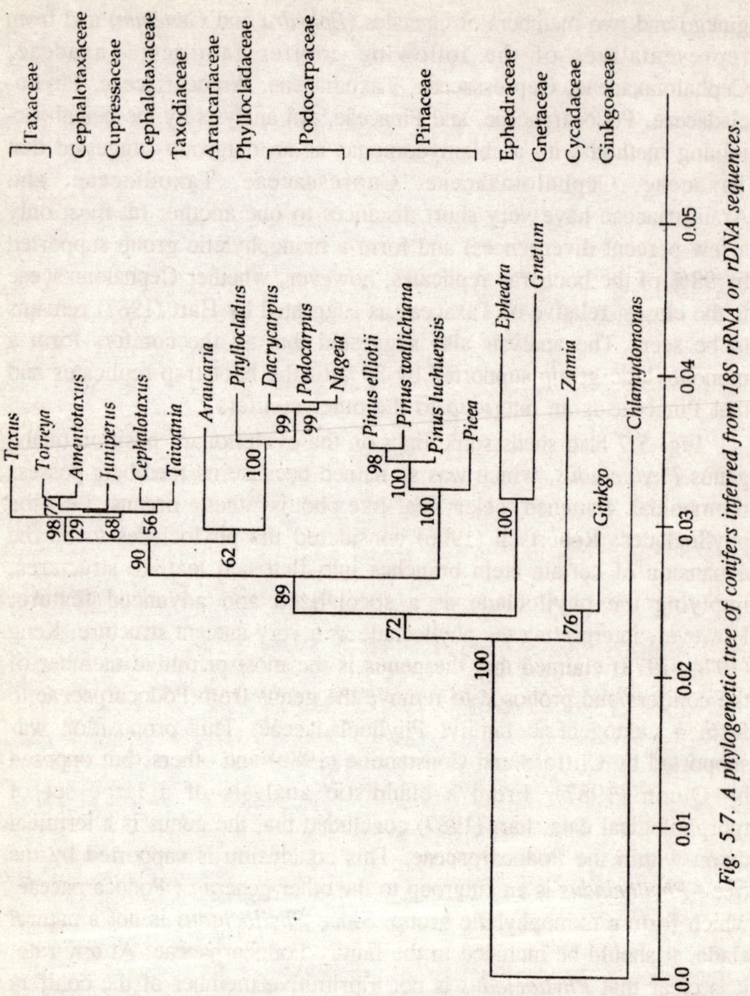

Fig. 5.7. A phylogenetic tree of conifers inferred from 18S rRNA or rDNA sequences.

of conifers and against the view that the terminal, uniovulate seed of Taxaceae is a primitive character. Moreover, the 18S rRNA sequence data do not support the view that the Taxaceae are direct descendants of the Podocarpaceae. Thus, it seems that the uniovulate seed in Taxaceae and that in some species of *Podocarpus* have different origins, probably all reduced from multiovulate cones.

ORIGIN OF ANGIOSPERMS

The origin of angiosperms was dubbed by Charles Darwin as "an abominable mystery" and remains a highly controversial issue in plant evolution. Angiosperms began to radiate rapidly about 115 Myr ago (the middle Cretaceous) and became the dominant group of earth's flora by about 90 Myrs ago. However, no reliable angiosperm fossils older than 120 Myr have been found, and there is little agreement as to when the angiosperms arose and from which branch of the gymnosperms they stem. Since the progymnosperm lineage extends back to at least 370 Myr ago, there is an enormous range of time during which angiosperms might have had their beginnings. Theories as to why there are no fossils of progenitor angiosperms fall into two basic types: either (1) angiosperms did not exist until the early Cretaceous and then radiated explosively or (2) pre-Cretaceous angiosperms lived in habitats so refractory to fossilization that they left no record.

One way to decide between these two views is to estimate the date of divergence between monocotyledons (monocots) and dicotyledons (dicots), the two major classes of angiosperms, because this provides a minimal estimate for the age of angiosperms. The first application of DNA sequence data to estimate this data was made by Martin et al. (1989). Their analysis was based on the comparison of the sequences of the nuclear gene encoding cytosolic glyceraldehyde-3-phosphate dehydrogenase (GAPDH or gapC) from plants, animals, and fungi. By using several divergence dates between animal taxa and between the plant, animal, and fungal kingdoms they estimated the rate of evolution of this gene. From this, they estimated that monocots and dicots diverged about 320 Myr ago. This date appears to be too ancient, because the earliest land plant fossils are only about 420 Myr old and so the estimate of Martin et al. would imply that the lineages leading to bryophytes, pteridophytes, gymnosperms, monocots, and dicots all appeared within the first 100 Myrs of land plant evolution. Nevertheless, their data provided evidence for pre-Cretaceous origins of angiosperms.

Wolfe et al. (1989b) obtained a different estimate by using three approaches. The first one was based on a calibration of the rate of synonymous substitution in chloroplast genes with the maize—wheat divergence as a reference. Using DNA sequence data, they first demonstrated that maize, wheat, and rice all originated at approximately the same time, i.e., represent approximately a trichotomy. Since fossils of rice leaf epidermis have been described from the upper Eocene (about 40 Myr ago), they took 50 Myr as a lower bound for the origin of maize, wheat, and rice. Further, they used 70 Myr as an upper bound for the maize-wheat divergence, as suggested by Stebbins. From the average number of synonymous substitutions per site between maize

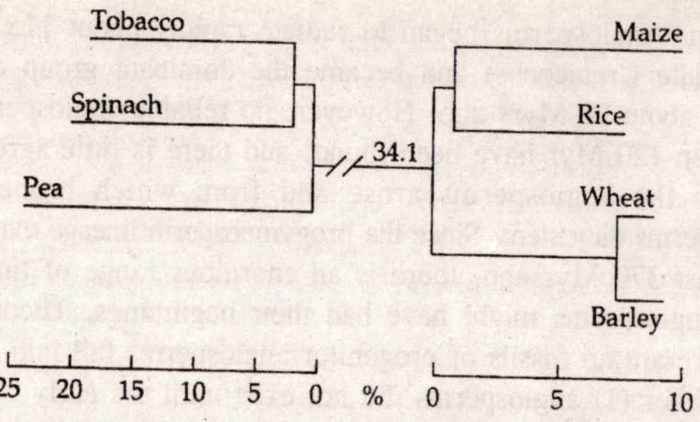

Fig. 5.8. Phylogenetic tree for three dicot and four monocot species.

and wheat chloroplast genes, one can readily estimate the rate of synonymous substitution to be $a=0.173/(2T)=1.73\times10^{-9}$ or 1.24×10^{-9} substitutions per site per year, depending on whether the lower bound of T=50 Myr or the upper bound of T=70 Myr is used. The average number of synonymous substitutions per site per year between monocot (maize and wheat) and dicot (tobacco) genes is $(0.578+0.576)/2=0.577$. Therefore, the date of the monocot-dicot divergence is estimated to be 0.577/(2a) or between 170 and 230 Myr. The second approach was based on a calibration of the rate of nonsynonymous substitution with the bryophyte-angiosperm divergence as a reference point, which was taken to be between 350 and 450 Myr ago; the nonsynonymous rate was used because synonymous substitutions have been saturated for this degree of divergence and therefore cannot be reliably estimated. This approach gave an estimate between 150 and 260 Myr. The third approach was based on a calibration of the substitution rate in nuclear ribosomal RNA genes with the plant-animal divergence as a reference point, which was assumed to be about 1000 Myr ago. The estimate obtained from the 26S rRNA sequence data was between 200 and 250 Myr and that obtained from the 18S rRNA sequence data was between 200 and 210 Myr. From the above estimates, Wolfe et al. suggested that monocots and dicots diverged about 200 Myr ago with an uncertainty of about 40 Myr. These results are supported by recent estimates from mitochondrial gene sequences and suggested that monocots and dicots could have diverged in the early Jurassic. Therefore, the molecular data strongly support the hypothesis that angiosperms existed long before they came to prominence.

There are other major issues concerning the origin and evolution of the angiosperms. First, what is the common ancestor of the angiosperms? One view is that the angiosperms are derived from a gymnosperm lineage, i.e., from the ancestor of Gnetales. This implies that Gnetales and angiosperms share a common ancestor not shared by other gymnosperms, so that the gymnosperms are paraphyletic. Since Haeckel (1894) this view has been espoused by many plant taxonomists because of the angiosperm-like features of the Gnetales, such as the dicotyledonous seeds, vessels in the secondary wood, and reduced female gametophytes. It has gained support from cladistic analysis of morphological data; parsimony analyses of *rbcL* and rRNA genes and parsimony analysis of combined morphological and molecular data. The opposing view is that no extant gymnosperm is a sister group of the angiosperms and the angiosperms are derived from the progymnosperm. This view is supported by the 5S RNA sequence data of Hori et al. (1985), the partial 18S rRNA sequence data of Troistky et al. (1991), the *rbc*L sequence data of Hasebe et al. (1992), and the chloroplast sequence data of Goremykin et al. (1996). This issue remains to be resolved. Second, do the dicots have multiple origins? The traditional view that the dicots are paraphyletic is supported by the partial rRNA sequence data of Hamby and Zimmer (1992) and the *rbc*L tree inferred by Chase et al. (1993), but it requires more extensive data (from different genes) to completely settle the issue. Third, are the monocots monophyletic? This may not be the case according to Hamby and Zimmer (1992), but further research is needed to resolve the question.

Origin of Eukaryotes

The living world has historically been dichotomously divided into eukaryotes and prokaryotes. Those organisms with a well-formed nucleus are eukaryotes whereas those without such a nucleus are prokaryotes. The eukaryotes include all protists such as ciliates and flagellates, as well as fungi, plants, and animals. The only prokaryotes are the bacteria, including the cyanobacteria, which were formerly known as blue-green algae.

C.R. Woese and coworkers have challenged this dichotomous view. Since the late 1960s they have been studying bacterial relations by comparing the sequences of the ribosomal RNAs from different species. In their earlier studies they used T1 ribonuclease to digest the 16S rRNA sequence into oligonucleotides (words), each of which ends with a single G, e.g., CUAAG and UCAUCG. The oligonucleotides thus produced were short enough to be sequenced by the techniques then available. To reduce the chance of recurrence of the same word in a 16S rRNA molecule, they used only words with six letters or more to form a dictionary characteristic of an organism. They then analyzed the data in terms of an association coefficient, S_{AB}, which is defined as twice the number of nucleotides in the words common to both of dictionaries A and B divided by the number of nucleotides in all words in the two dictionaries. They showed that the S_{AB} values can be used to study the evolutionary relationships among organisms.

Woese and coworkers came acrose a totally unexpected finding when examining the ribosomal RNA of the methanogenic bacteria. These unusual organisms live only in oxygen-free environments such as the intestinal tract of animals and sewage-treatment plants, and generate methane (CH_4) by the reduction of carbon dioxide. Methanogens are no doubt bacteria because they are the size of bacteria, have no nuclear membrane, and have a low DNA content. Thus, they would be expected to be more closely related to other bacteria than to eukaryotes. However, the S_{AB} values showed that methanogens are related as closely to the latter as to the former and that the group formed by methanogens is phylogenetically as deep as any other bacterial group. On the basis of these results and the argument that methanogenic metabolism (the reduction of carbon dioxide to methane) is ideally suited to the kind of atmosphere thought to have existed on the primitive earth (rich in carbon dioxide but virtually devoid of oxygen), Woese and Fox (1977) proposed to name methanogens and their relatives the archaebacteria, implying that this group of bacteria is at least as ancient as the true bacteria (eubacteria).

The archaebacterial group includes also extreme halophiles (halobacteria) and extreme thermophiles. Halobacteria require high concentration of salt and grow in salty habitats along the ocean borders and in inland waters such as the Great Salt Lake and the Dead Sea. They are the only photosynthetic archaebacteria; they transduce light into chemical energy by means of a proton pump based on bacteriorhodopsin. The extreme thermophiles grow at high temperatures, some near the boiling point of water. All species grow anaerobically, though a minority of species can also grow aerobically. Most species require sulfur as an energy source and live in hot sulfur springs.

Woese and Fox (1977a) and Fox et al. (1980) proposed that archaebacteria, eubacteria, and eukaryotes were derived from a common ancestor and represent the three primary lines of life because the S_{AB} values suggested that these three groups re about equally distant from one another. Under this hypothesis, the prokaryotes are not monophyletic as traditionally believed but are divided into two groups:

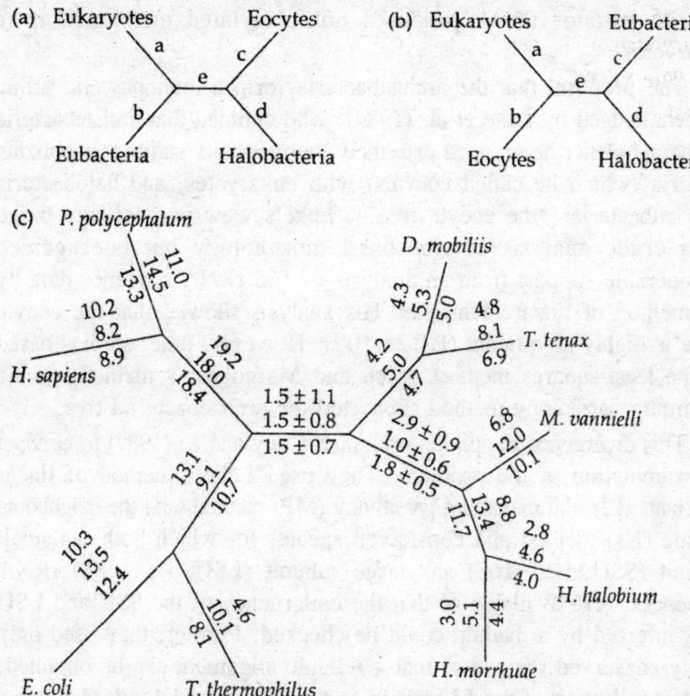

Fig. 5.9. (a) The archaebacterial tree; for simplicity, methanogens are not included. (b) The eocyte tree. (c) The unrooted tree inferred by the neighbour-joining and maximum parsimony methods.

the eubacteria and the archaebacteria. Moreover, the eukaryotic lineage is as deep as the two other lineages, contrary to the traditional view that the eukaryotic lineage arose long after the diversification of prokaryotes. The view of Woese and Fox can be represented by "the archaebacterial tree" if one neglects the root of the tree; for simplicity methanogens are not included.

Woese and Fox's proposal has stimulated a great deal of controversy. Van Valen and Maiorana (1980) argued that archaebacteria were derived from other bacteria and gave rise to the eukaryotes. They pointed out that archaebacteria and eukaryotes shared several specific and derived similarities. For example, the archaebacterium *Thermoplasma* has a histone-like protein that produces nucleosome-like condensations of DNA, has actin- and myosin-like proteins, and can invaginate its plasma membrane, an ability that is necessary for a cell that engulfs others. Moreover, *Halbacterium*, another archaebacterium, has a light-inducing pigment almost identical to rhodopsin, and its initiator tRNA produces nonformylated methione, as in eukaryotes.

The proposal that the archaebacteria form a monophyletic group was challenged by Lake et al. (1984), who claimed that archaebacteria are paraphyletic; he grouped extremely thermophilic, sulfur-metabolizing bacteria (which he called eocytes) with eukaryotes, and halobacteria with eubacteria ("the eocyte tree"). Lake's view was initially based on a crude analysis of ribosomal morphology but later gained considerable support from an analysis of 16S rRNA sequence data by the method of linear invariants. His analysis showed that the eocyte three is highly significant ($P<2\times10^{-5}$). However, other studies based on the least-squares method, Fitch and Margoliash's method, or the maximum parsimony method supported the archaebacterial tree.

This divergence in opinion prompted Gouy and Li (1989) to conduct an examination of the problem. They used Lake's method of linear invariants (LI), the maximum parsimony (MP) method, and the neighbour-joining (NJ) method and considered species for which both the small subunit (SSU, i.e., 16S) and large subunit (LSU, i.e., 28S) rRNA sequences were available so that the congruency of the SSU and LSU trees inferred by a method could be checked. Further, they used only highly conserved regions so that a reliable alignment can be obtained. While application of the LI method to the SSU data did indeed support the eocyte tree, application of this method to the LSU data strongly supported the archaebacterial tree ($P<5\times10^{-6}$). When the SSU and LSU data were combined, the LI method also supported the archaebacterial tree. Application of the NJ method to the SSU and LSU data either separately or jointly all gave the archaebacterial tree. The same was true for the MP method; for the SSU data, the archaebacterial, eocyte, and halobacterial trees were supported by 20, 10, and 9 informative sites, respectively, and for the LSU data, the corresponding numbers were 34, 27, and 14. In conclusion, their study supported the monophyly of archaebacteria.

The study of Rivera and Lake (1992) on elongation factor EF-1α sequences has reopened the controversy on the archaebacterial tree vs. the eocyte tree. Elongation factor EF-1α (termed EF-Tu in bacteria) is an ubiquitous protein that transports aminoacyl-tRNAs to the ribosome and participates in their selection by the ribosome. While the sequences of eubacteria and the archaebacteria *T. celer* (a methanogens species) and *Halobacteria volcanii* contain a 4-amino acid segment (GPMP or its variants) near the guanosine diphosphate binding domain, Matheson et al. (1990) found that the sequence of the eocyte *Sulfolobus acidocaldarius* contains an 11-amino acid segment similar to the eukaryotic one. Later, Rivera and Lake (1992) found that other eocyte species also contain the 11-amino acid segment. This difference in segment length can be explained by one change under the eocyte tree but by at least two changes under the archaebacterial tree. Rivera and Lake therefore concluded that the eocyte tree is the true tree. This conclusion assumes that gap events occur extremely rarely, so that one difference is sufficient to draw a conclusion on the tree topology. This assumption is, however, not supported by the fact that the sequence of *Methanogens vannielii* contains a 7-amino acid segment instead of the 4-amino acid segment in other methanogens. At any rate, the issue deserves further investigation, for Rivera and Lake's conclusion is contrary not only to that of Gouy and Li (1989) but also to that of Gogarten et al. (1989), who showed that the ATPase data support the monophyly of the archaebacteria. The monophyly was further supported by a recent study using aminoacyl-tRNA synthetase genes. However, the expanded database of elongation factor sequences tends to support the paraphyly of archaeons, with the *Sulfolobus/Desulfurococcus* clade as a sister group to the eukaryotes.

The phylogenetic analyses discussed above actually provided only unrooted trees—the placement of the root between the eubacterial lineages and the other two lineages was based on other molecular and cellular data. Gogarten et al. (1989) and Iwabe et al. (1989) have come up with an ingenious idea to infer the root of the tree. They

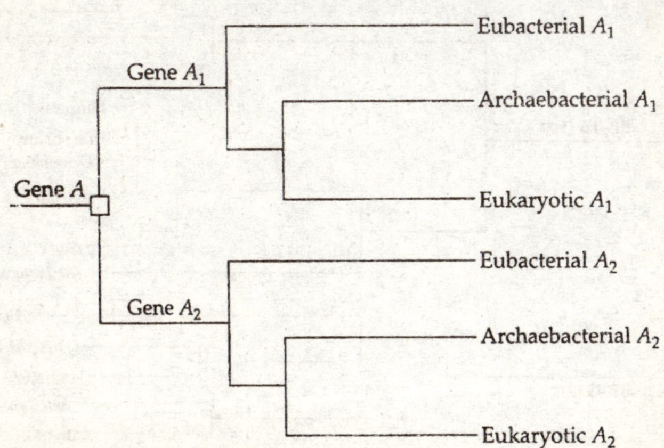

Fig. 5.10. Duplication of gene A into A_1 and A_2 prior to the divergence of the eubacterial, archaebacterial, and eukaryotic lineages.

used duplicate genes that were derived before the separation of the three lineages (domains), which are called Bacteria, Archaea, and Eucarya by Woese et al. (1990). Suppose that gene A was duplicated into A_1 and A_2 before the divergence of the three lineages. Subsequently, as the three lineages diverged, A_1 (and A_2) should also diverge in the same order. Therefore, A_2 sequences are outgroups to A_1 sequences and can be used to root the A_1 sequences, i.e., to root the three lineages. Similarly, A_1 sequences can be used to root the A_2 sequences and the three lineages.

Iwabe et al. (1989) applied this concept to the EF-Tu (EF-1α) and EF-G genes. Since these two elongator genes are duplicate genes present in all prokaryotes and eukaryotes, the duplication event should have occurred before the separation between eukaryotes and prokaryotes. For this reason, the EF-Tu sequences can be used as outgroups to infer the root of the tree for the EF-G sequences, and the tree in the lower part of Fig. 5.11 suggests that the eukaryote lineage (as represented by *Dictyostelium* and hamster) is derived from the archaebacterial lineage (as represented by *Methanococcus*) and that the eubacterial lineage is the most ancient. The same conclusion was reached when the EF-G sequences were used as outgroups to infer the tree for the EF-Tu sequences. The tree was further supported not only by a more extensive analysis of elongation factor sequences but also by other sets of duplicate genes, such as aminoacyl-tRNA synthetase genes, which are also derived from ancient gene duplications that

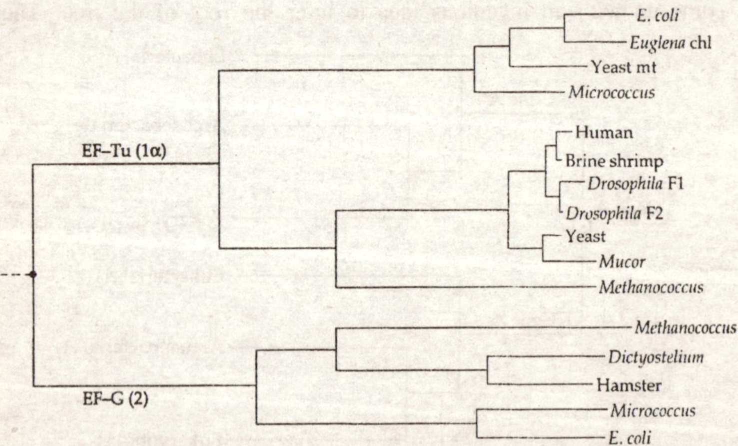

Fig. 5.11. Phylogenetic tree inferred from a simultaneous comparison of EF-Tu(1a) and EF-G(2) sequences from archaebacteria, eubacteria, and eukaryotes.

predated the last common ancestor of all living organisms. Therefore, it seems fairly certain that the eukaryotes originated from an archaebacterial ancestor and that the root of the tree of life is between the eubacterial lineage and the ancestral lineage of the archaebacterial and eukaryotic lineages.

There is, however, the proposal that the eukaryotic genome is a chimera derived from a fusion of a (gram-negative) eubacterium and an archaebacterium. In a maximum likelihood and parsimony analysis of 24 protein sequences from gram negative and gram positive bacteria, archaebacteria and eukaryotes, Golding and Gupta (1995) found that although nine proteins significantly supported the archaebacteria—eukaryote clade, seven significantly supported the gram negatives-eukaryote clade; the rest gave no significant support to any of the three alternative topologies. They argued that such a large proportion of cases supporting the gram negative-eukaryote clade is unlikely to be due to horizontal gene transfer, convergent evolution, or methodological errors, especially in view of the fact that none of the proteins significantly supported the gram positives-eukaryote clade. This chimeric origin hypothesis needs further investigation.

Ancient DNA

A novel approach to archeology and paleontology was initiated in 1984 when Higuchi et al. (1984) succeeded in extracting DNA from a 140-year old salt-preserved skin of a quagga, an extinct animal that looks like a cross between a zebra and a horse. They cloned DNA fragments isolated from dried skin and connective tissue and obtained mitochondrial DNA sequences totaling 229 bp. These sequences differed by 12 bp from the corresponding sequences of a mountain zebra. These

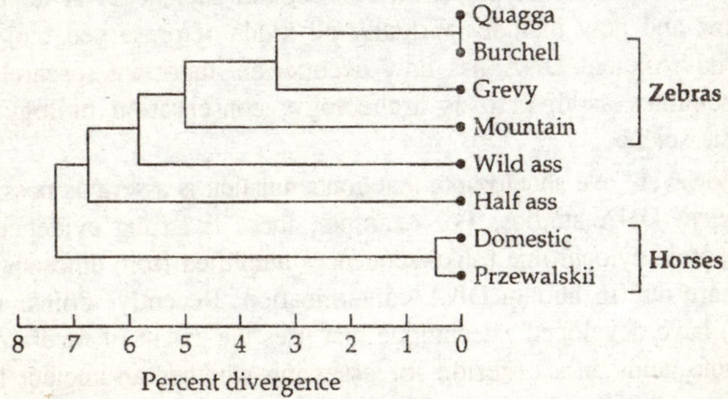

Fig. 5.12. Phylogenetic tree relating mitochondrial DNA from the extinct quagga to mtDNAs from other members of the horse genus.

and several sequences obtained in subsequent studies allowed the quagga to be placed into a phylogeny of the horse and its relatives. The results argue for a close relationship between the quagga and Burchell's zebra and against the view that the quagga is more closely related to the domestic horse.

The study of Higuchi et al. was soon followed by the cloning of DNA from the skin of an ancient Egyptian mummy dated ~2400 years before present (BP). However, application of cloning techniques to DNA extracted from ancient material is difficult because of degradation and chemical modification and because of the small amounts that can be extracted. The situation changed when the polymerase chain reaction (PCR) technique was introduced into the study of ancient DNA, because PCR requires minute amounts of DNA, theoretically just a single DNA molecule, and can function efficiently even if the DNA is damaged and degraded to some extent. The power of PCR to ancient DNA was first demonstrated in an application to a 7000-year-old brain from excavations in Florida and later in an application to human bone dated to ~5450 years BP. The latter was a significant step because it was thought that ancient DNA can be found only in artificially preserved material and because bones form the great bulk of ancient organic material. An even more remarkable achievement was the PCR amplification of an *rbc*L DNA segment from a fossilized magnolia leaf from the Miocene, between ~17 and 20 Myr BP succeeded in the amplification of an *rbc*L sequence from a Miocene *Taxodium* (bald cypress) from the same site. Because among all modes of fossilization preservation in amber may more consistently yield better quality DNA, even older fossilized DNA has now been amplified. The list of specimen types shown to contain ancient DNA continued to grow and now includes virtually all kinds of preserved biological material. Ancient DNA has now become an important research tool in disciplines as diverse as archeology, conservation biology, and forensic science.

However, one should note that contamination is a serious possibility in ancient DNA studies. For example, there is strong evidence that the reported cytochrome b-like sequences amplified from dinosaur bone DNA are due to human DNA contamination. Recently, Poinar et al. (1996) have developed a technique that uses the extent of racemization of amino acids as a criterion for assessing whether an ancient tissue sample contains endogenous DNA. Their study suggested that many paleontological finds from which DNA sequences purportedly millions of years old have been reported probably did not contain intact DNA because they showed extensive racemization. Fortunately, specimens bound in amber have a high probability to yield ancient DNA.

Another sensational event is the report of revival of bacterial spores in a 25- to 40-million-year-old amber. This bacterium is identified to be most closely related to *Bacillus sphaericus*, based on the sequencing of 530 bp of the 16S rRNA. Reviving an ancient organism enables scientists to study evolution through a true living fossil. Good quality DNA can be obtained in large amounts, and a reliable rate of nucleotide substitution can be estimated, if the ancient organism can be reliably dated. However, there remains considerable reservation about the report because the experiment has not been repeated.

6

SUBSTITUTIONAL EVOLUTION

The vast majority of mutations that occur are either neutral with respect to *fitness* (defined as the individual's ability to survive and reproduce) or are disadvantageous. If they are disadvantageous, they will tend to be removed from the population since their bearers will be less likely to survive and/or reduce (*negative* or *purifying selection*). Occasionally, a new mutation confers a selective advantage and increases the fitness of individuals bearing it so that it will eventually reach *fixation* (the point at which the allele frequency in the population becomes 100%). This is termed *positive selection*. Selection is thus nothing more than the differential and nonrandom reproduction of genotypes resulting from the superior or inferior fitness of their associated phenotypes. In vertebrates, direct evidence at the molecular level for the occurrence of selection, whether positive or negative, has however often been hard to obtain and there are as yet relatively few good practical examples.

Gene evolution does not however invariably require selection since changes in allele frequency can also occur by chance owing to random sampling of gametes (*genetic drift*). Whereas selection implies directed change, genetic drift may be viewed as a stochastic process of undirected change. Genetic drift can cause rapid changes in small populations but its effect will be fairly minimal in large ones. This is one of the central conclusions of the *neutral theory of molecular evolution*. The importance of this theory to our understanding of the evolutionary process cannot be understated. Its majority points may be summarized as follows:

1. Most polymorphic variation within a species and most amino acid substitutions between species are likely to be neutral with respect

to selection. This is not to say that variant alleles have no biological function or phenotypic effect, merely that the alleles are indistinguishable (neutral) in terms of their function.

2. The amino acid substitution rate can vary between genes and this is likely to be related to the nature and extent of structural and functional constraints upon the different protein products.
3. Those parts of a protein with the fewest functional constraints will evolve most rapidly. Conversely, those regions of a protein which are functionally important will be those that are most highly conserved evolutionarily.
4. Genes whose protein products have a stable function manifest a *molecular clock* i.e. the amino acid substitution rate is similar fora given gene in different evolutionary lineages.
5. A high level of genetic polymorphism will exist and this is predicted to increase with population size since the probability of fixation of a specific allele is lower in large populations.
6. In coding sequences, nucleotides in the third codon position evolve faster than the first two positions. The proportion of substitutions in the third position should be higher for proteins that are evolving more slowly.
7. Any change in the function of a gene sequence will serve to alter the proportion of amino acid substitutions that are neutral. Genes that are no longer capable of expression (pseudogenes) will accumulate mutations rapidly and are destined eventually to lose recognizable homology to the original sequence.

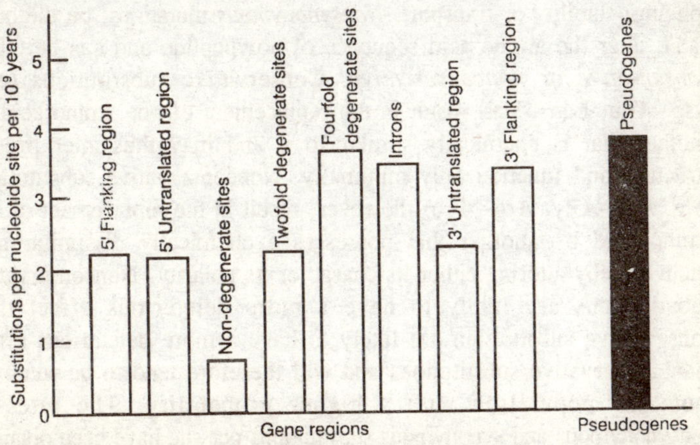

Fig. 6.1. Average rates of substitution in different parts of genes and in pseudogenes.

8. Sequences of portions of sequences that are evolving faster than the neutral rate for the species are likely to have been subject to positive selection.
9. Genes manifesting a high rate of variation should also evolve at a higher rate.

Synonymous and Nonsynonymous Substitutions

Rates of nucleotide substitution vary between different gene regions, tending to be higher in introns and flanking regions than in coding sequences. Single base-pair changes in noncoding regions do not usually affect gene expression unless they occur in a promoter or regulatory region or alternatively impair mRNA splicing efficiency. Such changes are not usually subject to the effects of negative selection and may therefore accumulate in the genome. Using anonymous DNA segments, Copper et al. (1985) estimated the unique DNA sequence heterozygosity in the human genome to be 0.0037, indicating that ~1/250 bases vary polymorphically in human populations and that most of these variants may be expected to occur in noncoding regions. More recently, and employing rather larger sample sizes, other workers have estimated that polymorphisms in the human genome occur with a frequency of between 1/200 and 1/1000 bp.

In coding regions, single base-pair substitutions that are silent are termed *synonymous* in that they do not change the amino acid sequence of the gene product. Most of these changes occur at the third base positions of codons. Such mutations are likely to be neutral with respect to fitness, assuming that they do not alter gene expression or mRNA splicing, stability or transport. *Nonsynonymous* mutations, on the other hand, alter the amino acid sequence of polypeptide and can be either *conservative* or *nonconservative*. Conservative substitutions (e.g. Asp→Glu, Leu→Val) result in the replacement of one amino acid by another that is chemically similar to it and may thus alter protein structure and function only minimally. Nonconservative substitutions (e.g. Arg→Gly, Arg→Pro), however, result in the replacement of one amino acid by another that possesses a chemically dissimilar side chain thereby altering either its charge or its polarity. Nonconservative substitutions are likely to have a more deleterious effect than conservative substitution are likely to have a more deleterious effect than conservative substitutions and will therefore tend to be removed from the population with a higher probability. The rates of nonsynonymous and synonymous substitutions per site have been obtained for a variety of eukaryotic genes and are of the order of 0-2 $\times$ 10^{-9}

and 2-12 × 10^{-9}, respectively. The study of 1880 human-rodent orthologue pairs yielded average estimates of 0.52 × 10^{-9} and 2.92 × 10^{-9} for nonsynonymous and synonymous substitutions, respectively. In general, eukaryotic pseudogenes evolve at approximately the same rate as the synonymous ties in functional genes, a rate that can therefor be considered to represent the *neutral rate of substitution.*

Owing to the nonrandom nature of the genetic code, different amino acids are encoded with different degrees of degeneracy. Base positions in codons can thus belong to any one of three classes :

(i) *Nondegenerates sites* (~65% of base positions in human codons): all possible substitutions are nonsynonymous. The base substitutions rate at these sites is very low on account of the strong selection pressure to avoid amino acid exchanges.

(ii) *Two-fold degenerate sites* (~19% of base position in human codons): base positions at which only one of the three possible substitutions is synonymous.

(iii) *Four-fold degenerate sites* (~16% of base positions in human codons): base positions in which all three possible substitutions are synonymous. The base substitution rate at these sites is higher than those noted for non-degenerate sites and two-fold degenerate sites. The comparison of human and rodent genes has allowed the rates of transitional and transversional silent substitutions in four-fold degenerate sites to be estimated (1.71 ×10^{-9}) and 1.22 × 10^{-9} per site per year, respectively) since the human-rodent divergence.

Positive and Negative Selection in Protein Evolution

Protein coding genes exhibit considerable variation in their rates of synonymous (k_s) and nonsynonymous (k_a) substitutions. The k_s/k_a ratio serves as a rough guide to the degree of evolutionary conservation and therefore the functional constraints placed upon a protein by natural selection (Li et al., 1985). Thus, some human gene sequences are very highly conserved (k_s/k_a ration high), for example ubiquitins, histones H3 and H4 ribosomal proteins, H19, calmodulin, and the G protein α-subunits indicating that these proteins have evolved under negative or purifying selection. By contrast, other human/primate genes have evolved very rapidly (k_s/k_a less than unity), for example the sex determining locus, SRY (Yp11.3) and those genes encoding apolipoprotein C-I, (APOC1; 19q13.2) protamines P1 and P2 (PRM1, PRM2; 16p13.3), myelin proteolipid protein pregnancy-specific glycoprotein 1 and carcinoembryonic antigen (PSG1, CEA; 19q13.2), eosinophil cationic protein (RNASE3; 14q24-q31), the rhesus blood group genes (RHD, 1pp34-p36.2; RHAG, 6p11-p21.1) and β-microseminoprotein (MSMB, 10q11.2). Such examples of a significantly higher rate of nonsynonymous nucleotide substitution than synonymous substitution provide strong evidence for the action of positive selection.

Lysozyme, originally a bacteriolytic enzyme whose origin preceded the emergence of the vertebrates, has been independently recruited as a digestive enzyme by two groups of mammals, the ruminant artiodactyls and the leaf-eating colobine monkeys. Both groups are able to ferment plant material in the foregut and possess stomachs that contain high levels of lysozyme. Advanced ruminants such as cows, sheep and deer possess ~10 lysozyme genes as a result of gene amplification events which occurred after the divergence from the pig lineages. Sequence comparison of the lysozyme genes from human (LYZ; chromosome 12) and other primates has indicated a k_s/k_a ratio significantly less than unity in both the colobine and hominoid lineages, indicative of the action of positive selection for amino acid replacements noted in the colobine lineage also occurred in parallel or convergently in the ruminant artiodactyls. Comparison of the nucleotide sequences of the coding regions of the lysozyme genes of advanced ruminants and pigs revealed no difference in the rate of synonymous substitution consistent with the view that it was a change in selective pressure rather than the mutation rate that was responsible for changes in the rate of stomach lysozyme evolution. The reasons for positive selection for lysozyme amino acid replacements in hominoids are at present unclear but Messier and Stewart (1997) suggested that it might have been associated with the increased neutrophil expression of lysozyme in hominoids as compared with other catarrhines.

Since synonymous substitutions are likely to be neutral with respect to selection, they have been employed in numerous attempts to calibrate *molecular clocks*. However substitutions rates can vary quite widely between orthologous gene sequences in different taxonomic groups or lineages, as well as between different genes in the same species. The speed of the molecular clock appears to vary according to the lineages leading to extant species A and B, a *relative rate test* is employed. This involves the use of a third (reference) species, C, which is known to have branched off prior to the divergence of A and B. Pairwise comparisons of orthologues in A and C, and in B and C, are used to calculate the k value, the number of substitutions per 100 sites. The k_{AC} and k_{BC} values then provide a measure of the relative rates of mutation in the lineages leading to species A and B, respectively. Such calculations have suggested that the substitution rates in the lineages leading to mouse and rat are approximately equal [~7.9 × 10^{-9} (3.9 × 11.8)], whereas comparable estimates for humans and Old World monkeys [~ 2.2 × 10^{-9} (1.8-2.8)] and humans and chimpanzees [-13 × 10^{-9} (0.9–1.9)] are considerably lower. This is held to be indicative of a slowdown in the substitution rate in primates which appears to be at its greatest in hominoids. Despite these data, the *hominoid slowdown* is not a universal phenomenon in primate evolution since it is not apparent with some gene sequences.

Some genes/proteins show a rather different, discontinuous pattern of evolution. Thus, for example, in both primates and artiodactyls, the growth hormone (GH1; 17q23) gene exhibits a pattern of near stasis punctuated by bursts of rapid evolution during which the rate of evolutionary change has increased at least 25-fold. Interestingly, during mammalian evolution, only the coding region of the growth hormone gene corresponding to the mature protein has exhibited rapid change whereas other regions of the gene (signal peptide, nonsynonymous substitutions and 5' and 3' untranslated regions) have remained relatively unchanged. This would be consistent with the rapid bursts of evolution being of adaptive significance and Wallis (1997) suggested that this may have involved what he termed *function switching*. Briefly, if primate GH were to have had biological functions other than growth promotion viz. lactogenic (prolactin-like) activity, GH could then also have played a role in maintaining the nutritional balance between mother and young. Acquisition of its lactogenic function could have involved changes in GH structure away form the best adapted to growth regulation to one that represented an evolutionary compromise which optimized the protein for its dual function. If the secondary function of GH had then changed (e.g. the adoption of a pattern of seasonal breeding following migration or climatic change, in which gestation and suckling no longer overlapped), selection for dual function would no longer have operated and the selection pressure on GH structure would have been rated to its growth promoting function alone. If several amino acid substitutions had already occurred, then the process of functional reversal involving simply the reversal of each amino acid change would have been very unlikely. Rather, GH would have adopted a new structural form, slightly different from the original, but nevertheless adapted to its primary function. Switching back to dual function might then have led to a new cycle of structural change, driven by selection but without any overall change in function. This 'pushme-pullyou' mechanism of *function switching* can in principle be resolved by gene duplication with subsequent divergence of the paralogues to perform different functions. Indeed, the gene duplications giving rise to the GH cluster occurred after the rapid evolution of the pituitary expressed GH1 gene would probably have ceased although the relatively high evolutionary rate would still have been maintained by the placentally expressed genes. Whilst this idea is very appealing, it should be pointed out that the evolution of the placental lactogens early on in the mammalian radiation might reasonably be expected to have removed any selective pressure on the lactogenic properties of GH.

Another example of the intermittent acceleration and deceleration of the nucleotide substitution rate is provided by the cytochrome c oxidase subunit IV (COX4; 16q22-qter) gene at different stages of primate evolution. In passing, it is interesting to note that conceptually, the idea of function switching had its origin in the days of the early evolutionists:

"It is an error to imagine that evolution signifies a constant tendency to increased perfection. That process undoubtedly involves a constant remodeling of the organisms in adaptation to new conditions; but it depends on the nature of those conditions whether the direction of the modifications effected shall be upward or downward."

Intriguingly, Miyata et al. (1994) have claimed that the rate of evolutionary change of tissue-specific genes varies with the site of expression. Thus, brain-specific genes were reported to evolve at a slower rate than immune system-specific genes. However, since the identity of the compared genes was not revealed and the sample size small, it is hard to comment on the validity or otherwise of the authors' conclusions.

Mutation Rates and their Evolution

Although very much dependent upon the type of mutation being considered and the identity of the gene in question, studies of disease pathology have suggested that the mutation rate in males is higher than that in females. This has been held to reflect the rather higher number of cell divisions required from zygote to mature germ cell in male as compared to the female. In an evolutionary context, Miyata et al. (1987) developed a method for estimating the male-to-female ratio of mutation rates (α_m) from rates of nucleotide substitution in sex-linked and autosomal sequences. They predicted that if α_m were very large, the rate of synonymous substitution in X-linked genes would be ~2/3 of that in autosomal genes on the basis that the X chromosome is twice as likely to be present in a female as in a male. This prediction was borne out by their analysis of human and rodent gene sequences. Further, the rate at which the Y chromosome accumulated substitutions was found to be twice that of the autosomes (the Y chromosome mutates at relative rate α_m as mutation always occurs in a male). The comparison of rodent *Ubel* genes and pseudogenes allowed am to be estimated to be of the order of 2.0. Similarly, comparison of SMCX/SMCY (Xp11.21-p11.22Yq) genes from mouse, horse and human yielded an estimate for α_m of 3.0. By contrast, comparison of intronic sequences of the homologous ZFX (Xp21.3-p22.3) and ZFY (Yp11.32) loci in humans, orangutans, baboons and squirrel monkeys suggested that α_m may be as high as 6.0 in primates. The actual value of α_m may however be very much dependent upon the species compared. Thus, η-globin pseudogene sequence data were used to derive estimates of α_m between 3 and 6 in higher primates but only ~2 in mice and rats. If studies of many different systems indicate that the mutation rate is consistently higher in males than in females, then it may indeed be possible to view evolution as being 'male-driven'.

That some sequence are hypermutable in the human genome is clear from studies of pathological lesions responsible for human genetic disease. Hotspots for somatic mutation are also apparent from studies of mammalian immunoglobulin genes (consensus sequences, G C/T A/T and TAA). Such hotspots can be interpreted in terms of nearest neighbour effects on nucleotide substitution rates. In an evolutionary context, these effects are also apparent from sequence comparison studies of human genes and pseudogenes.

The mutation rate in higher organisms has long been assumed to be compromise between keeping the frequency of deleterious substitutions low at the same time as not completely abolishing the potential for generating adaptive variation. One prediction of this *trade-off theory* is that a lower equilibrium mutation rate should evolve if the deleterious effect of mutation is increased. To examine the validity of this postulate, McVean and Hurst (1997) compared rates of synonymous nucleotide substitution in 33 X-linked genes and 238 autosomal genes in mouse and rate. Since the X chromosome is hemizygous in male mammals, deleterious recessive mutations arising on it might reasonably be expected to have a greater effect on fitness than those arising on the autosomes. If, however, synonymous substitutions were completely neutral with respect of fitness, then they should accumulate at a rate equal to the mutation rate and could be used directly to estimate mutation rates. McVean and Hurst (1997) found that the X-linked genes exhibited significantly lower rates of synonymous substitution than their autosomal counterparts and this was held to be explicable in terms of the X-chromosomal mutation rate being reduced by natural selection.

Mutation pressure may have influenced the evolution of the eukaryotic genetic code in that the code's organization at least appears to provide at least some protection against the deleterious consequences of single base-pair substitutions. However the converse argument is more persuasive: that the design of the genetic code and the functional similarity (or dissimilarity) of amino acids to one another may have affected the relative mutabilities of individual codons during evolution. Codon usage varies quite widely between different vertebrates, a finding which may be related to the influence of the codon frequencies of highly expressed genes on translation efficiency via tRNA pools. Codon usage may also vary between different genes in the same species. In humans, such a finding has been suggested to be related to chromosomal location.

Deleterious Mutation Rate in Humans

It has been suggested that humans may experience a high deleterious mutation rate. If a significant proportion of these mutations were even mildly deleterious, such lesions would tend to accumulate in populations with small effective sizes or in which selection had been relaxed. Eyre-Walker and Keightley (1999) estimated the human deleterious mutation rate per diploid genome per generation, U, by comparing the expected and observed rates of nonsynonymous substitution in 46 orthologous gene pairs from human and chimpanzee. Under conservative assumptions of 60,000 genes in the human genome, an average length of protein coding sequence of 1.52 kb, a human-chimpanzee divergence time of 6 Myrs ago, and an average generation time of 25 years in the human lineage, Eyre-Waler and Keightley (1999) estimated the rate of nonsynonymous substitutions, M, to be 4.2 $\pm$ 0.5 mutations per diploid genome per generation, and the deleterious mutation rate, U, to be 1.6 $\pm$ 0.8 mutations per diploid genome per generation. Estimates of U for chimpanzee (1.7 $\pm$ 0.8) and gorilla (1.2 $\pm$ 0.6) were found to be similar to that obtained for humans. Eyre-Walker and Keightley (1999) concluded that the human deleterious mutation rate is close to the upper limit tolerable by a species with a low reproductive rate. This implies that in hominids, synergistic epistasis may have occurred between deleterious mutations. Further, the level of selective constraint (U/M) inhuman protein coding sequences (0.38 ± 0.17) was judged to be atypically low as were estimates from chimpanzee (0.53 ± 0.16) and gorilla (0.38 ± 0.17). A large number of slightly deleterious mutations may therefore have become fixed in the hominids and the most likely explanation for this, is small long-term effective population size.

Mutation in Pathology and Evolution; Two Sides of the Same Coin

In the early days of genetics, many thinkers saw spontaneous variation purely and simply in terms of its role as the evolutionary fuel for speciation. The first to draw parallels with disease was probably the British geneticist William Bateson who, at the turn of the century, maintained that disease presented 'a discontinuity closely comparable with that of many variations'. Indeed, he speculated that the problem of species (may well) be solved by the study of pathology; for the likeness between variation and disease goes far to support the view which Virchow has forcibly expressed, that 'every deviation from the type of the parent animal must have its foundation on a pathological accident'.

Couched in modern terms, single base-pair substitutions in human gene pathology and evolution may be viewed as to sides of the same coin. This appealing supposition has recently been corroborated by an in-depth comparison of mutations causing inherited disease with mutations in noncoding DNA that have become fixed during the evolutionary divergence of human and other mammalian genome. Mutations in noncoding DNA may date back some millions of years and their survival has been independent of natural selection. By contrast, disease-associated substitutions are of fairly recent origin by comparison with the evolutionary timescale. This difference notwithstanding, under the assumption that the underlying molecular mechanisms of mutation have not changed substantially during mammalian evolution, some resemblance between the two mutational spectra was to be expected. Consistent with this explanation, the relative base substitutions rates observed in the context of human genetic disease were found to be remarkably similar to those derived by Hess et al. (1994) in an extensive analysis of human gene/pseudogene alignments. This pattern of similarity was still apparent after allowance had been made for the DNA sequence environment at the site of mutation by the use of nearest neighbour-dependent mutation rates; the only notable difference was slight under-representation among pseudogene mutations of C→T and G→A substitutions. This was held to be explicable in terms of either a lower level of germline methylation or, perhaps more likely, a deficiency of 5 mC-containing CG dinucleotides in non-coding DNA sequences.

The claim that relative mutation rates exhibit a strong positive correlation between pathological mutations in coding sequences and evolutionarily fixed mutations in non-coding sequences might appear surprising in the light of other experimental results. For example, a study of murine *aprt* gene sequences has claimed that silent mutations accumulate more slowly in transcribed sequences, possibly due to preferential DNA repair. Further, Hanawalt (1990) has shown that in vitro-induced pyrimidine dimers and interstrand DNA crosslinks are repaired with a substantially higher efficiency in active genes than in noncoding regions. Although this type of lesion is specific to the action of particular exogenous chemical mutagens and irradiation, the idea of a system which is generally more effective at removing endogenous mutations from coding DNA as opposed to noncoding DNA is appealing. This is because efficient DNA repair should only have conferred a substantial selective advantage in coding regions. By contrast, the results of Krawczak and Cooper (1996a) suggest that the relative contribution (via variable efficiency) of different DNA repair pathways to the generation of mutations is unlikely to differ substantially between intragenic and intergenic sequences.

To what extent is the likelihood of generation of a mutation related to its phenotypic consequences? When codon substitutions causing genetic disease were categorized according to whether they were neutral or whether they changed the hydrophobicity or polarity of the encoded amino acid residue, it emerged that neutral changes were characterized by larger likelihoods of generation via mutation than nonneutral substitutions. This disparity suggested that selection has operated on the cellular DNA repair machinery in such a way as to optimize the removal of the latter type of mutation. If nonneutral changes were more likely to result in a disadvantageous (disease) phenotype than neutral substitutions, then any repair bias operating against these changes at the DNA level would have had a selective advantage.

The hypothesis of a mutational repair bias was further supported by the finding that the likelihood of generation of an amino acid substitution in humans is negatively correlated with its likelihood of coming to clinical attention. The extent of the correlation was, however, found to differ dramatically between different types of substitutions. Thus, a significant decrease in mutation generation likelihood was only associated with an increased likelihood of clinical observation for substitutions which affected hydrophilic and polar residues. Various explanations may be considered to account for these findings. First, substitutions of hydrophilic and polar amino acid residues could have resulted in more severe and/or variable consequences for the phenotype than other types of substitution during most of the time period that the DNA repair mechanisms were evolving. Second, it has been observed that, during evolution, the effect of an amino acid change in the hydrophobic protein core is often compensated for by another change in the immediate vicinity. Thus, any selective pressure acting upon the organism to avoid substitutions at hydrophobic core residues may have been balanced by the requirement for evolutionary fixation of a second compensatory mutation. Finally, the majority of adaptive events shaping the eukaryotic DNA repair process will have occurred in organisms other than human. It therefore follows that, were current clinical observation likelihoods to be specific to humans, this could have obscured any correlation between mutation generation likelihoods and the phenotypic consequences of substitutions at hydrophobic or nonpolar residues.

If mutations in human genes are biased against amino acid replacements with a high present day probability of resulting in a disease phenotype, the question arises as to whether such a bias might also be reflected in the evolutionary history of protein sequences. To explore this possibility, Krawczak and Cooper (1996a) calculated a quantity termed the *relative evolutionary acceptability* for each possible amino acid substitution from the data of Collins and Jukes (1994) who had reported the numbers and types of amino acid mismatches deduced from the alignment of 337 pairs of human-rodent cDNA sequences. Krawczak and Cooper (1996a) found that clinical observation likelihoods were negatively correlated with evolutionary acceptability values. This implied that the more likely a given amino acid substitution was to result in a disease phenotype (at least in contemporary humans), the less often has it been tolerated during the evolution of mammalian protein sequences.

In summary, it is evident that the evolutionary requirement to avoid a deleterious phenotype has left its footprints in the mechanisms of mutation generation. In this context, the most promising target for selection would appear to be the intracellular DNA repair mechanism.

Although the effect of a given amino acid replacement upon protein structure is known to be heavily dependent upon its precise location within the tertiary structure of the molecule, some basic rules which relate local causes and consequences may nevertheless be perceived. If the efficiency of mutation removal were directed by the immediate DNA sequence context of lesion, it may be that this has facilitated the avoidance during evolution of hazardous amino acid replacements by consideration of the genetic code.

Importance of Evolutionary Conservation in the Study of Pathological Mutations at the Protein Level using Human Factor IX as a Model

The nature, frequency and location of gene lesions causing human genetic disease are highly specific and, are determined in part by the local DNA sequence environment. Once a given mutation has arisen, however, the likelihood that it will come to clinical attention is a complex function of the nature of the resulting amino acid substitution, its precise locations and immediate environment within the protein molecule, and its effects upon protein structure and function. Although in-depth investigations of the *in vivo* effects of missense mutations upon specific human proteins are generally rare, two such studies have nevertheless been performed for human factor IX the liver-expressed zymogen of a vitamin K-dependent serine protease that activates factor X in the presence of factor VIIIa. These studies will be discussed in some detail.

The vast majority of known lesions in the F9 (Xq26-q27) gene causing hemophilia B are missense mutations, causing ~59 of severe (<1% FIX:C) and moderate (1-5% FIX: C) hemophilia, and perhaps as much as 97% of mild (>5% FIX: C) hemophilia. On the basis of 95 independent missense mutations, Bottema et al. (1991) concluded that substitutions of 'generic' factor IX residues (conserved in factor IX of other mammals and in related human serine proteases) almost invariably cause hemophilia B. Mutations at factor IX-specific residues (conserved only in mammalian factor IX) and nonconserved residues were, by contrast, found to be some six to 33-fold less likely to result in a disease phenotype. Even though the study of Bottema et al. (1991) provided new insights into the identity of amino acid residues of structural or functional importance to factor IX, the authors did not employ models of the tertiary structure of the protein or its constituent domains. The significance of the location of specific amino acid residues within the structure of the factor IX molecule to the consequences of mutation could therefore not be assessed. Moreover, neither the variable propensity of different regions of the F9 gene to mutate nor the nature of the resulting amino acid exchanges were considered.

In many ways, factor IX represents an ideal system in which to assess the influence of positional determinants upon the disease-associated mutations spectrum of a single protein. Firstly, the number of known F9 missense mutations is among the highest of all human genes (*Human Gene Mutation Database*). Secondly, the amino acid sequences of numerous other vertebrate factor IX proteins and evolutionarily related serine proteases are available for direct comparison. Finally, the structure of factor IX has been determined by X-ray crystallography and the three-dimensional structures of a number of homologues serine proteases are also known. Wacey et al. (1994) constructed by comparative methods, a multidomain model of the quaternary structure of activated factor IX (FIXa) and used this model to study the expression pathway of F9 gene lesions from genotype to clinical phenotype: a total of 277 different single base-pair substitutions in the F9 gene, comprising 241 missense mutations and 36 nonsense mutations, were analysed. Comparison of the relative nearest neighbour-dependent single base pair substitution rates in the F9 gene with estimates derived from a wide range of other human genes revealed similar profiles (with CpG dinucleotides representing hotspots for mutation), suggesting that similar mutational mechanisms were operating at the DNA level.

Wacey et al. (1994) classified F9 missense mutations as either *conservative* or *nonconservative* on the basis of the chemical difference between the wild-type and the mutant amino acid residue. *Chemically difference* is a measure, originally devised by Grantham (1974), that combined the three interdependent properties of chemical composition, polarity and molecular volume of an amino acid residue. When the magnitude of biochemical change upon substitution was measured in this way and related to the clinical severity of the resulting disease phenotype, nonconservative substitutions (i.e. those characterized by large chemical difference) were found to result in severe rather than mild or moderate hemophilia B approximately 1.7 times more often than conservative substitutions. Conversely, conservative substitutions were 3- to 4- fold more likely to be associated with a moderate or mild phenotype than their nonconservative counterparts. The possibility was considered that conservative amino acid substitutions might be more likely to come to clinical attention in the tightly packed core of the protein as opposed to the surface of the molecule where they might be more readily tolerated. However, since mutations from all chemical difference classes appeared to be scattered over all domains of the protein and since no one chemical difference class was found to be associated solely with surface or buried residues, there appeared to be no relationship between the magnitude of the amino acid exchange and the location of the affected residue.

The extent of evolutionary sequence conservation exhibited by amino acid residues in factor IX was also found to correlate with disease severity. Whilst 71% of mutations at 'highly conserved' residues (residues conserved in all mammalian factor IX proteins and in three other serine proteases) caused severe rather than mild or moderate hemophilia B, this was the case for only 50% of mutations at less conserved residues. Furthermore, Wacey et al. (1994) estimated that missense mutations at non-conserved residues were 15-20 times less likely than mutations at conserved residues to result in a disease phenotype at all. Although this implies that many missense mutations at evolutionarily unconserved residues are tolerated by the molecule and do not come to clinical attention, the relative importance of such residues was considered to be greater than previously claimed by Bottema et al. (1991). Secondly, Wacey et al. (1994) allowed for determinants neglected by Bottema et al. (1991) viz. the actual F9 gene coding sequence and the redundancy of the genetic code. Finally, the two studies were not directly comparable since Wacey et al. (1994) confined their estimation of clinical observation likelihoods to severe cases of hemophilia B in order to cope with the problem of identical-by-descent mutations which are likely to be more prevalent in cases of mild or moderate disease.

Amino acid residues which are sequence conserved both in mammalian factor IX proteins and four different human serine proteases are likely to be critical for functions common to all serine proteases. These residues are located predominantly in the interior of factor IX, within α-helices or β-turns, and Wacey et al. (1994) found that substitutions tended to cluster in the Gla domain, the EGF domains and the serine protease domain (around the reactive site and oxyanion hole). All but one of the Cys residues known to be involved in disulfide bonding were affected by mutation as were the reactive site residues, residues involved in carboxylase recognition and activation peptide cleavage site, and residues which contribute to factor IX binding. By contrast, mutations at factor IX residues which are sequence conserved in mammals and one or two other serine proteases were found to cluster in the serine protease domain and also at domain boundaries within the protein structure. Presumably, docking of the constituent domains of factor IX and the other homologous serine proteases during protein folding requires amino acid conservation at these boundaries. Finally, factor IX residues which are sequence conserved between mammals but not between other serine proteases may be exclusively important for the structure and function of factor IX. When the 3D structure of the factor IX protein was considered, spatially clustered groups of mutations at such residues became apparent on the surface of the EGF domains, in regions implicated in the binding of factors Va and VIIIa, and of the serine protease domain.

Regions which exhibit similar functions in different homologous proteins may not only be sequence conserved but may also exhibit structural conservation. Although *sequence conserved regions* (SeqCRs) of mammalian serine proteases are invariably structurally conserved, *structurally conserved regions* (SCRs) may differ markedly with respect to their amino acid sequences. SCRs were defined by Greer (1990) as those portions of known protein structures that overlap very well when superimposed. In serine proteases, SCRs usually comprise secondary structure elements, the active site and other essential structural framework residues of the molecule.

The locations of structurally conserved regions inhuman factor IXa were determined by Wacey et al. (1994) employing their homology model of the factor IX protein. No clear relationship was noted between the severity of the hemophilia B phenotype and the level of structural conservation of mutated factor IX amino acid residue, although substitutions at structurally conserved residues were estimated to have an approximately two-fold higher likelihood of resulting in a disease state than mutations at nonconserved residues. Interestingly, mutated sites which were not sequence-conserved were nearly all structurally conserved. The only exception involved two missense mutations at Gly59. However, Gly59 lies immediately adjacent to a type β hairpin SCR and would be predicted to be critical in defining this structural element. Some SCRs, although not sequence-conserved, may thus serve as structural supports through their backbone interactions and should therefore be regarded as 'scaffolding' residues rather than 'spacers'.

The topological properties of a mutated factor IX amino acid residue are also important for determining clinical severity. Mutations at residues with their side chains pointing away from the solvent ('buried residues') were found to cause severe hemophilia 1.5 times more often than mutations at residues with solvent-accessible side chains. Finally, the likelihood of a mutation resulting in a severe disease phenotype was higher for substitutions in hydrophobic as opposed to polar regions, probably because of the critical importance of these residues for correct protein folding. This is consistent with the conclusions of other workers that amino acid substitutions occurring in the protein core give rise to a 'continuum of increasingly non-native properties' affecting the stability and/or the folding dynamics of the protein.

Equilibrium of Synonymous Codon Substitutions

Single base-pair substitutions in coding regions that do not change the encoded amino acid sequence can be assumed to be comparatively free of selectional constraints. Although it cannot be entirely excluded that these synonymous (silent) changes might influence gene expression via effects on mRNA translation (e.g. via alterations in RNA secondary structure or local imbalances in the tRNA reservoir of a cell), the probability of survival and ultimate population fixation should in general be much higher for silent mutations than for missense mutations. This view is consistent with the fact that the vast majority of evolutionarily stable base substitutions in coding regions of human genes have taken place at the wobble positions of degenerate codons.

There are 19 groups of triplets which encode the same amino acid such that the constituent triplets of each group can be replaced by each other via single base-pair substitution. If mutations that cause an amino acid exchange are ignored, the mutation dynamics within each group of codons can be modelled by a simple system of linear equations involving the relative rates of different single base pair substitutions. With evolutionary time, this system will approach an equilibrium state and the equilibrium codon frequencies within each group can be determined by solving this system of equations. The actual frequencies within degenerate codons are still some distance from equilibrium in humans.

When a similar analysis is performed for other vertebrates species with known codon usage, it turns out that humans are not closest to their own equilibrium. In 17/19 cases, *Xenopus laevis* ranks first whereas humans and rodents form a second groups of species, all ranking equally low. This class is followed by chicken, dog, cow, pig, rabbit and sheep in order of distance from equilibrium. There is thus some correlation between distance from equilibrium and generation time and, with the exception of humans, the ranking of species reflects the total synonymous substitution rates estimated by Bulmer et al. (1991). These

Table 6.1. Euclidean distance between the vectors of current and equilibrium frequencies within degenerate codons of human genes

Encoded amino acid	*Codon group*	*Euclidean distance*
Glu	GAA GAG	0.083
Lys	AAA AAG	0.088
Asp	GAT GAC	0.164
Asn	AAT AAC	0.173
Tyr	TAT TAC	0.196
His	CAT CAC	0.222
Pro	CCT CCC CCA CCG	0.240
Thr	ACT ACC ACA ACG	0.275
Gln	CAA CAG	0.294
Ser	TCT TCC TCA TCG	0.295
Cys	TGT TGC	0.311
Gly	GGT GGC GGA GGG	0.314
Ala	GCT GCC GCA GCG	0.318
Phe	TTT TTC	0.367
Arg	CGT CGC CGA CGG AGA AGG	0.371
Ser	AGT AGC	0.416
Leu	TTA TTG CTT CTC CTA CTG	0.431
Val	GTT GTC GTA GTG	0.444
Ile	ATT ATC ATA	0.540

data are therefore consistent with a model of DNA sequence evolution which, after species divergence, allows ancestral general sequence to approach equilibrium codon usage faster in one species than in another.

If current codon usage in different species were indeed the result of the divergent evolution of common ancestral sequences progressing at different absolute (albeit equal relative) substitution rates, than one species should always be closer to equilibrium than the other in all codon groups. This, however, is not the case. For example, hamster is closer to equilibrium than dog for the tyrosine (TAT, TAC) and histidine (CAT, CAC) encoding triplets, but not for glutamine (CAA, CAG). Furthermore, the codon frequencies for lysine (AAA, AAG), aspartic acid (GAT, GAC) and glutamic acid (GAA, GAG) in *Xenopus* are on opposite sites of the equilibrium when compared to all other vertebrates.

One may therefore surmise that codon usage has not evolved in a strictly uniform way. Although it cannot e excluded that relative mutation rates differ between species thereby resulting in different equilibria, the fact that *Xenopus* and rodents are very close to an equilibrium which is itself based upon human genetic disease data, argues strongly against this objection. It is thus more likely that species divergence has been accompanied by substantial changes in codon usage, allowing some species to manifest sudden changes with respect to their distance from equilibrium. This would also be consistent with the finding that differences in synonymous codon usage between vertebrates is not explicable merely by differential absolute DNA repair efficiency.

Ikemura and Wada (1991) were able to demonstrate through an analysis of approximately 2000 human gene sequences that codon usage differs dramatically between different genomic regions. The major proportion of GC-rich genes was observed in T-bands whereas AT-rich genes were located mainly in G-bands. Further, the average G+C percentage at the third position of codons was found to be related to the quinacrine dullness and the mitotic chiasma density of a particular chromosome. Since species divergence has almost always been characterized by gross structural chromosomal rearrangements, and since different genes are known to evolve at different absolute rates even in one and the same species, the piece-wise reconstitution of new genomes from their common ancestors is likely to have altered codon frequencies substantially.

Single Base-pair Substitutions in Gene Regions

Neighbouring-Nucleotide Effects on the Rate of Germline Single Base-pair Substitutions in Human Genes

In terms of their relative frequency of occurrence, the most important category of single base-pair substitution causing human genetic disease is represented by C→T and G→A transitions within CpG dinucleotides; some 23% of all pathological single base-pair substitutions found within the coding regions of human genes are of this type. Allowing for the confounding effects of codon usage and differential clinical observation likelihoods through consideration of relative mutabilities, this proportion translates into a mean transition rate for either CG→TG or CG→CA that is five times higher than the base mutation rate.

Krawczak et al. (1998) demonstrated that th proportion of pathological CG→(TG, CA) transitions is significantly higher for autosomal gens (25.0%) than for X-linked genes (17.7%). These proportions are a direct reflection of the significantly lower frequency of CpG in the coding sequences of X-linked genes (2.9%) as compared to autosomal genes (3.7%). Krawczak et al. (1998) speculated that the lower CpG frequency in X-chromosomal genes may be a consequence of a generally increased level of DNA methylation resulting from the evolutionary recruitment of this post-synthetic modification to play a role in X-inactivation.

For CpG dinucleotides to be hypermutable in the context either of genetic disease or evolution, they must be methylated in the germline. Since it cannot be excluded that the efficiency of both DNA methyltransferase action and G : T mismatch repair may be influenced by sequence motifs flanking the CpG dinucleotide, the question arises as to whether some CpGs may be intrinsically more mutable than others by virtue of their DNA sequence context. Significant differences in the relative mutation rate of CpG dinucleotides depending upon their flanking nucleotides were indeed noted by Krawczak et al. (1988). These results were consistent with those of Ollila et al. (1996) who noted a preference for 5' pyrimidines and 3' purines flanking mutated CpG dinucleotides.

Comparison of the sequences flanking the human and chimpanzee β-globin genes has shown that the CpG dinucleotide is hypermutable over evolutionary time and subject to high frequency C→T or G→A transitions. Indeed, some 40% of the CpG dinucleotides present in either the human or chimpanzee sequences were found to be affected by nucleotide sequence changes. Similar conclusions have been drawn by Perrin-Pecontal et al. (1992). Comparison of the CpG mutation rates exhibited by globin gene and pseudogene sequences from human, chimpanzee and macaque yielded an estimate of the rate of 5mC deamination of $\sim 1 \times 10^{-16}$. The absence of any significant difference between deamination rate estimates derived from gene and pseudogene sequence data suggested that the action of selection has had a negligible effect on the transition rate at CpG dinucleotides in primate β-globin genes. Indeed, this constancy in the CpG deamination rate is consistent with a neutralist view of gene evolution. Moreover, the successful use of evolutionary comparisons of DNA sequences to derive consistent values of the CpG deamination rate has demonstrated the feasibility of using the CpG deamination rate as a 'molecular clock', at least over relatively short periods of evolutionary times.

Although the nonrandomness of mutation is at its most dramatic for CpG mutations, non-CpG mutations also appear to be distributed nonrandomly. A subtle neighbouring nucleotide effect reminiscent of misalignment models of mutagenesis was noted by Krawczak et al. (1998) in their study of pathological mutations; a substantial proportion of the observed single base-pair substitutions exhibited identity of the newly introduced base to one of the bases immediately flanking the site of mutation. Since this effect occurred only at a distance of one base-pair and in the absence of surrounding repetitive sequences, it was potentially explicable in terms of misalignment mutagenesis involving highly localized DNA slippage, misincorporation and realignment events between template and primer at the replication fork. Intriguingly, this next neighbour mutational bias only occurred at specific codon positions and exhibited polarity. Since such a phenomenon is unlikely to be explicable in terms of the primary mutational event, it is probably associated with the DNA repair process. Implicit in this assumption, however, is that the DNA repair machinery is able to recognize the reading frame and to utilize this information as a cue in effecting DNA repair. Consistent with such a relationship, the observed correction bias would operate in such a way as to remove newly introduced termination codons. The ability of a DNA repair mechanism to take the reading frame into account, thereby minimizing the effects of mutation, would have had positive selective value owing to the relatively deleterious nature of in-frame termination codons. Evidence for reading frame sensitivity in the DNA repair process first came from the observation that relative single base-pair substitution rates are biased toward the avoidance of those replacements that (i) change the chemical characteristics of the encoded amino acid residue substantially and (ii) have a high likelihood of coming to clinical attention. Krawczak et al. (1998) concluded that, by consideration of the genetic code, selection has optimized the DNA repair mechanism in such a way as to avoid the most hazardous of amino acid replacements, a category which would certainly include nonsense mutations.

Although the same substitution types appear to be subject to next neighbour effects on the coding and noncoding DNA strands, the quantitative differences in non-CpG single base-pair substitution rates observed by Krawczak et al. (1998) confirmed that the two DNA strands are not fully equivalent in terms of their rates and patterns of mutation. Nucleotide substitutions that have accumulated in noncoding sequences during evolution are also asymmetric between DNA strands. There are several possible (and nonmutually exclusive) reasons for this strand asymmetry. Firstly, the four nuclear DNA polymerases, each associated with its own distinctive mutational spectrum, may be differentially involved in the synthesis of the leading and lagging strands during DNA replication. Secondly since the transcriptional elongation complex is asymmetric, mutation rates may differ between transcribed and non-transcribed strands on account of either unequal exposure to DNA damage or differential repair. Not only may the transiently single-stranded nontranscribed DNA strand be particularly vulnerable to mutation (e.g. by methylation-mediated deamination) but transcription-coupled repair, a process that corrects lesions specifically on the transcribed DNA strand, could also account for different mutation rates between transcribed and nontranscribed strands. Both these mechanisms would predict a higher mutation rate for the nontranscribed as opposed to the transcribed DNA strand.

Finally, in their dataset of pathological substitutions, Krawczak et al. (1998) found that thermodynamic stability of DNA triplets was positively correlated with the average relative rate at which the central nucleotide of a triplet underwent substitutions. At first sight, this finding might appear to be counterintuitive since it implies that higher rather than lower DNA duplex stability would render a gene region more prone to single base-pair substitution. However, the consistent absence of flanking repeat elements noted for the mutations analyzed here suggests that extensive strand slippage (which would require the DNA to be single-stranded) is unlikely to play an important role in the generation of single base-pair substitutions. Nevertheless, a high degree of thermodynamic stability could in principle impair DNA replication in various ways. First, the likelihood that DNA helicases would be incapable of unwinding the two DNA strands correctly or efficiently may be expected to be higher in more stable regions. Second, temporary reannealing of the two native DNA strands during replication might be favoured and could be more enduring in such regions. In both cases, DNA polymerase activity would be seriously impeded by localized double-stranded DNA structures, which could result in either a cessation of polymerization or the skipping of one or more nucleotides, leaving a gap in the nascent DNA strand. Miscorrection during the post-replicative repair of such nicks would then introduce a single base-pair substitution. Alternatively, the observed correlation could reflect the increased stability of at least some slippage-mediated misalignments during replication of the native and nascent DNA strands, allowing enough time for misincorporation of a noncomplementary nucleotide. In this case, however, the thermodynamic stabilities of the misaligned structures must be comparable to those of the wild type triplets, an assumption for which there is currently no evidence.

Single Base-pair Substitutions in Evolution which have Altered the Function of Specific Amino Acid Residues

Some single base-pair substitutions occurring during gene evolution have introduced nonsense mutations into protein coding regions thereby either prematurely truncating the protein product or abolishing the expression of that product altogether. However, other single base-pair substitutions have introduced missense mutations that have served to alter the function of specific amino acid residues and these are the topic of this Section.

Nucleotide substitution that have occurred and become fixed through the action of genetic drift are readily apparent in any comparative analysis of orthologous genes and provide abundant evidence for a neutralist model of evolution. Nucleotide substitutions resulting in human genetic disease may be taken as representative examples of negative (purifying) selection since many of the lesions that have come to clinical attention are likely to reduce (or would once have reduced) survival and/or reproductive fitness. By contrast, unequivocal example of single base pair substitutions that have occurred within the coding regions of mammalian genes during evolution which have been sufficiently well characterized for us to be able to identify the consequent change in protein structure and function that has been subject to positive selection. Several illustrative examples are discussed below. Since these systems have not so far been fully characterized, further studies are eagerly awaited.

Visual pigments

The visual pigments, which comprise an integral membrane protein (opsin) coupled to a light-sensitive chromophore, are spectrally tuned to a particular wavelength of maximal absorption, λmax. The visual pigments in rods (photoreceptors cells that function in dim light) are rhodopsins which have λmax 495 nm. In codes, the photoreceptor cells that mediate colour vision, there are three are three types of visual pigment which in humans have λmax values of 420 nm (blue/short wavelength-sensitive), 530 nm (green/middle wave-length-sensitive) and 560 nm (red/long wavelength-sensitive). The human genes encoding these opsins are: rhodopsin (RHO; 3q21-24), the blue cone pigment (BCP; 7q31-35), the green cone pigment (GCP; Xq28), and red cone pigment (RCP; Xq28). These genes have evolved by a process of duplication and divergence from a common ancestor ~500 Myrs ago and encode proteins that harbour specific amino acid changes that are directly responsible for the shifts in λmax values between the different visual pigments.

To understand the molecular basis of spectral tuning of visual pigments, it is necessary to correlate the sequences of the visual pigments with their λmax values. Such an analysis was first performed by Yokoyama and Yokoyama (1990) who compared the red and green visual pigments from human and the Mexican cave-fish, *Astyanax fasciatus*. The red pigments in humans and cavefish were found to have evolved from the green pigments *independently* by three specific amino acid changes (Ala180Ser, Phe277Tyr and Ala285Thr, i.e. AFA SYT). A similar study in primates also concluded that the spectral difference between red and green pigments could be accounted for by the difference between AFA (green) and SYT (red). These three critical amino acid residues are located near the chromophore and their experimental substitution has been shown to alter λmax values not only in human red and green pigments but also in bovine rhodopsin.

Most mammals have dichromatic vision, possessing blue visual pigments together with either red or green pigments. Phylogenetic analysis of a range of vertebrates has suggested that the vertebrate ancestor of the opsin gene was a green pigment gene encoding AFA at the three critical sites and that the common ancestor of tetrapods acquired a red pigment by two amino acid substitutions. The SYT of the red pigments of extant tetrapods was acquired by a further amino acid substitution, Ala180Ser.

Old world monkeys have both red and green pigments and therefore possess trichromatic vision. Phylogenetic analysis of the primate opsins has indicated that the common ancestor of hominoids, Old World monkeys and new World monkeys possessed a red visual pigment with AYT at the three critical sites but no green pigment. The ancestor of the green pigment gene is thought to have arisen by gene duplication ~35 Myrs ago in the Old World monkey lineage and the green pigment was then derived in two distinct steps (AFAβ→AYT→AFA).

New World monkeys have only one X-linked opsin gene but exhibit multiple alleles at this locus. Thus, in the squirrel monkey (*Saimiri sciureus*), the visual pigments derived from three alternative alleles have λmax values of 532, 547, and 561 nm and manifest amino acids AFA, AFT, and SYT respectively at the critical sites. Whilst all male and homozygous female squirrel monkeys are dichromatic, heterozygous females that happen to have pigments with lmax values of 532 nm and 561 nm possess trichromatic vision. Tovee (1994) suggested that colour vision in New World monkeys might be an adaptation to allow a wide variety of colour vision types within a single family group. However, an alternative explanation, which does not invoke group selection, would be for the alternative alleles to be maintained in the population by overdominant selection. This postulate appears to be supported by the independent evolution of triallelic systems in several other New World monkey lineages represented by the marmoset (*Callithrix jacchus*), the saki monkey (*Pithecia irroata*), the capuchin (*Cebus nigrivittatus*) and the tamarin (*Saguinus mystax*).

Zhou et al. (1997) studies the X-linked opsin gene of two nocturnal prosimians, the bushbaby species *Galago senegalensis* and *Otolemur garnettii*. At those amino acid positions known to cause spectral differences, however, the cone pigment possessed identical residues to those of the marmoset protein. This suggests that in spite of the bushbaby's nocturnal existence, its X-linked opsin gene is under functional constraint. Consistent with this view, Zhou et. al. (1997) noted two amino acid substitutions which may be important for maximizing dim light sensitivity.

Kringle domains of apolipoprotein(a)

The human apolipoprotein(a) (LPA; 6q27) gene emerged from the plasminogen (PLG; 6q27) gene by a process of gene duplication followed by intragenic amplification of exons within the LPA gene. Apolipoprotein (a) is a component of the cholesteryl ester-rich particle lipoprotein(a) within which it is covalently linked to apolipoprotein B100. Lipoprotein(a) has been postulated to play a role in fibrinolysis by competing with plasminogen for binding to fibrin, thereby interfering with clot lysis. Fibrin binding appears to be potentiated by lysine-binding sites in some of the many kringle domains of apolipoprotein(a). Kringle IV-10 of human apolipoprotein(a) most closely resembles that of plasminogen kringle IV and appears to be critical for the fibrin-binding potential of lipoprotein(a). The lysine binding sites of apolipoprotein(a) consist of a hydrophobic trough containing three aromatic amino acids (Trp62, Phe64, and Trp72), an anionic centre composed of two aspartic acid residues (Asp55 and Asp57), and a cationic centre comprising two residues, Lys35 and Arg71. Kringles IV-1, IV-2, IV-3, and IV-4 of apolipoprotein(a) do not bind to lysine and in each case, this is associated with the absence of Asp57 within the kringle. Two different substitutions in kringle IV-10 have occurred during primate evolution which are associated with the loss of the lysine-binding properties of the lipoprotein(a) particle; a Trp-72→Arg substitution in rhesus monkey and an Asp57→Asn substitution in chimpanzee. The physiological consequences of these two substitutions are however as yet unknown.

DNA-binding specificity of steroid receptors

Gene regulation by steroid hormones is mediated by binding of the hormone ligand to its cognate receptor. Upon ligand binding, most nuclear receptors then interact as dimers with their response elements, each monomer binding to a half-site sequence. These response elements comprise two 6 bp half-site sequences organized as palindromic repeats with 3-5 bp separating the half sites. Thus the thyroid hormone, retinoic acid and estrogen receptors recognize the half-site sequence TGACCT whereas the androgen, progesterone and glucocorticoid receptors recognize the half-site sequence TGTTCT. More specifically, the consensus estrogen response element (ERE) is AGGTCANNNTGACCT whilst the consensus glucocorticoid response element (GRE) is GGTACANNNT-GTTCT. These response elements exhibit a 3 bp spacer between half-sits. By contrast, the thyroid hormone receptor recognizes a 4 bp spacer and the retinoic acid receptor a 5 bp spacer. Thus, both the sequence and the spacing between the half-sites determines the specificity.

Most nuclear receptors have two or three amino acids (P box residues) in their DNA-binding domains that serve to specificity recognition of the responses elements within the promoter of the target gene. Some residues (e.g. Val443 in the glucocorticoid receptor or Glu439 in the estrogen receptor) can contribute to specificity both by forming a positive interaction with a base in the cognate response element and by forming and negative interaction with a base in a noncognate response element. In the glucocorticoid receptor, Ser440 inhibits the interaction of the receptor with the ERE but at the cost of also reducing affinity for the GRE. Thus, the diversification of steroid receptor specificity was probably achieved during evolution by a relatively small number of single base-pair substitutions either in the P-box encoding residues of steroid receptors or in the response elements of their target genes.

Calcium-dependent and-independent synaptotagmins

Synaptotagmins constitute a large family of proteins involved in membrane trafficking. At least five different synaptotagmin genes have been characterized in the human genome (SYT1, 12cen-q21; SYT2, 1q; SYT3, 19q; SYT4, 5q; SYT5, 11p). Most synaptotagmins are capable of binding calcium through their calcium-binding C_2A domains. Synaptotagmins IV and IX are unique in their inability to bind calcium and Von Poser et al. (1997) have shown that this inability is caused by the substitution of Ser for Asp at residue 230 in the C_2A domain. This substitution is evolutionarily conserved in synaptotagmin IV and it seems likely that synaptotagmins IV and XI shared a common ancestor in which the substitution originated. Von Poser et al. (1997) postulated that evolution selected for the loss of calcium binding in two different synaptotagmins while leaving the remainder of the protein structures intact. This would of course imply that these synaptotagmins also possess calcium-independent properties and that these properties have been retained by selection through evolutionary time.

Olfactory receptor-ligand interactions

Certain residues in the α-helical sixth transmembrane domain of human olfactory receptors have been implicated interactions with their odorant ligands and these residues appear to have been subject to positive selection. The highly variable residues 6-22 and 6-25 are thought to constitute a receptor sub-site that binds hydroxyl groups on odorant ligands; substitutions at these critical positions may therefore help to determine the odor specificity of olfactory receptor sub-types. A Val/Ile difference at residue 206 of the orthologous rat and mouse I7 olfactory receptor proteins has been shown to be responsible for a species-specific odorant response; rats prefering octanal to heptanal, mice the reverse.

Single Base-pair Substitutions in Evolution which have Affected mRNA Splicing

Single base-pair substitutions affecting mRNA splicing make up between 8% and 15% of all mutations causing human genetic disease. These lesions occur disproportionately at the most evolutionarily conserved positions within the splice and fall into three main categories: (i) mutations within 5' or 3' splice sites which reduce the amount of correctly processed mature RNA and/or activate alternative ('cryptic') splice sites in the vicinity, (ii) mutations outwith actual splice sites which create cryptic splice sites, and (iii) mutations in the branch-point sequence. The vast majority of the pathological lesions affecting mRNA splicing so far reported have been single base-pair substitutions within splice sites. This is not only because these are comparatively frequent but also because they are both readily detectable and highly likely to result in a severe clinical phenotype. Disease-associated mutations affecting 5' splice sites are approximately twice as frequent as mutations at 3' splice sites. This discrepancy coincides with a much higher level of sequence conservation at 5' splice sites and is likely to reflect the strong requirement for U1 snRNA binding at 5' splice sites to promote alignment and cleavage. Regarding the phenotypic consequences of pathological mutations affecting mRNA splicing, the exclusion of one or more exons from the end-product (*exon skipping*) is observed more frequently than *cryptic splice site utilization*. Some evidence exists, at least for 5' splice sites, that cryptic splice site usage is favoured under conditions where a number of potential sites are present in the vicinity of the mutated splice site and where these potential splices sites exhibit sufficient homology to the consensus sequence. In most such cases, the activating mutation improves the similarity between the cryptic site and the splice site consensus sequence. At the 3' splice sites, the amount of mRNA product consequent to the utilization of the cryptic splice site appears to be correlated with the level of similarity to the consensus sequences; at 5' splice sites, the distance to the nearest wild-type splice site may also play a role.

Splicing mutations and gene inactivation

The alteration of mRNA processing as a result of mutations in splice sites also occurs during evolution. One dramatic example is that of the once active human L-gulono-λ-lactone oxidase (*GULOP*) gene (Chromosome 8p21.1) which was inactivated at least in part by the introduction of single base-pair substitutions at the invariant bass of splice sites. Another similar example is provided by the CSHL1 gene, long regarded as a pseudogene of the human placentally expressed growth hormone/chorionic somatomammotropin gene family which is clustered on chromosome 17q22-q24. The CSHL1 'pseudogene' was originally thought to have been inactivated by the introduction of a G→A substitution in the first position of the intron 2 donor splice site. However, *in vitro* expression studies have shown that it may only have been partially inactivated. Although five alternatively spliced forms of CSHL1 mRNA are produced from the 'pseudogene', the majority of these transcripts lack exon 2 which encodes the signal peptide necessary for normal secretion. Some low abundance CSHL1 mRNAs are nevertheless produced which possess the leader peptide and these could in principle encode novel hormones of physiological significance.

Splice site differences between orthologous genes

In an evolutionary context, gene inactivation is likely to be a relatively rare occurrence and alternations in splice junction sequences are more often to be found associated with more subtle changes in mRNA processing such as alternative splicing. A number of examples have been noted in orthologous genes. Thus, in the erythroid 5-aminolevulinate synthase (ALAS1; 3p21.1) gene, exon 4 is involved in alternative splicing in the human but not in the dog or mouse. Conversely, the alternative splicing of the 45 bp exon 3 of the murine *Alas1* gene utilizes a major upstream splice site (85% of mRNAs) and a minor downstream site (15% of mRNAs). This is not found in human owing to an A G transition which abolishes the consensus sequence of the 3' splice site thereby preventing the possibility of alternative splicing. Finally, the alternative splicing pathway involving the mutually exclusive exons 6A and 6B of the β-tropomyosin (TPM2; 9q13) gene differs between chicken, rat and *Xenopus*. The chicken β-tropomyosin exon 6A is flanked by stronger splicing signals than its rat counterpart and this has been related to inter-specific differences in both the donor and acceptor splice sites. A chicken-specific pyrimidine-rich splicing enhancer present upstream of exon 6A may also play a role.

Splice site differences between paralogous genes

Evidence for splice site mutations having occurred during evolution has also come from the study of paralogous genes. The human glycophorin B (GYPB; 4q28.2-q31) gene lacks exon 3 by comparison with the related glycophorin A (GYPA; 4q28.2-q31) gene owing to a G→T transversion at the exon 3 donor splice site. Comparison with the GYPB gene has identified an A→T substitution in the first base of exon 5 of the GYPA gene which leads to the alternative use of an acceptor splice site 9 bp upstream and the incorporation of nine extra bases into the GYPA coding sequence.

The human genes encoding interleukin-1α, (IL1A; 2q13-q21), interleukin-1β (1L1B: 2q13-q21) and interleukin-l receptor antagonist (IL1RN; 2q14.2) are evolutionarily related members of the interleukin family which remain closely linked on the long arm of chromosome 2. The first exon of IL1RN (encoding a leader peptide) is homologous to the untranslated first exon of IL1B but the IL1RN gene lacks the exons corresponding to the first three expressed exons of the IL1A and IL1B genes. Hughes (1994) suggested that the common ancestor of the IL1B and IL1RN genes was an alternatively spliced gene: one transcript could have included exons 1-7 encoding the ancestral IL1B protein whereas the other transcript may have included exons 1 and 5-7 encoding the IL1RN protein. The duplication of this ancestral gene would have freed one copy from functional constraints so that it could encode IL1RN only. Selection would no longer have been able to conserve the intron-exon junctions involved in the splicing of exons 2-4 of the IL1RN gene. Consistent with this view of events, the region of the IL1B gene between exons 1 and 5 is more than twice the length of the analogous region of the IL1RN gene.

Finally, the pituitary-expressed growth hormone (GH1; 17q22-q24) gene differs from the closely linked placentally-expressed growth hormone (GH2) gene in its pattern of splice site selection. Whereas 90% of GH1 mRNA transcripts contain exon 2 spliced to an alternative acceptor site located 45 bp into exon 3, the GH2 gene does not utilize this alternative splicing pathway. Three single base substitutions located between the two alternative acceptor splices sites have been shown to be both necessary and sufficient to define the GH1 alternative splicing event. One of these changes is thought to specify a lariat branchpoint essential for alternative acceptor site usage whereas the other two bases may serve to modulate the frequency with which the site is selected.

7

Pattern of Nucleotide Substitution

The mathematical theory can be used to study the rate of nucleotide substitution, which is a basic quantity in the study of molecular evolution. In order to characterize the evolution of a DNA sequence, we need to know how fast it evolves. It is also interesting to compare the substitution rates among genes and among different DNA regions of a gene, because this can help us understand the mechanism of nucleotide substitution in evolution.

Knowing the rate of nucleotide substitution may also enable us to date evolutionary events such as the divergence between species. However, to do this we need to know whether the rate estimated from one group of species is applicable to another group. This raises the issue of how variable the rate is among different evolutionary lineages or, in other words, whether there exists a molecular clock.

In this chapter we present data on the rates of nucleotide substitution and discuss the factors that may cause differences in rates and patterns of nucleotide substitution.

Estimation of Substitution Rates

The *rate of nucleotide substitution* is commonly defined as the number of substitutions per site per year and can be calculated by dividing the number of substitutions per site between two homologous sequences (K) by 2T, where T is the time of divergence between the two sequences. This is,

$$r = K/(2T) \qquad ...(1)$$

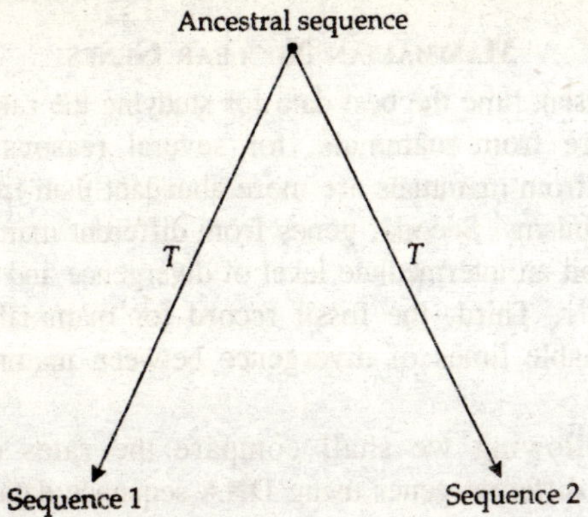

Fig. 7.1. Divergence of two homologous sequences from a common ancestral sequence T years ago.

The divergence time, T, is usually assumed to be the same as the time of divergence between the two species from which the two sequences were taken and is usually inferred from paleontological data.

From Equation 1 it is easy to see that the accuracy of r depends on the accuracy of the estimates of K and T. When K is small, estimates of K have large standard errors relative to its expected value (i.e., large coefficients of variation), unless the sequence is long. Moreover, if T is small, then even a small error in the estimation of T can lead to a substantial error in r, because T is in the denominator. For these reasons, DNA sequences from closely related species are not suitable for estimating r. On the other hand, if K is large, it is difficult to obtain a reliable estimate of K owing to difficulties in correcting for multiple substitutions at the same site. Thus, highly divergent sequences are also not suitable for estimating r.

In this chapter we will deal with the issue of rate variation among genes and among different regions in a gene. For this purpose, it is advisable to use the same species pair for all the genes under consideration. The reason is twofold. First, there are usually considerable uncertainties about paleontological estimates of divergence times. By using the same pair of species we can compare rates of substitution among genes without knowledge of the divergence time. Second, the rate of substitution may vary considerably among lineages. In this case, differences in rate between two genes may be due to differences between lineages (i.e., lineage effects) rather than due to differences that are attributable to the genes themselves.

MAMMALIAN NUCLEAR GENES

At the present time the best data for studying the rate of nucleotide substitution are from mammals, for several reasons. First, DNA sequence data from mammals are more abundant than from most other groups of organisms. Second, genes from different mammalian orders generally exhibit an intermediate level of divergence and so are suitable for estimating r. Third, the fossil record for mammals is relatively good, and reliable times of divergence between mammalian species are available.

In the following we shall compare the rates of nucleotide substitution in different genes using DNA sequence data from humans and mice or rats; if both the mouse and rat genes have been sequenced, the mouse gene is used. The human and rodent lineages are chosen because many gene airs are available for comparison. As will be seen, the rodent lineage appears to have evolved considerably faster than other mammalian lineages, so the average rate obtained below may be higher than the average for mammalian lineages. For noncoding regions, it is not easy to obtain reliable alignments between human and rodent sequences, and other pairs of species will be used.

As different regions of a gene serve different functions and are apparently subject to different stringencies of functional constraints, it has been customary to treat different regions separately.

Coding Regions

The protein-coding regions of genes have attracted the most attention from both molecular and evolutionary biologists because of their functional importance. As a consequence, a large amount of sequence data has become available for these regions and many comparative studies of nucleotide sequences in these regions have been published. These studies have shown that synonymous substitution usually occurs much faster than nonsynonymous substitution.

Table 7.1 lists the rates of synonymous and nonsynonymous substitutions for 47 genes. The rates were obtained by using the method of Li (1993) and Pamilo and Bianchi (1993) and by assuming that the divergence time between the human and rodent lineages is 80 million years.

It is clear from Table 7.1 that the rate of nonsynonymous substitution is extremely variable among different kinds of genes, ranging from effectively zero for the histones to about 3×10^{-9} substitutions per nonsynonymous site per year for interferon γ. Nonsynonymous substitution rates are, of course, reflected in the rates of protein

Table 7.1. Rates of synonymous and nonsynonymous substitution in various mammalian protein-coding genes

Gene	*L*	*Nonsynonymous rate*	*Synonymous rate*
Histones			
Histone 4	102	0.00±0.00	3.94±0.81
Histone 3	135	0.00±0.00	4.52±0.87
Ribosomal proteins			
S17	134	0.06±0.04	2.69±0.53
S14	150	0.02±0.02	2.16±0.42
Contractile system protein			
Actin α	376	0.01±0.01	2.92±0.34
Myosin β heavy chain	1933	0.10±0.01	2.15±0.13
Hormones and other active peptides			
Somatostatin-28	28	0.00±0.00	3.10±1.98
Insulin	51	0.20±0.10	3.03±1.02
Insulin C-peptide	31	1.07±0.37	4.78±2.14
Insulin-like growth factor II	179	0.57±0.11	2.01±0.37
Erythropoietin	191	0.77±0.12	3.56±0.53
Parathyroid hormone	90	1.00±0.20	3.47±0.87
Luteinizing hormone	140	1.05±0.17	2.90±0.54
Growth hormone	189	1.34±0.17	3.79±0.63
Urokinase-plasminogen activator	430	1.34±0.11	3.11±0.35
Interleukin-1	265	1.50±0.15	3.27±0.46
Relaxin	53	2.59±0.51	6.39±3.75
Hemoglobins and myoglobin			
α-globin	141	0.56±0.11	4.38±0.77
β-globin	146	0.78±0.14	2.58±0.49
Myoglobin	153	0.57±0.11	4.10±0.85
Apolipoproteins			
β (partial)	273	1.32±0.14	3.64±0.50
E	291	1.10±0.12	3.72±0.51
A-I	235	1.64±0.17	3.97±0.63
Imunoglobulins			
Ig k	106	2.03±0.30	5.56±1.18
Ig V_H	100	1.10±0.20	4.76±1.12
Interferons			
γ	136	3.06±0.37	5.50±1.45
α1	166	1.47±0.19	3.24±0.66
β1	159	2.38±0.27	5.33±1.24

Other proteins			
Aldolase A	363	0.09±0.03	2.78±0.33
APRT	179	0.68±0.11	3.56±0.59
Albumin	590	0.92±0.07	5.16±0.48
Hydroxanthine phosphoribosyl-transferase	217	0.12±0.04	1.57±0.31
Creatine kinase M	380	0.15±0.03	2.72±0.34
Glyceraldehyde-3-phosphate dehydrogenase	332	0.20±0.04	2.30±0.30
Lactate dehydrogenase A	331	0.19±0.04	4.06±0.49
Fibrinogen γ	411	0.58±0.07	4.13±0.46
Amylase	506	0.63±0.06	3.42±0.38
ANF	149	0.72±0.13	3.38±0.60
BADR	412	0.45±0.06	3.07±0.35
CA I	260	0.84±0.11	3.22±0.47
EF 2	857	0.02±0.01	4.37±0.36
Glucagon	29	0.00±0.00	2.36±1.08
Glutamine synthetase	371	0.23±0.04	2.44±0.30
Lipoprotein lipase	437	0.19±0.04	2.95±0.33
Prion protein	224	0.29±0.06	3.89±0.63
TNF	231	0.76±0.11	2.91±0.45
Thymidine kinase	232	0.43±0.08	3.93±0.59
Average		0.74(0.67)	3.51(1.01)

evolution, which may vary by over two orders of magnitude. As is well known, certain structural proteins (e.g., histones, actins and ribosomal proteins) are extremely conservative. Some peptide hormones, e.g., somatostatin-28 and glucagon, are also extremely conservative, but others have evolved at intermediate rates (e.g., erythropoietin and parathyroid hormone) or at high rates (e.g., relaxin). The insulin C-peptide has often been used as an example of rapid evolution, but it actually has evolved considerably more slowly than relaxin. Most enzymes have evolved at low rates. Hemoglobins, myoglobin, and some carbonic anhydrates have evolved at intermediate rates, while apolipoproteins, immunoglobulins, interferons, and interleukins have evolved rapidly. Apolipoprotein B is a huge protein (4536 amino acids) and the relatively high rate in the region included in Table 7.1, which is actually the most conservative part of the protein (the low density lipoprotein receptor-binding domain), implies that the whole protein has evolved at a high rate. In contrast, the entire sequence of myosin β heavy chain has evolved at a low rate, though it is also a large protein (1933 amino acids). The average rate of nonsynonymous substitution for all the genes in Table 7.1 is 0.74×10^{-9} substitutions per nonsynonymous site per year; the method of Li et al. (1985b) gives a slightly higher average rate, i.e., 0.78×10^{-9}. It is worth noting that the rate in the β globin is close to the average rate.

The rate of synonymous substitution also varies considerably from gene to gene, although not so much as the rate of nonsynonymous substitution. Part of the variation is probably due to statistical fluctuations, because the number of synonymous sites compared is small for many genes. For example, the estimate for the synonymous rate for relaxin is very high, but it has a large standard error. The average synonymous rate over all genes is 3.51×10^{-9} substitutions per synonymous site per year. This value is considerably smaller than that given in Li and Graur (1991) because they used the earlier method Li et al. (1985b); when the latter is applied to the data in Table 7.1, it gives a value of 4.31×10^{-9}, which is close to that of Li and Graur (1991).

The sites in a coding region can also be divided into nondegenerate, twofold degenerate, and fourfold degenerate sites. Table 7.2 shows that at fourfold degenerate sites the rate of transitional substitution (2.2×10^{-9}) is higher than the rate of transversional substitution (1.5×10^{-9}), although at each site two types of transversional change and only one type of transitional change can occur. This observation can be largely explained by the fact that transitional mutation occurs more frequently than transversional mutation. At twofold degenerate sites, the rate of transitional substitution (1.9×10^{-9}) is only slightly lower than that at fourfold degenerate sites, but the rate of transversional substitution (0.4×10^{-9}) is considerably lower than the corresponding rate at fourfold degenerate sites, owing to the fact that all transversional changes at twofold degenerate sites are nonsynonymous. At nondegenerate sites, the rates of transitional and transversional substitution are about the same ($\sim 0.4 \times 10^{-9}$) and are considerably lower than the corresponding values at fourfold degenerate sites because all changes at nondegenerate sites are nonsynonymous. Thus, the (total) rate of nucleotide substitution is lowest at nondegenerate sites, intermediate at twofold degenerate sites, and highest at fourfold degenerate sites.

Noncoding Regions

Data from noncoding regions are much less abundant than data from coding regions, and so only a limited comparative analysis can be done at the present time. (Note that in order to estimate the rate

Table 7.2. Rates of transitional and transversional substitution at nondegenerate, twofold degenerate, and fourfold degenerate sites of codons

Type of substitution	*Nondegenerate*	*Twofold degenerate*	*Fourfold degenerate*
Transition	0.40	1.86	2.24
Transversion	0.38	0.38	1.47
Total	0.78	2.24	3.71

of substitution in a sequence we must have data from at least two species.) Since most published sequences are mRNAs, which do not include introns and flanking regions, the 5' and 3' untranslated regions are the only noncoding regions that can be studied in detail. Table 7.3 shows the substitution rates in these two regions for 16 genes from humans and rodents. In both regions the rates vary greatly among

Table 7.3. Rates of nucleotide substitution in 5' and 3' untranslated regions and at fourfold degenerate sites of protein-coding genes, based on comparison between human and mouse or rat genes

	5' untranslated		*3' untranslated*		*Fourfold degenerates*	
Gene	*L*	*Rate*	*L*	*Rate*	*L*	*Rate*
ACTH	99	1.87±0.41	97	2.32±0.49	275	2.78±0.34
Aldolase A	124	1.08±0.26	154	1.73±0.32	195	3.16±0.48
Apolipoprotein A-IV	83	3.06±0.68	134	1.73±0.33	160	3.38±0.50
Apolipoprotein E	23	1.27±0.69	84	1.70±0.42	153	4.00±0.60
Na,K-ATPase β	118	2.45±0.45	1,117	0.57±0.06	118	2.87±0.54
Creatine kinase M	70	1.71±0.46	168	1.79±0.30	178	2.81±0.41
α-fetoprotein	47	3.64±1.13	144	2.79±0.49	225	4.14±0.54
α-globin	34	1.56±0.65	90	2.21±0.50	81	4.47±0.98
β-globin	50	1.30±0.46	126	2.85±0.49	78	2.42±0.56
Glyceraldehyde-3-phosphage dehydrogenase	70	1.34±0.38	121	1.74±0.36	170	2.43±0.39
Growth hormone	21	1.79±0.85	91	1.83±0.41	83	3.82±0.78
Insulin	56	2.92±0.80	53	3.09±0.81	62	4.19±1.00
Interleukin-1	59	1.09±0.38	1,046	2.02±0.14	105	2.97±0.60
Lactate dehydrogenase A	95	2.79±0.55	470	2.48±0.23	152	3.64±0.60
Metallothionin-II	61	1.88±0.52	111	2.57±0.48	23	2.37±1.00
Parathyroid hormone	84	1.79±0.43	228	2.21±0.30	38	3.85±1.21
Average		1.96 (0.78)		2.10 (0.61)		3.33 (0.69)

genes, but this variation may largely represent sampling effects due to the fact that both of these regions are usually very short. In almost all genes, the rates in the 5' and 3' untranslated regions are lower than the rate of substitution at fourfold degenerate sites (i.e., site at which all possible nucleotide substitutions are synonymous). The average rates for the 5' and 3' untranslated regions are 1.96×10^{-9} and 2.10×10^{-9} substitutions per year, respectively, which are both about 55% of the average rate of 3.71 substitutions per year at fourfold degenerate sites.

Pseudogenes are DNA sequences that were derived from functional genes but have been rendered nonfunctional by mutations that prevent their proper expression. Since they are subject to no functional constraints, they are expected to evolve at a high rate. Table 7.4 shows a comparison between the rate of substitution in cow and goat $\psi\beta^X$ and $\psi\beta^Z$ globin pseudogenes and the rates in the noncoding regions and fourfold degenerate sites in the β- and γ-globin genes. The rate in these pseudogenes is indeed slightly higher than that in the other regions. This result seems to be generally true for pseudogenes, though currently pseudogene data are limited.

Table 7.4. Divergence (±standard error) in 5' and 3' flanking (FL) regions between cow and goat β- and γ-globin genes and between cow and goat β-globin pseudogenes.

Statistic	*5'FL*	*5'UT*	*Fourfold*	*Introns*	*3'UT*	*3'FL*	*Pseudogenes*
Present divergence	5.3	4.0	8.6	8.1	8.8	8.0	9.1
Standard error	1.2	2.0	2.5	0.7	2.2	1.5	0.9

Figure 7.2 presents a comparison of the rates of substitution in different regions of the gene, as well as in pseudogenes. The rates for the 5' and 3' untranslated regions, nondegenerate sites, twofold degenerate sites, and fourfold degenerate sites are the average rates for the genes in Table 7.2 and 7.3. The rate for the 5' flanking region was computed by assuming that the ratio of this rate to that at fourfold degenerate sites is 5.3/8.6 and that the average rate at fourfold degenerate sites is 3.71×10^{-9} substitutions per year. The rates for introns, the 3' flanking region, and pseudogenes were computed in the same manner. Since these are rough estimates based on limited data and since the rate in a region varies from gene to gene, the rates may not be applicable to any particular gene but are meant to provide a rough general pattern of the relative substitution rates in different

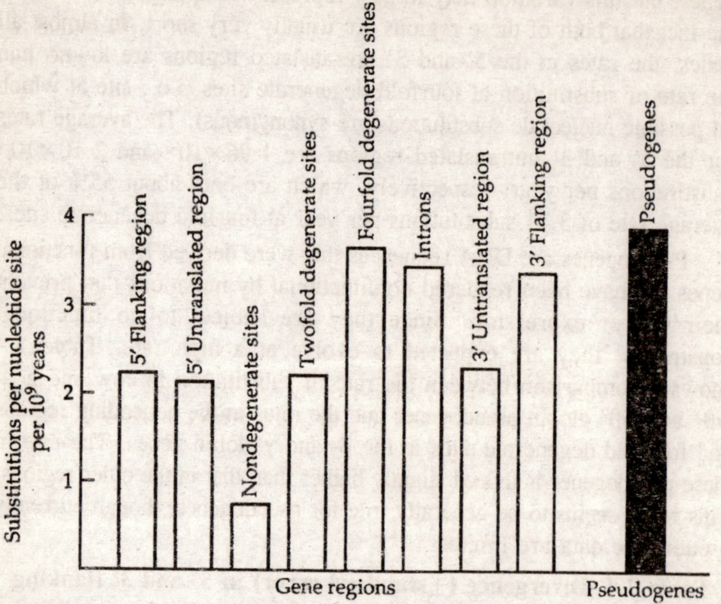

Fig. 7.2. Average rates of substitution in different parts of genes and in pseudogenes.

DNA regions. With this precaution, we note that the substitution rate in a gene is highest at fourfold degenerate sites; slightly lower in introns and the 3' flanking region; intermediate in the 3' untranslated region, the 5' flanking and untranslated regions, and twofold degenerate sites; and lowest at nondegenerate sites. Pseudogenes have the highest rate of substitution, although only slightly higher than that at the fourfold degenerate sites of a functional gene.

Causes of Variation in Substitution Rates

We have seen in the preceding section that the rate of nucleotide substitution varies greatly among genes and among different regions of a gene. We now discuss possible causes of these variations. For this purpose, we note that the rate of substitution is determined by two factors: (1) the rate of mutation and (2) the probability of fixation of a mutation. The probability of fixation depends on whether the mutation is advantageous, neutral, or deleterious and may vary enormously from gene to gene. Since the rate of mutation is unlikely to vary much within a gene but may vary among genes, it is simpler to discuss the rate variation among different regions of a gene and the variation among genes separately.

Variation among Different Gene Regions

Let us first consider the large difference between the synonymous and nonsynonymous rates in a gene. Since the rate of mutation at synonymous and nonsynonymous sites within a gene should be the same, or at least very similar, the difference in substitution rates is usually attributed to differences in the intensity of purifying selection between the two types of sites. This is understandable in light of the neutral theory of molecular evolution. Mutations that result in an amino acid substitution have a higher chance of causing deleterious effects on the function of the protein than do synonymous changes. Consequently, the majority of nonsynonymous mutations will be eliminated from the population by purifying selection. The result will be a reduction in the rate of substitution at nonsynonymous sites. In contrast, synonymous changes have a better chance of being neutral or nearly neutral, and more of them can become fixed in a population.

Of course, nonsynonymous substitutions may have a better chance of improving the function of a protein. However, if advantageous selection plays a major role in the evolution of proteins, the rate of nonsynonymous substitution should exceed that of synonymous substitution. Indeed, in some immunoglobulin genes, the nonsynonymous rate in the complementarity-determining regions (CDRs; also known as hypervariable regions) is higher than the synonymous rate. The higher rate has been attributed to overdominant selection for antibody diversity. However, when the entire immunoglobulin gene is considered, the nonsynonymous rate is still considerably lower than the synonymous rate. This result indicates that, even in immunoglobulins, most nonsynonymous mutations are disadvantageous and are eliminated from the population. In a case similar to that of immunoglobulins, Hughes and Nei (1989) reported that in certain regions of the major histocompatibility complex genes, the nonsynonymous rate of substitution exceeds the rate of synonymous substitution. They attributed the higher rates of nonsynonymous substitution to overdominant selection.

The contrast between synonymous and nonsynonymous rates in a gene demonstrates a well-known principle in molecular evolution, namely that the stronger the functional constraints on a macromolecule, the slower the rate of evolution. This may be explained under the selectionist view by using Fisher's (1930) argument that the larger the effect a mutation has, the smaller the chance of being advantageous it will have. However, neutralists have provided a more systematic explanation. In particular, Kimura (1977) has formulated the following model. Suppose that a certain fraction, f_0, of all mutations in a certain molecule are selectively neutral or nearly neutral and that the rest are deleterious. (Advantageous mutations are assumed to occur only rarely, such that their relative frequency is effectively zero, and they do not contribute significantly to the overall rate of molecular evolution.) If we denote by υ_T the total mutation rate per site per unit time, then the rate of neutral mutation is $\upsilon_0=\upsilon_T f_0$. According to the neutral theory of molecular evolution, the rate of substitution is $k=\upsilon_0$. Therefore,

$$k=\upsilon_T f_0 \qquad \text{...(2)}$$

Within any given gene, the υ_T value can be assumed to be the same for synonymous sites and nonsynonymous sites. However, the f_0 value is higher for synonymous sites than for nonsynonymous sites and so the former should evolve faster than the latter. Thus, although the model is oversimplified, it is helpful for explaining the rate differences among different DNA regions.

According to the above model, the highest rate is expected to occur in a sequence that does not have any function, so that all mutations in it are neutral (i.e., $f_0=1$). In fact, pseudogenes do seem to have the highest rate of nucleotide substitution. The observation that 5' and 3' untranslated regions have lower substitution rates than the rate of synonymous substitutions in coding regions further supports the neutral line of reasoning, because these untranslated regions contain important signals for transcription initiation, termination, and regulation.

Within a protein, different structural or functional domains are likely to be subject to differential functional constraints and to evolve at different rates. A good example is provided by proinsulin, which consists of three segments: A, B, and C. Segment C residues in the middle of the proinsulin molecule and is removed during the formation of the active hormone (insulin), which consists of the two remaining segments, A and B. Segment C does not take part in the hormonal activity of insulin and is thought only to facilitate the creation of the proper tertiary structure of the hormone. Consequently, the nonsynonymous substitution rate for the region coding for the regions coding for the A and B chains. Nevertheless, considerable constraints must still operate on the C segment, because the nonsynonymous substitution rate in this region is rather low, comparable to that in β globin.

There are many proteins that contain clearly defined functional or structural domains that are highly conserved in evolution. The steroid

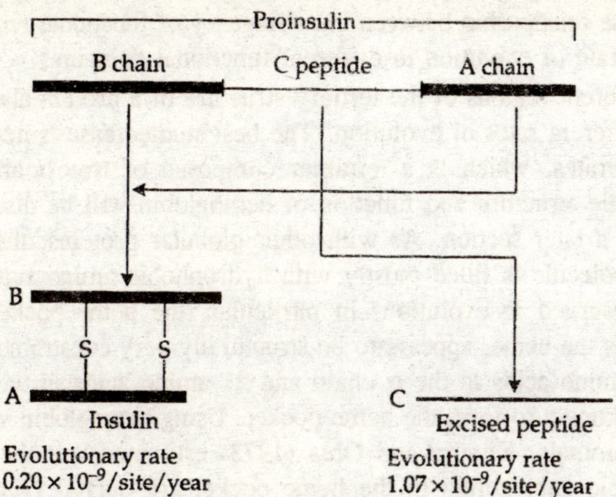

Fig. 7.3. Comparison between the rates of nucleotide substitution for DNA regions coding for functional insulin and the C peptide.

and thyroid hormone receptors are such examples. These receptors are factors that regulate the transcription of eukaryotic protein-coding genes in target tissues. The basic structural requirements for the receptors to control gene expression are the ability to bind the hormone with high affinity and specificity and the ability to bind to specific regions of DNA or chromatin. The DNA binding domain is extremely well conserved; e.g., 100% identity at the amino acid level between human and chicken estrogen receptors. The hormone binding domain is also well conserved, though not so strongly as the DNA binding domain. Domain D is a hydrophilic region that may serve as a hinge between the DNA and hormone binding domains. This domain and the other domains, whose function is unknown, are considerably less well conserved than the DNA and hormone binding domains. Thus, there is

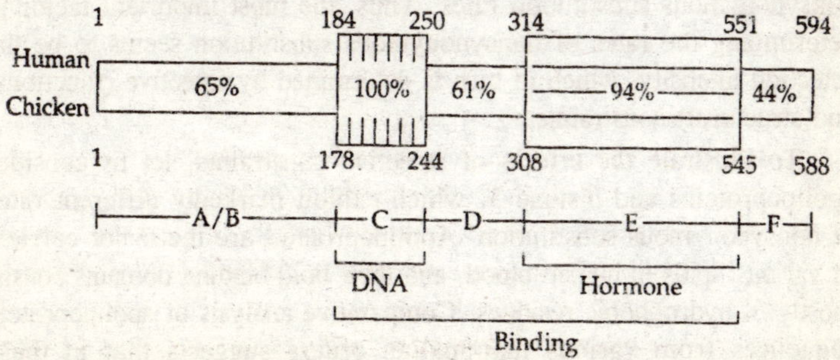

Fig. 7.4. Schematic comparison of the human and chicken estrogen receptor sequences.

an inverse relationship between the stringency of functional requirement and the rate of evolution in different functional domains.

Different regions of the tertiary structure of a protein also usually show different rates of evolution. The best studied case is hemoglobin in vertebrates, which is a tetramer composed of two α and two β chains; the structure and function of hemoglobin will be discussed in detail in a later section. As with other globular proteins, the interior of the molecule is filled mostly with hydrophobic amino acids and is well conserved in evolution. In particular, the heme pocket, which surrounds the heme, appears to be structurally very constrained. There are 19 amino acids in the α chain and 21 amino acids in the β chain that are known to form the heme pocket. Using hemoglobin sequences from mammals, Kimura and Ohta (1973) estimated that the rates of amino acid substitution in the heme pocket are only 0.17×10^{-9} and 0.24×10^{-9} substitutions per amino acid site per year for the α and β chains, respectively. In contrast, for those sites on the surface of hemoglobin, the rates of amino acid substitution in the α- and β-globin are 1.4×10^{-9} and 2.7×10^{-9}, which are about ten times higher than the corresponding rates in the heme pocket.

Variation among Genes

To explain the large variation in the rates of nonsynonymous substitution among genes, we again consider the two possible factors: the rate of mutation and the intensity of selection. The assumption of equal rates of mutation for different genes may not hold in this case, since different regions of the genome may have different propensities to mutate. Wolfe et al. (1989c) suggested that different regions of the mammalian nuclear genome may differ from each other by a factor of two in their rates of mutation. However, a twofold difference in mutation rates is too small to account for the thousandfold range in nonsynonymous substitution rates. Thus, the most important factor in determining the rates of nonsynonymous substitution seems to be the selection intensity, which in turn is determined by selective (functional and structural) constraints.

To illustrate the effects of selective constraints, let us consider apolipoproteins and histone 3, which exhibit markedly different rates of nonsynonymous substitution. Apolipoproteins are the major carriers of various lipids in human blood, and their lipid-binding domains consist mostly of hydrophobic residues. Comparative analysis of apolipoprotein sequences from various mammalian orders suggests that in these domains the substitution of a hydrophobic amino acid (e.g., valine, leucine) for another hydrophobic amino acid is acceptable at many sites. This lax structural requirement may explain why the nonsynonymous rates in these genes are fairly high.

Table 7.5. Comparison of evolutionary rates between the heme pocket and the surface in hemoglobin chains

Region	*α-globin*	*β-globin*
Heme pocket	0.17	0.24
Surface	1.35	2.73

At the other extreme, we have histone 3. Since most amino acid in histone 3 interact directly with either the DNA or other core histones in the formation of the nucleosome, it is reasonable to assume that there are very few possible substitutions that can occur without impeding the function of this protein. In addition, histone 3 must retain its strict compactness and its high alkalinity, which are necessary for interaction with an acidic molecule, the DNA. As a consequence, histone 3 is very intolerant of most molecular changes. Indeed, this protein is one of the slowest evolving protein known, more than one thousand times slower than the apolipoproteins.

Why the rate of synonymous substitution also varies from gene to gene is less clear. There may be two reasons for this variation. First, the rate of mutation may differ among different regions of the genome and the variation in rates of synonymous substitution may simply reflect the chromosomal position of the gene. Since there is currently little data on the variation of mutation rate among different regions of the mammalian genome, the importance of this factor cannot be evaluated. Wolfe et a. (1989c) hypothesized that the mutation rate of a sequence may depend on its GC content. However, analysis of DNA sequences of mammalian genes revealed only minor effects of GC content on the rate of synonymous substitution.

The second reason may be that not all synonymous codons are equivalent in fitness, so that some synonymous changes are favoured whereas others are selected against. Such selection will create codon usage bias, a subject that will be discussed later. If the intensity of selective contraints on codon usage varies among genes, so will the rate of synonymous substitution. In this connection, it is interesting to note the existence of a positive correlation between the synonymous and nonsynonymous substitution rates. In fact, for the genes used in Table 7.1 (excluding the histones), the correlation coefficient between the synonymous rate and the nonsynonymous rate is about 70%. There are several possible causes for this correlation. First, if the rate of mutation varies among genes, then there will be a correlation between synonymous and nonsynonymous rates of substitution. Since little data exist on the variation of mutation rate, the importance of this factor is difficult to evaluate. Second, there might be selection for translational accuracy in mammalian genes, as in the case of *Drosophila* genes. Because highly constrained proteins are less tolerant to translational errors, their genes may be more constrained to the use of optimal codons and so will have lower rates of synonymous substitution as well as lower rates of nonsynonymous substitution. Akashi (1994) showed that this is true for *Drosophila* genes and proposed that this might also be the case for mammalian genes. Third, in the presence of codon usage bias, a nonsynonymous change may cause a shift from an optimal codon to a nonoptimal codon. As a hypothetic example, suppose that GUG and GCC are the optimal codons for valine and alanine, respectively. Then, even if the change from valine to alanine is neutral at the protein level, the change from GUG to GCG is not neutral, because GCG is not the optimal codon for alanine. Thus, codon preference can slow down the rate of nonsynonymous substitution, depending on the degree of usage preference. On the other hand, if the nonsynonymous substitution of GCG for GUG has already occurred, then the change from GCG to GCC is advantageous and will be accelerated. Thus, a higher nonsynonymous rate can result in a faster synonymous rate. For these reasons, a positive correlation may exist between synonymous and nonsynonymous rates. However, Akashi's (1994) analysis of *Drosophila* genes suggested that this factor is less important than the selection for translational accuracy, because the majority of nonsynonymous changes do not cause a shift in optimal codons. Fourth, the extent of selection at synonymous positions might be affected by the nucleotide composition at adjacent nonsynonymous positions. Fifth, there might be an excess of tandem (doublet) substitutions between adjacent bases, because doublets can cause both synonymous and nonsynonymous substitutions at the same time and thus can create a positive correlation between the two.

Drosophila Nuclear Genes

Some *Drosophila* species are also well sequenced, though not as extensively as humans and rodents. Table 7.6 provides a list of substitution rates between the *D. melanogaster* and *D. obscura* groups, which are assumed to have diverged about 30 million years ago. The number of genes compared is smaller than that in Table 7.1, and so is

Table 7.6. Rates of synonymous and nonsynonymous substitution in various *Drosophila* genes

Gene	*L*	*Nonsynonymous*	*Synonymous*
Arrestin B	401	0.23±0.07	11.62±1.48
Heat shock protein 83	374	0.30±0.08	9.65±1.27
Myosin alkali light chain 1	155	0.32±0.12	3.67±0.93
Neither-inactivation-nor-after potential E	353	0.33±0.08	9.48±1.15
Antennapedia	378	0.48±0.10	12.23±1.53
Rhodopsin 3	382	0.78±0.13	15.38±1.90
Adh related gene (Adhr-Dup)	272	0.77±0.15	19.22±2.77
Rhodopsin 2	381	0.87±0.13	17.18±2.10
Alcohol dehydrogenase	254	0.90±0.17	9.50±1.30
Glucose dehydrogenase	612	0.88±0.10	16.90±1.55
Rhodopsin 4	378	0.93±0.13	15.58±1.82
Amylase, distal	494	1.13±0.13	6.80±0.78
Rosy	1334	1.18±0.08	18.78±1.22
Ultrabithorax	249	1.28±0.20	11.25±1.58
Adenine phosphoribosyltransferase	181	1.47±0.27	16.63±2.62
Adenosine 3 (Gart)	1351	1.43±0.10	19.48±1.23
Minute-(3)-99D (rp49)	131	1.52±0.30	11.02±2.15
Urate oxidase	345	1.57±0.20	13.43±1.72
Bicoid	474	1.80±0.18	16.22±1.68
Lethal (2) giant larvae	1161	2.05±0.12	24.82±1.87
Serendipity delta	419	2.23±0.22	29.78±5.13
Pupal cuticle protein	184	2.47±0.35	18.50±3.65
Serendipity beta	342	2.60±0.25	21.77±3.25
Larval cuticle protein 4	111	2.57±0.45	10.30±2.13
Period	1065	2.82±0.15	17.92±1.52
Esterase 6	542	3.03±0.23	21.52±2.37
Chorion protein 16	137	3.05±0.47	14.57±2.58
Toll	1095	3.28±0.17	16.08±1.08
Chorion protein 19	162	3.53±0.45	14.35±2.88
Serendipity alpha	523	4.10±0.28	24.13±2.78
Chorion protein 15	111	5.57±0.75	13.18±2.55
Cp18 Chorion protein 4	169	5.63±0.60	18.38±3.40
Average		1.91 (±1.42)	15.60 (±5.50)

the range of nonsynonymous rates. Nevertheless, a 25-fold difference is seen between the lowest and the highest rates (0.23×10^{-9} vs.

5.68×10^{-9}). The range of synonymous rates in Table 7.6 is from 3.7×10^{-9} to 29.7×10^{-9}, an 8-fold difference, larger than the mammalian range. As will be discussed later, *Drosophila* genes appear to be subject to codon usage constraints, which are likely to vary among genes, thus causing variation in synonymous rates. This might explain why the range of synonymous rates is larger in *Drosophila* genes than in mammalian genes. The estimated average synonymous rate, 15.4×10^{-9}, is about four to five times higher than that in mammalian genes. Of course, the exact magnitude of difference is not certain, because both estimates involved assumptions about the divergence times.

Rates of Substitution in Organelle DNA

Organelle genomes are much smaller and easier to investigate experimentally than nuclear genomes. Moreover, the discovery of an exceptionally high rate of substitution in the mammalian mitochondrial genome stimulated further interest in the evolution of organelle DNA.

The mammalian mitochondrial genome consists of a circular, double-stranded DNA about 15,000-17,000 bp long, approximately 1/10,000 of the smallest animal nuclear genome. It contains mostly unique (i.e., nonrepetitive) sequences: 13 protein-coding genes, two rRNA genes, 22 tRNA genes, and a control region that contains sites for replication and transcription initiation. The genome is structurally very stable, as is evident from the small variation in genome size and gene arrangement among mammalian species.

In sharp contrast, plant mitochondrial genomes exhibit a high structural variability. They undergo frequent rearrangements, duplications, and deletions. For this reason, the genome size varies from 40,000 to 2,500,000 bp. The mitochondrial genome of the liverwort (*Marchantia polymorpha*) is 186,608 bp in length. It contains 3 rRNA genes, 29 genes for 27 species of tRNA, and 30 open reading frames (ORFs) for functionally known proteins (16 ribosomal proteins, 3 subunits of H^+-ATPase, 3 subunits of cytochrome c oxidase, apocytochrome b protein, and 7 subunits of NADH ubiquinone oxidoreductase). In addition, 32 ORFs of unknown function were predicted. Therefore, this genome contains 94 possible genes.

The chloroplast DNA (cpDNA) or genome in vascular plants is circular and varies in size from about 120,000 to 220,000 bo, with an average size of 150,000 bp. Much of the variation in size is the result of variation in the length of an inverted repeat, and the genome is known to be structurally stable. The completely sequenced cpDNA from tobacco (a dicot) is 155,844 bp long, including an inverted repeat of 25,339 bp. The genome contains about 113 genes in total: 21 ribosomal proteins, 4 rRNAs, 30 tRNAs, 4 RNA polymerase subunits, 1 translation initiation factor, 1 putative intron maturase, 29 genes for photosynthesis functions, 11 genes for chlororespiration (NADH dehydrogenase subunits), 1 gene for a subunit of C1p portease, 1 gene for a subunit of acetyl-CoA carboxylase, and 10 conserved ORFs of unknown function. The gene contents in the cpDNAs of the rice and the liverwort are similar to that of the tobacco cpDNA.

The synonymous rate of substitution in mammalian mitochondrial genes has been estimated to be 5.7×10^{-8} substitutions per synonymous site per year. This is about 10 times the value for synonymous substitutions in nuclear protein-coding genes. The rate of nonsynonymous substitution varies greatly among the 13 protein-coding genes, but it is always much higher than the average nonsynonymous rate for nuclear genes. The reason for these high rates of substitution in mammalian mitochondria seems to be a high rate of mutation relative to the nuclear rate. The high mutation rate might be due to (a) a high turnover rate of mt genome and (b) a high concentration of mutagens resulting from the metabolic functions performed by the mitochondria. On the other hand, the stringency of purifying selection against nonsynonymous mutations seems to be of the same order of magnitude as that operating on nuclear genes.

Early studies based on a few gene sequences or on restriction-enzyme mapping suggested that chloroplast genes have lower rates of nucleotide substitution than mammalian nuclear genes and that plant mitochondrial DNA evolves slowly in terms of nucleotide substitution, though it undergoes frequent sequence rearrangements. These results have been confirmed by more extensive analyses of DNA sequences.

Table 7.7 shows a comparison of the substitution rates in the three genomes of higher plants. The average numbers of substitutions per nonsynonymous site (K_A) in the chloroplast and mitochondrial genomes are similar, but the average number of substitutions per synonymous site (K_S) in the chloroplast genome is almost three times that in the mitochondrial genome for the comparison between monocot and dicot species, and it is two to three times higher for the comparisons between maize and rice. The average synonymous substitution rate in plant nuclear genes is four to five times that in chloroplast genes. Thus, the synonymous substitution rates in plant mitochondrial, chloroplast, and nuclear genes are in the approximate ratio of 1:3:12.

Table 7.7. Comparison of the rates of nucleotide substitution in plant chloroplast, mitochondrial, and nuclear genes

Genomes	K_S	L_S	K_A	L_A
A. Comparison between monocot and dicot species				
Chloroplast genes	0.58±0.02	4,177	0.05±0.00	14,421
Mitochondrial genes	0.21±0.01	1,219	0.04±0.00	4,380
B. Comparison between maize and rice				
Nuclear genes	0.57±0.01	8,898	0.07±0.00	3,702
Chloroplast genes	0.12±0.00	7,872	0.02±0.00	28,518
Mitochondrial genes	0.05±0.01	1,845	0.02±0.00	6,357

If we take the divergence time between maize and rice to be 50-70 million years, the nuclear data in Table 7.7 indicate an average synonymous rate of $4.1\text{-}5.7 \times 10^{-9}$ substitutions per site per year. This is similar to the rates of synonymous substitution seen in mammalian nuclear genes. Interestingly, the rate of nucleotide substitution does not correlate with the rate of structural changes in the genome of organelles. In mammals, the mitochondrial DNA evolves very rapidly in terms of nucleotide substitutions, but the spatial arrangement of genes and the size of the genome are fairly constant among species. In contrast, the mitochondrial genome of plants undergoes frequent structural changes, but the rate of nucleotide substitution is extremely low. In chloroplast DNA, both the rates of nucleotide substitution and structural evolution are very low. The lack of correlation between the rates of nucleotide substitution and the rates of structural evolution suggests that the two processes occur independently.

VIRAL GENES

Many viral genomes evolve rapidly because of a high mutation rate. The rate of mutation in a virus depends on the polymerase or polymerases it uses to synthesize its genome. Polymerases are classified into four classes, depending on whether they use DNA or RNA templates and on whether they produce DNA or RNA products. RNA-dependent RNA polymerases use RNA as a template and catalyze the synthesis of RNA polymers. RNA-dependent DNA polymerases are known as reverse transcriptases because they use RNA templates but produce, in a reverse sense, DNA polymers. The two other classes, DNA-dependent DNA and RNA polymerases, use DNA templates and produce, respectively, DNA and RNA polymers. RNA-dependent polymerases are more error prone than DNA-dependent polymerases because they lack the proofreading (3'→5' exonuclease) activity. For this reason, the mutation rates are in general much higher in RNA viruses than in DNA viruses; note, however, that some DNA viruses, e.g., the hepatitis B virus, go through an RNA stage and reverse transcription, so that their mutation rates are also high Table 7.8 shows a compilation of estimates of mutation rates for (1) RNA viruses, which are here defined as viruses that exclusively replicate by RNA and do not go through a DNA stage in their life cycle, (2) retroviruses, which are RNA viruses but go through a DNA stage by reverse transcription, and (3) DNA viruses. Although the estimates are rough, it is clear that the mutation rates in RNA viruses are much higher than those in DNA viruses (e.g., bacteriophages). For example, the mutation rate for phage Qβ, which is an RNA virus, is about 10,000 times higher than that in bacteriophage M13 benefits from proofreading and mismatch repair, two antimutagenic mechanisms that are apparently absent for phage Qβ. The mutation rates for retroviruses are approximately one order of magnitude lower than those for RNA viruses. The mutation rates for *E. coli* and fungi are 10^6 to 10^7 times lower than those for RNA viruses.

Table 7.8. Rates of spontaneous mutation per replication

Organism	*Mutation rate*
A. RNA viruses	
Phage QB	1.5×10^{-3}
Poliovirus	$0.1\text{-}1.6 \times 10^{-4}$
Vesicular stomatis virus	3.2×10^{-4}
Influenza virus A	$>7.3 \times 10^{-5}$
B. Retroviruses	
Spleen necrosis virus	$0.5\text{-}3.6 \times 10^{-5}$
Moloney murine leukemia virus	$0.4\text{-}>6.6 \times 10^{-6}$
Rous sacoma virus	4.6×10^{-5}
C. DNA-based microbes	
Bacteriophage M13	7.2×10^{-7}
Bacteriophage λ	7.7×10^{-8}
Bacteriophage T2	2.7×10^{-8}
Bacteriophage T4	2.0×10^{-8}
Escherichia coli	5.4×10^{-10}
Saccharomyces cerevisiae	2.3×10^{-10}
Neurospora crassa	2.0×10^{-11}

Table 7.9. Rates of synonymous and nonsynonymous substitutions of viral genes and nuclear genes

Organism	Gene	Substitutions per site per year	
		Synonymous	*Nonsynonymous*
A. RNA viral genes			
Influenza A virus	Hemagglutinin	13.10×10^{-3}	3.59×10^{-3}
Hepatitis C virus	E	6.29×10^{-3}	0.32×10^{-3}
HIV-1	*gag*	9.70×10^{-3}	1.70×10^{-3}
Moloney murine sarcoma virus	v-*mos*	2.75×10^{-3}	0.82×10^{-3}
Moloney murine leukemia virus	*gag*	1.16×10^{-3}	0.54×10^{-3}
B. DNA viral genes			
Hepatitis B virus	P	4.57×10^{-5}	1.45×10^{-5}
Herpes simplex virus type-1	Genome	3.50×10^{-8}	
C. Nuclear genes			
Mammals	c-*mos*	5.23×10^{-9}	0.93×10^{-9}
Mammals	α globin	3.94×10^{-9}	0.56×10^{-9}

Table 7.9 presents a list of the rates of nucleotide substitution per year in viral gene. Both influenza A and hepatitis C viruses are RNA viruses and their rates of synonymous substitution are very high. The retroviruses HIV-1 (the human immunodeficiency virus type 1), the Moloney murine sarcoma virus and the Moloney murine leukemia virus also have high rates of synonymous substitution. In particular, the rate in HIV-1 is comparable to that in influenza A virus. Since the mutation rates in retroviruses are in general one order of magnitude lower than those in RNA viruses, the high substitution rate in HIV-1 might be due to a very short "generation time" (replication cycle of the viral genome); note that the rate of substitution depends not only on the rate of mutation per replication but also on the generation time of the virus. Although hepatitis B virus (HBV) is a DNA virus, it goes through a stage of reverse transcription. However, its rate of synonymous substitution is 100 times lower than those of retroviruses. It is possible that the replication of the HBV genome does not always depend on reverse transcription or that the replication frequency of the HBV genome is not as high as that of retroviruses. Herpes simplex virus is a DNA virus. The synonymous rate in this virus was estimated from restriction analysis of the entire genome and so might be 3 to 4

times that shown in Table 7.9. However, it is clear that its rate of substitution is about 100,000 times lower than those in RNA viruses. As is well known, the rate in mammalian nuclear genes is at least one million times lower than those in RNA viruses.

NONRANDOM USAGE OF SYNONYMOUS CODONS

Because of the degeneracy of the genetic code, most of the 20 amino acids are encoded by more than one codon. Since synonymous mutations do not cause any change in amino acid sequences and since natural selection was thought to operate predominantly at the protein level, synonymous mutations were proposed as candidates for selectively neutral mutations. However, if all synonymous mutations are indeed selectively neutral, the synonymous codons for an amino acid should be used with more or less equal frequencies, unless the base composition is very skewed. As DNA sequence data accumulated, however, it became evident that the usage of synonymous codons is distinctly nonrandom in both prokaryotic and eukaryotic genes. In fact, in many yeast and *E. coli* genes the bias in usage is highly conspicuous. For example, 21 of the 23 leucine residues in the *E. coli* outer membrane protein II (*omp*A) are encoded by the codon CUG, though there exist five other codons for leucine. Such a large bias cannot be explained by nonrandom mutation. How to explain the widespread phenomenon of nonrandom codon usage became a controversial issue, to which, fortunately, some clear answers seem to be emerging.

An observation that has been helpful for understanding the phenomenon of nonrandom usage is that genes in an organism or in related species generally show the same pattern of choices among synonymous codons. Thus, mammalian, *E. coli*, and yeast genes fall into distinct classes of codon usage. Grantham et al. (1980) therefore proposed the *genome hypothesis*, postulating that the genes in any given genome use the same coding strategy with respect to choices among synonymous codons, i.e., the bias in codon usage is species-specific. The genome hypothesis turns out to be true in general, though there is considerable heterogeneity in codon usage among genes in a genome.

A simple measure of nonrandom usage of synonymous codons in a gene is the Relative Synonymous Codon Usage (RSCU). The RSCU value for a codon for a particular amino acid is the observed number of occurrences of that codon in the gene divided by the number of occurrences expected under the assumption of equal usage of all codons for that amino acid. For a given amino acid, let n be the number (from 1 to 6) of synonymous codons for that amino acid and X_i be the number of occurrence of the ith codon for that amino acid. Then, under the assumption of equal usage, the expected occurrences of a codon for that amino acid is $(1/n) \sum X_i$ and so the $RSCU_i$ for the ith codon is given by

$$RSCU_i = \frac{X_i}{\frac{1}{n}\sum_{i=1}^{n} X_i} \qquad ...(3)$$

If the synonymous codons of an amino acid are equally used in a gene, their RSCU values are equal to 1. Deviations from equal usage can arise if mutation among the four nucleotides is not random or if natural selection occurs among synonymous codons.

Codon Usage in Unicellular Organisms

Studies of codon usage in *E. coli* and yeast have greatly increased our understanding of the factors that affect the choice of synonymous codons. Post et al. (1979) found that *E. coli* ribosomal-protein genes preferentially use synonymous codons that are recognized by the most abundant tRNA species. They suggested that the preference resulted from natural selection because using a codon that is translated by an abundant tRNA species will increase translational efficiency and accuracy. Their finding prompted Ikemura (1981, 1982) to gather data on the relative abundances of tRNA species in *E. coli* and the yeast *Saccharomyces cerevisiae*. He showed that, in both species, a positive correlation exists between the relative frequencies of the synonymous codons in a gene and the relative abundances of their cognate tRNA species. The correlation is very strong for highly expressed genes. For instance, in *E. coli* the most abundant of the four leucine tRNAs is $tRNA_1^{Leu}$, which recognizes the CUG codon, and the *omp*A gene, which is very highly expressed, uses predominately this codon for leucine.

Schematically the correspondence between the frequencies of the six leucine codons and the relative abundances of their cognate tRNAs. In *E. coli*, $tRNA_1^{Leu}$ is the most abundant leucine tRNA species, and in highly expressed genes, CUG (the codon recognized by this tRNA) is indeed much more frequently used than the other five codons. On the other hand, in yeast, the most abundant leucine tRNA species is $tRNA_3^{Leu}$, and the codon recognized by this tRNA (UUG) is the predominant codon. In contrast, in genes with low levels of expression, the correspondence between tRNA abundance and the use of the respective codon is much weaker in both species.

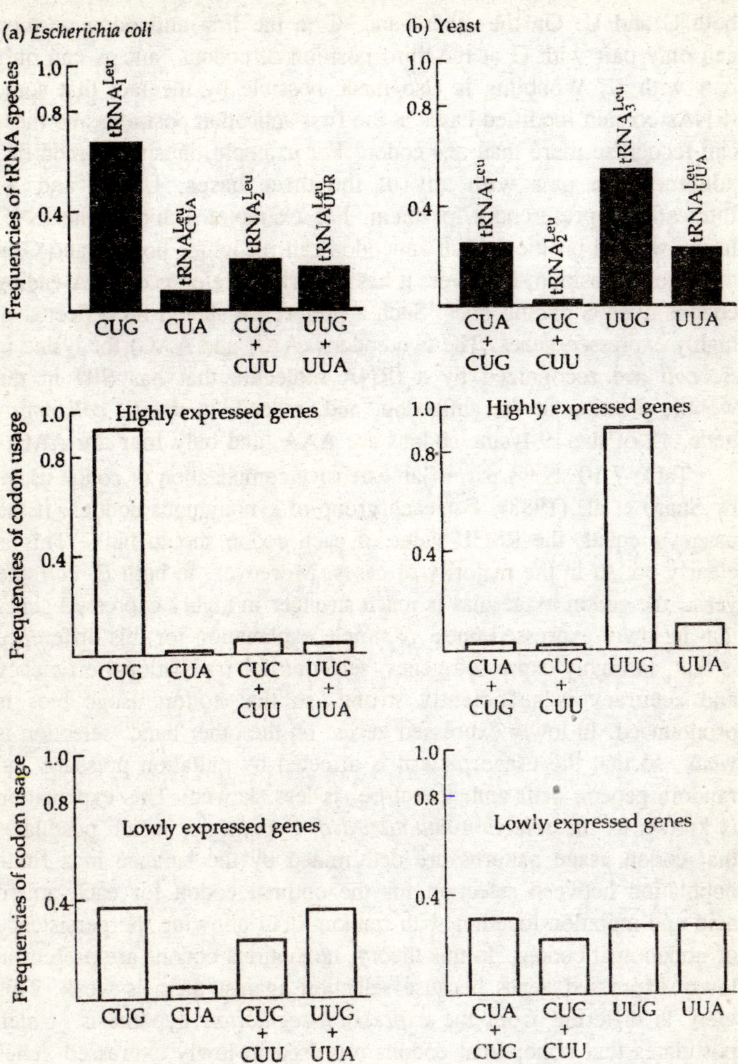

Fig. 7.5. Diagram illustrating the relationship between the relative frequency of codon usage for leucine and the relative abundance of the corresponding cognate tRNA species in (a) Escherichia coli and (b) Saccharomyces cerevisiae.

The importance of translational efficiency in determining the codon-usage pattern in highly expressed genes is further supported by the following observation. It is known that codon-anticodon pairing involves *wobbling* at the third position. For example, U in the first position of anticodons can pair with both A and G. Similarly, G can pair with both C and U. On the other hand, C in the first anticodon position can only pair with G at the third position of codons, and A can only pair with U. Wobbling is also made possible by the fact that some tRNAs contain modified bases at the first anticodon position, and these can recognize more than one codon. For example, inosine (a modified adenine) can pair with any of the three bases, U, C, and A. Interestingly preferences for them. For example, 4-thiouridine (S^4U) in the wobble position of an anticodon can recognize both A and G in the wobble position; however, it has a marked preference for A-ending codons over G-ending ones. Such a preference should be reflected in highly expressed genes. The two codons (AAA and AAG) for lysine in *E. coli* are recognized by a tRNA molecule that has S^4U in the wobble position of the anticodon, and, indeed, in the *E. coli ompA* gene, 15 of the 19 lysine codons are AAA, and only four are AAG.

Table 7.10 shows part of an extensive complication of codon usage by Sharp et al. (1988). For each group of synonymous codons, if the usage is equal, the RSCU value of each codon should be 1. This is clearly not so in the majority of cases. Moreover, in both *E. coli* and yeast, the codon usage bias is much stronger in highly expressed genes that in lowly expressed ones. A simple explanation for this difference is that, in highly expressed genes, selection for translational efficiency and accuracy is sufficiently strong, so that codon usage bias is pronounced. In lowly expressed genes, on the other hand, selection is weak, so that the usage pattern is affected by mutation pressure and random genetic drift and, therefore, is less skewed. This explanation is known as the *selection-mutation-drift* hypothesis, which postulates that codon usage patterns are determined by the balance in a finite population between selection for the optimal codon for each amino acid and mutation together with random drift allowing the persistence of nonoptimal codons. In this theory, nonoptimal codons are present in lowly expressed genes because selection against them is weak. This view is different from the *expression-regulation* hypothesis, which postulates that nonoptimal codons are used in lowly expressed genes as a mechanism for keeping their expression low. An argument against the latter hypothesis is that it would be more efficient to modulate the level of gene expression by changing the strength of the promoter or the ribosome binding site.

A theoretical study shows that strong usage bias occurs in a gene if the selective disadvantage(s) against nonoptimal codons is larger than 2/N, where N is the effective size of a haploid population. Since the effective size for most unicellular organisms is likely to be larger than 10^6, s may need to be only of the order of 10^{-6} or even smaller for strongly biased codon usage to occur; indeed, Hartl et al. (1994) obtained an estimate of $s=7\times10^{-9}$ for the *gnd* gene in *E. coli*. Thus, the strong codon usage bias observed in highly expressed genes in unicellular organisms may actually require only very small selective advantages for optimal codons. On the other hand, the fact that codon usage is not strongly biased in lowly expressed genes suggests that the selective differences between optimal and nonoptimal codons in these genes are extremely small.

Table 7.10. Codon usage in four species

Amino		*Escherichia coli*		*Saccharomyces cerevisiae*		*Drosophila*		*Human*	
acid	*Codon*	*High*	*Low*	*High*	*Low*	*High*	*Low*	*G+C*	*A+T*
Leu	UUA	0.06	1.24	0.49	1.49	0.03	0.62	0.05	0.99
	UUG	0.07	0.87	5.34	1.48	0.69	1.05	0.31	1.01
	CUU	0.13	0.72	0.02	0.73	0.25	0.80	0.20	1.26
	CUC	0.17	0.65	0.00	0.51	0.72	0.90	1.42	0.80
	CUA	0.04	0.31	0.15	0.95	0.06	0.60	0.15	0.57
	CUG	5.54	2.20	0.02	0.84	4.25	2.04	3.88	1.38
Val	GUU	2.41	1.09	2.07	1.13	0.56	0.74	0.09	1.32
	GUC	0.08	0.99	1.91	0.76	1.59	0.93	1.03	0.69
	GUA	1.12	0.63	0.00	1.18	0.06	0.53	0.11	0.80
	GUG	0.40	1.29	0.02	0.93	1.79	1.80	2.78	1.19
Ile	AUU	0.48	1.38	1.26	1.29	0.74	1.27	0.45	1.60
	AUC	2.51	1.12	1.74	0.66	2.26	0.95	2.43	0.76
	AUA	0.01	0.50	0.00	1.05	0.00	0.78	0.12	0.64
Phe	UUU	0.34	1.33	0.19	1.38	0.12	0.86	0.27	1.20
	UUC	1.66	0.67	1.81	0.62	1.88	1.14	1.73	0.80
Met	AUG	1.00	1.00	1.00	1.00	1.00	1.00	1.00	1.00

In conclusion, in *E. coli* and yeast, the choice of synonymous codons is constrained by tRNA availability and other factors related to translational efficiency. These constraints will result in purifying selection, thus slowing down the rate of synonymous substitution. In fact, it has been shown that the rate of synonymous substitution in enterobacterial genes is negatively correlated with the degree of codon usage bias. Therefore, the phenomenon of nonrandom usage of synonymous codons may not be taken as evidence against the neutral theory of molecular evolution, since it can be explained in terms of the principle that stronger selective constraints result in lower rates of evolution.

Codon Usage in Multicellular Organisms

Codon usage bias also varies widely among *Drosophila* genes. The bias is strongly associated with the GC content at third codon positions, which, in turn, has been found to be positively correlated with the GC content in their adjacent introns, though the latter correlation was not detected in earlier studies. Since most parts of introns are apparently subject to only weak or no selective contraints, their GC content should be largely determined by the mutation pressure. Thus, a simple explanation for the positive correlation between the GC contents at third codon positions and in their adjacent introns is that the GC content at third codon positions (and therefore also codon usage) are

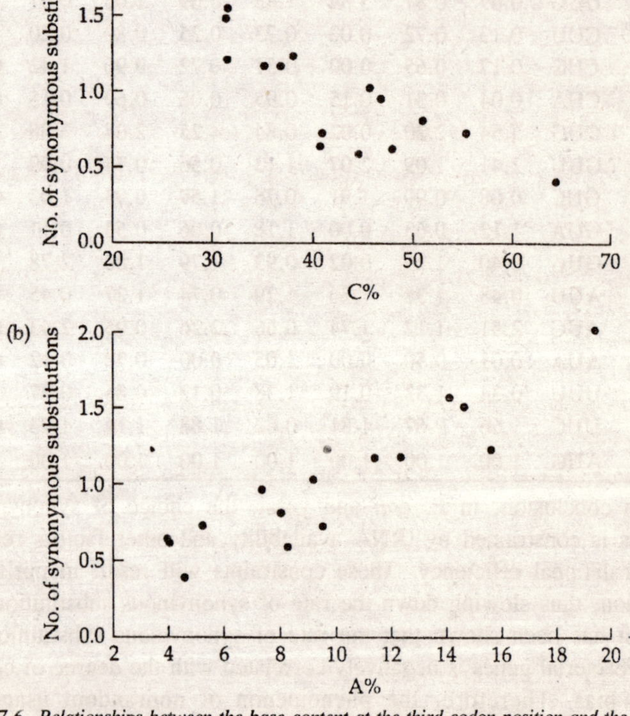

Fig. 7.6. Relationships between the base content at the third codon position and the number of synonymous substitutions per site for the comparisons between Drosophila melanogaster and D. pseudoobscura genes.

affected by the mutation pressure. The effect is, however, weak, especially in genes with high codon usage bias. For fourfold degenerate sites, Spearman's rank-order correlation is 0.139 for the high-bias class of genes and 0.396 ($P<0.001$) for the low-bias class. Thus, variation in mutation pattern cannot explain the GC variation at fourfold degenerate sites among high-bias genes and can account for only 16% of this variation among low-bias gene. For twofold degenerate sites the correlations are only slightly higher: $r=0.272$ ($P<0.02$) for high-bias gene and $r=0.399$ ($P<0.001$) for low-bias genes. In conclusion, the large variation in the GC content at degenerate sites appears to be mainly due to variation in the strength of natural selection, though the contribution of variation in mutation pattern is not negligible. Indeed, in high-bias genes there appears to be a strong preference for codons ending with C. Available data indicate that those genes with high GC contents at degenerate sites are more highly expressed than those with low GC contents. Other interesting observations are that the GC content in pseudogenes is reduced compared with their functional counterparts and that codon usage bias is less strong in *Drosophila melanogaster* than in *D. simulans*, as a result of reduction in population size in the former species. These observations suggest that, as in the case of unicellular organisms, selection for translational efficiency plays an important role in determining the choice of synonymous codons in *Drosophila*. In fact, Akashi (1994) has provided evidence for selection for translational accuracy. Apparently, the effective population size in *Drosophila* is large enough for selection to be effective in producing strong codon usage bias in highly expressed genes. Moreover, as in the case of bacteria, the rate of synonymous substitution has been shown to be negatively correlated with codon usage bias. Since the high GC content at degenerate sites is largely due to preference for C-ending codons at the expense of G-ending codons, the rate of synonymous substitution is negatively correlated with the C content but positively correlated with G (A) content at the third codon positions.

In many human genes, codons tend to end in either G or C (i.e., to have a high GC content at the third position), whereas other genes have a low third-position GC content. However, there are several reasons why this bias may not be strongly related to the level of gene expression. First, the α- and β-globin genes have different GC contents at the third position of codons (high and low, respectively), although they are both expressed in the same tissues (erythrocytes) in approximately equal quantities and, therefore, should have the same level of expression. Second, in chicken, the frequency with which codons are used does not correlate well with tRNA availability. Finally, the GC content at the third codon position is strongly correlated with the GC level in introns and in both flanking regions. For example, the α-globin gene is high in GC and resides in a high-GC region, whereas the β-globin gene is low in GC and resides in a low-GC region. Thus, it appears that the codon-usage bias in a human gene is to a large extent determined by the GC content in the region that contains the gene. As will be discussed, whether the GC content in a region is determined by natural selection or mutational bias is a controversial issue. However, since the GC content at the third position of codons in a gene tends to be higher than in introns and in surrounding regions, it is possible that the codon usage pattern in human genes is affected to some extent by natural selection. Furthermore, as mentioned earlier, a simple explanation for the positive correlation between nonsynonymous and synonymous rates in mammalian genes is that the degree of codon usage constraints varies among genes, i.e., genes subjected to strong codon usage constraints tend to have both low synonymous and nonsynonymous rates.

Adaptive Evolution of Lysozyme in Cow, Langur, and Hoatzin

It is usually very difficult to provide experimental evidence that a certain molecular change is advantageous or has an adaptive value. Instead, evolutionists often rely on sequence comparison to infer whether convergent or parallel changes have occurred under similar conditions. Lysozyme has been a favourite protein for this purpose.

Lysozyme is a ubiquitous bacteriolytic enzyme found in virtually all animals. Its catalytic function is to cleave the β(1-4) glycosidic bonds between N-acetyl glucosamine and N-acetyl muramic acid in the cell walls of eubacteria. By virtue of its catalytic function and its expression in body fluids, such as saliva, serum, and tears, lysozyme usually serves as a first-line defense against bacterial invasion. In foregut fermenters, i.e., in animals in which the anterior part of the stomach functions as a chamber for bacterial fermentation of ingested plant matter, lysozyme is also secreted in the posterior parts of the digestive system and is used to free nutrients from within the bacterial cell.

Foregut fermentation has independently arisen twice in the evolution of placental mammals, once in the ruminants (e.g., cows) and once in leaf-eating colobine monkeys (e.g., langurs). In both cases, lysozyme, which in other mammals is not normally secreted in the stomach, has been recruited to degrade the cell walls of bacteria, which carry on the fermentation in the foregut. Stewart and Wilson (1987) compared the amino acid sequences of lysozyme (!130 residues) from cows, langurs, baboons, humans, rats, and horses. They noted that there are five uniquely shared amino acids between cows and langurs, compared to only one amino acid uniquely shared by cows and horses. Since cows and langurs diverged much earlier than the separation of the langur lineages from the human and baboon lineages, the uniquely shared amino acids in these two species were suggested to be the result of a series of adaptive parallel substitutions that occurred independently in both lineages.

Table 7.11. Pairwise comparison of lysozyme sequences among different species

	Species					
Species	Langur	Baboon	Human	Rat	Cow	Horse
Langur		14	18	38	32	65
Baboon	0		14	33	39	65
Human	0	1		37	41	64
Rat	0	1	0		55	64
Cow	5	0	0	0		71
Horse	0	0	0	0	1	

Recently, Kornegay et al. (1994) sequenced the cDNA of the hoatzin lysozyme to see whether parallel substitutions similar to those in the cow and langur lysozymes have occurred in this lysozyme, because hoatzin is the only known avian foregut fermenter and has separated from the mammals about 300 million years ago. Since the hoatzin lysozyme has a deletion of a proline at position 50 (depending on the alignment), the glutamic acid (E) residue at this position uniquely shared by the cow and langur lysozymes may not represent an adaptive substitution. What was unexpected is that the residues at positions 75 and 87 in the hoatzin lysozyme are the same as those [aspartic acid (D) and asparagine (N)] in the cow and langur lysozymes but different from those in the pigeon lysozyme. In addition, the hoatzin lysozyme has E instead of lysine (K) at positions 14 and 21. Since these two positions are at the surface of the enzyme, Kornegay et al. (1994) argued that having E in these positions may be more suitable for the less basic nature and lower pH profile of the stomach environment. In fact, position 14 is occupied by E in most ruminant stomach lysozymes other than those of the cow. Kornegay et al. made a similar argument for the E residue instead of K at position 126.

Table 7.12. Structural adaptations evident in hoatzin and mammalian stomach lysozymes

	Lysozyme type			
Characteristic	*Hoatzin stomach*	*Mammalian stomach*	*Chicken egg-white*	*Pigeon egg-white*
Low pH optimum	+	+	–	–
Isoelectric point	~6	6.2-7.7	11.2	~10.6
Total arginines	5	3-6	11	10
Adaptive residues				
E/K14	+	+	–	–
E/K21	+	+	–	–
D75	+	+	–	–
N87	+	+	–	–
E/K126	+	+	–	–

The adaptive structural changes proposed by Stewart and Wilson (1987) and Kornegay et al. (1994) are summarized in Table 7.12. They proposed that the residues at positions 14, 21, 75, 87, and 126 in the cow, langur, and hoatzin lysozymes represent adaptive substitutions. They provided evidence that some of these substitutions contribute to a better performance of lysozyme at low pH values. Furthermore, they suggested that the low arginine content confers increased resistance to inactivation by fermentation products such as diacetyl and to proteolytic digestion by trypsin.

Adaptation of Hemoglobin to High Altitudes

Hemoglobin has played a prominent role in the study of molecular evolution for several reasons. First, hemoglobin and its cousin myoglobin were the first proteins whose three-dimensional structures were worked out by X-ray crystallography. This knowledge has greatly facilitated the study of the relationship between structural constraint and rate of amino acid substitution. Second, its abundance in the blood of animals makes it easy to purify sufficient quantities for protein sequencing work, so it was one of the first proteins that had been sequenced in many species. The abundance of sequence data makes detailed comparative analyses feasible. Third, it is a protein that everyone is familiar with. For these reasons, hemoglobin has always been a favourite example for illustrating the principles of molecular evolution. In Table

7.5 it was used to illustrate the inverse relationship between stringency of structural constraints and rate of amino acid substitution in different regions of a protein. In this section it will be used as an example of adaptive changes to particular environments. Hemoglobin and myoglobin will be used as an example for the evolution of duplicate genes. For these purposes, it is helpful to have a summary of the structure and function of hemoglobin.

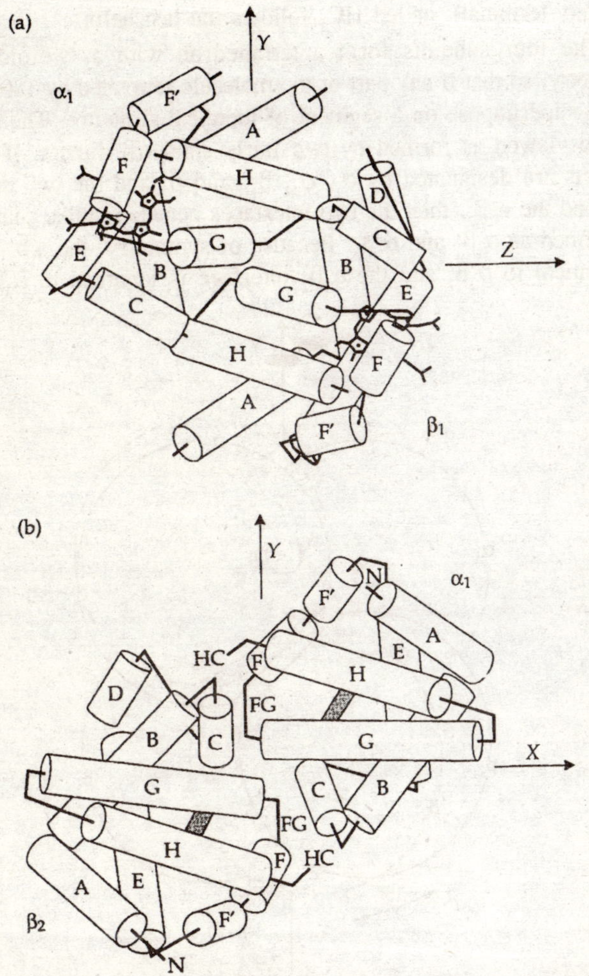

Fig. 7.7. Three-dimensional orientation of the subunits in carboxyhemoglobin. The top panel (a) shows the $\alpha_1\beta_1$ dimer and the heme groups of the α and β chains. The bottom panel (b) shows the $\alpha_1\beta_1$ interface.

Structure and Function of Hemoglobin

Hemoglobin consists of two α and β chains. In humans and almost all other mammals the α and β chains are 141 and 146 amino acids long, respectively. The α chain contains seven and the β chain eight helical segments, designated A to H; they are interrupted by nonhelical segments designated AB, BC, and so on. A short nonhelical amino-terminal segment, called NA, precedes the first helix, and another (carboxy-terminal), called HC, follows the last helix.

The four subunits form a tetrahedron with a twofold axis of symmetry, so that if any part of the molecule is rotated by 180 degrees, it will superimpose on a segment of identical structure. The molecule can be viewed as formed by two fairly rigid αβ dimers. If the four subunits are designated by α_1, α_2, β_1, and β_2 and the two dimers by $\alpha_1\beta_1$ and the $\alpha_2\beta_2$, then the two interfaces between unlike subunits can be defined as $\alpha_1\beta_1$ and $\alpha_1\beta_2$. Because of symmetry, the a_1b_2 interface is identical to $\alpha_2\beta_2$ and the $\alpha_1\beta_2$ interface is identical to $\alpha_2\beta_1$.

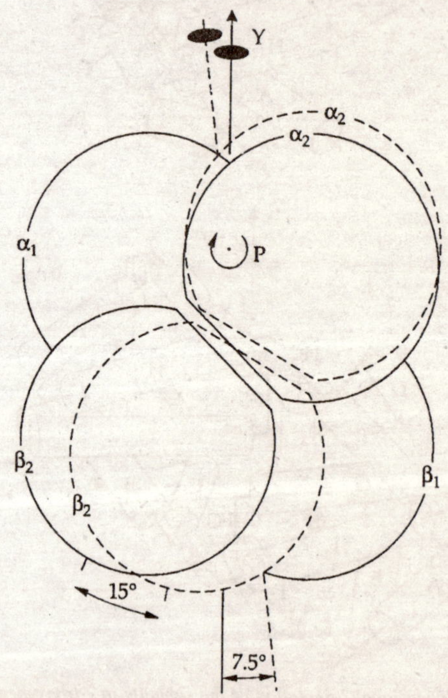

Fig. 7.8. Change in the position of one dimer ($\alpha_2\beta_2$) relevant to its partner dimer ($\alpha_1\beta_1$) as a result of ligation of the heme groups.

Uptake and release of oxygen are accompanied by small changes in tertiary structure of the segments surrounding the hemes and by a large change in quaternary structure involving a rotation of one αβ-dimer relative to the other by 15 degrees. The rotation that occurs on dissociation of oxygen widens the cavity between the two β chains so that the cationic groups that form its lining can bind organic phosphates. The two quaternary structures are referred to as the oxy or relaxed (R) and the deoxy or tense (T) structures.

As the oxygen carrier in human blood, hemoglobin must be able to load and unload oxygen at appropriate partial oxygen pressures (PO_2). Hemoglobin suits this physiologic role because it has the sigmoid oxygen-dissociation (saturation) curve. This sigmoid curve reflects the fact that hemoglobin binds oxygen in a cooperative fashion, that is, the binding of the first oxygen molecule facilitates the binding of subsequent molecules. In the air sacs of the lung, PO_2 is about 95 mm Hg and the hemoglobin in red cells becomes 97% saturated with oxygen.

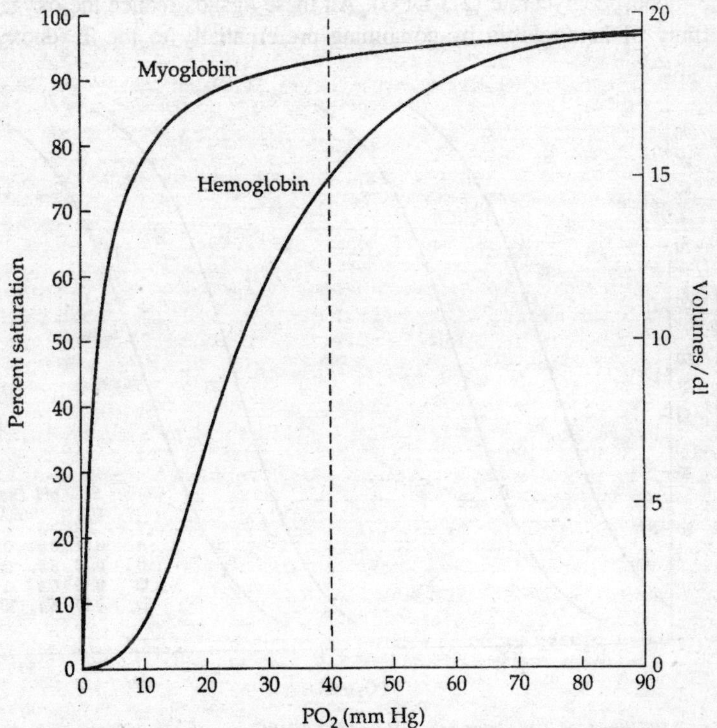

Fig. 7.9. Oxygen-binding curves of hemoglobin and myoglobin at 37°C and pH 7.4.

Following circulation through capillary beds, the mixed venous PO_2 is about 40 mm Hg and hemoglobin becomes only about 75% saturated with oxygen, so that 97%–75%=22% of the oxygen would have been unloaded. Should each of the four heme groups of hemoglobin bind oxygen independently of the others, the oxygen dissociation curve would have a hyperbolic shape, like that of myoglobin. Such a curve would be most unsuitable for oxygen transport, for in going from an arterial PO_2 of 95 mm Hg to a mixed venous PO_2 of 40 mm Hg, the hemoglobin is still 93% saturated, so that only 97%–93%-4% of the oxygen would be unloaded.

The oxygen affinity of hemoglobin decreases with decreasing pH, so that the oxygen-dissociation curve for hemoglobin is shifted to the right with increasing acidity. This is known as the (alkaline) Bohr effect, which facilitates the release of oxygen in tissue where acidity indicates the need for oxygen. Oxygen binding is influenced by chemical factors, known as heterotropic ligands, such as H^+, CO_2, Cl^-, and D-2,3-biphosphoglycerate (2,3-DPG). All these ligands reduce the oxygen affinity of hemoglobin by combining preferentially to the T (deoxy)

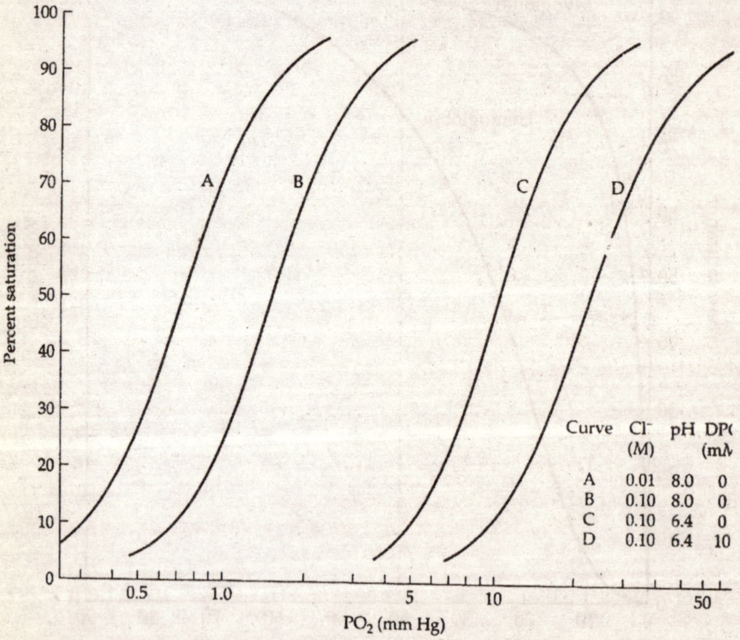

Fig. 7.10. Effect of ionic strength, pH, and 2,3-DPG on oxygen affinity of human hemoglobin A at 20°C.

structure, which has a high affinity for ligands but a low affinity for oxygen. The oxygen affinity of hemoglobin also decreases with rising temperature.

The oxygen affinity of hemoglobin can be conveniently expressed by the term P_{50}, the oxygen tension at which hemoglobin is half saturated. The higher the affinity of hemoglobin for oxygen, the lower the P_{50} value and vice versa. Thus, P_{50} is inversely related to oxygen affinity. The P_{50} value for human blood at physiologic pH (7.4) and temperature (37°C) is normally 26±1 mm Hg.

Llama Hemoglobin

Some animals living at high altitudes have adapted to life under hypoxic conditions (i.e., a low oxygen pressure) owing to changes in their hemoglobin sequences. The adaptation of llama (*Lama glama*), alpaca (*L. pacos*), guanaco (*L. guanacoe*), and vicuna (*L. vicugna*) to life in the Andes, South America, is such an example. The blood of these mammals has a high affinity for oxygen compared to their lowland relatives of the genus *Camelus*. This oxygen affinity is achieved by a low affinity for 2,3-DPG. In the deoxy form of most mammalian hemoglobins, 2,3-DPG is bound to the central cavity between the two β chains. The amino acids involved in these contacts are: β1Val, β2His, β82Lys, and β143His in human hemoglobin, where β refers to the β chain and the number to the residue position in the chain. A suitable substitution in one of these positions can cause a reduced binding of 2-3-DPG and thus increase the oxygen affinity by destabilizing the deoxy (T) hemoglobin. In llama, alpaca, guanaco, and vicuna, β2His is replaced by Asn. The side chain of Asn is shorter than that of His, and because of this shortening Asn cannot form contact with the phosphates of 2,3-DPG. The resulting higher oxygen affinity enables these animals to live in regions up to 4000 m. Interestingly, β2Asn is also found in the Indian and African elephant, allowing them to climb up to 4500 m. The vicuna, the largest wild animal permanently residing in altitudes between 4000 and 5000 m, has an additional substitution (α130Ala→Thr) affecting chloride binding, which in combination with β2Asn causes the highest blood oxygen affinity of all mammals investigated so far.

A few other differences are also observed among the *Camelus* and *Lama* sequences. Kleinschmidt et al. (1986) suggested that the internal nonpolar isoleucine at α10 in llama, alpaca, and guanaco does not affect the hemoglobin structure. They also suggested that although position α122 is involved in the $\alpha_1\beta_1$ contact, the substitution His→Asp in llama does not seem to influence the $\alpha_1\beta_1$ contact, for the carboxyl group of aspartic acid might form a salt bridge with the amino group of β30Arg. The substitutions α8Thr→Ala, α23Glu→Asp, α120Ser→Ala, and β76Asn→Ser between the genera *Camelus* and *Lama* concern external residues that normally do not affect the structure and the oxygen binding properties of the hemoglobin molecule.

Table 7.13. Amino acid substitution in Camelidae hemoglobins

	α chain						β chain		
Species	8	10	23	120	122	130	2	76	135
Lama glama	Ala	Ile	Asp	Ala	Asp	Ala	Asn	Ser	Ala
L. pacos	Ala	Ile	Asp	Ala	His	Ala	Asn	Ser	Ala
L. guanacoe	Ala	Ile	Asp	Ala	His	Ala	Asn	Ser	Ala
L. vicugna	Ala	Val	Asp	Ala	His	Thr	Asn	Ser	Ala
Camelus ferus	Thr	Val	Glu	Ser	His	Ala	His	Asn	Ala
C. dromedarius	Thr	Val	Glu	Ser	His	Ala	His	Asn	Ala

Bird Hemoglobins

The extensive protein sequencing work of Braunitzer and colleagues has provided many excellent examples of adaptation of bird hemoglobins to high altitudes. Three examples are discussed below.

The first example is the bar-headed goose (*Anser indicus*) hemoglobin. These geese live and hatch their young at the Tibetan lakes (altitudes 4000-6000 m) but winter in the plains of northwest India. They migrate semiannually across the Himalayan Mountains without obvious stops en route for acclimatization. During migration flocks have been observed crossing directly over Mt. Everest at an altitude of 9200 m. The hemoglobin sequence of the bar-headed goose was compared with those of the greylag goose, the Canadian goose, and the mute swan, which are its closest lowland relatives. The bar-headed goose hemoglobin differs from the greylag goose hemoglobin by only four amino acids, the smallest number of replacements found among all the pairwise comparisons:

	α18	α63	α119	β125
Greylag goose	Gly	Ala	Pro	Glu
Bar-headed goose	Ser	Val	Ala	Asp

The substitution at α18 affects an external residue, the site α63 lies in a surface crevice, and β125Glu is placed so that the hydrophilic side chain would protrude into the surrounding water. None of these replacements is likely to affect the function of hemoglobin. In contrast,

α119Pro touches leucine β55 at the $\alpha_1\beta_1$ contact and its replacement by alanine leaves a two-carbon gap so that the interaction between α119 and β55 can no longer occur. This loss of interatomic stabilization of T structure favours the existence of the high-affinity R structure. For this reason, Perutz (1983) suggested that the substitution α119Pro $\rightarrow$ Ala alone is responsible for the high affinity of the Bar-headed goose hemoglobin.

The second example is the hemoglobin of the Andean goose (*Chleophaga melanoptera*), which lives in the Andes at altitudes of 5000-6000 m. This hemoglobin differs from that of the greyleg goose by nine replacements in the α chain and seen in the β chain. Nevertheless, Hiebl et al (1987) postulated that its high oxygen affinity is caused by the single replacement β55Leu$\rightarrow$Ser, which leaves a two-carbon gap at the contact with α119Pro at the $\alpha_1\beta_1$ interface, the same proline that is replaced by alanine in the Bar-headed goose hemoglobin.

To test the above hypothesis that in both cases a single mutation was responsible for the high-oxygen affinity of the mutant hemoglobin (Hb), Jessen et al. (1991) used site-directed mutagenesis to produce two human Hb mutants: α119Pro$\rightarrow$Ala and β55Met$\rightarrow$Ser. They compared the oxygen affinities of these two mutant Hbs with those of native human Hb and goose Hbs. The increase in oxygen affinity (Δlog P_{50}) produced by either substitution was about 0.22, which was actually greater than the difference in oxygen affinity (0.15) between the Hbs of the Bar-headed and the greylag geese. They therefore concluded that the gap left by either substitution at the $\alpha_1\beta_1$ contact was a sufficient cause for the increase in oxygen affinity observed in nature. They also crystallized the mutant with β55Ser and determined its three-dimensional structure by X-ray analysis. They found that the major disturbances were restricted to the immediate neighbourhood of the amino acid substitution. This observation explained why the mutant Hb was fully functional and strengthened the view that the increase in oxygen affinity was caused exclusively by the gap introduced at the $\alpha_1\beta_1$ contact.

The Bar-headed and Andean goose hemoglobins represent two types of adaptation: transitory and permanent hypoxic stress. The blood of these geese contains two hemoglobin components, one major, Hb A, and one minor, Hb D. Both hemoglobins have the same β chain but differ in the α chains, denoted by α^A and α^B, which are encoded by two duplicated α loci. In the Andean goose the substitution β55Met$\rightarrow$Ser

Table 7.14. Oxygen affinity of human and goose hemoglobins

Hemoglobin	P_{50}*(mm Hg)*	*log* P_{50}	Δ *log* P_{50}
Human HbA	5.8	0.76	
Human HbPα119A	3.3	0.53	0.23
Human HbMβ55S	3.4	0.54	0.22
Greylag goose	2.8	0.45	
Bar-headed goose	2.0	0.30	0.15

is in the β chain and so affects both Hbs. Thus, this single substitution results in an effective increase of oxygen affinity of whole blood, which seems to suit well the Andean goose, for this species lives permanently in the High Andes. On the other hand, in the Bar-headed goose hemoglobin the substitution α119Pro→Ala, being in the α^A chain, affects only Hb A, so that Hb A has a high oxygen affinity compared to Hb D. Since Bar-headed geese have to cover altitudes from near sea-level to elevations as high as 9200 m on their migration flight over Mt. Everest, they need hemoglobins to adjust relatively quickly to different environmental conditions. Blood with hemoglobins of graded affinities can meet these requirement more effectively due to the larger variability in the fine tuning of oxygen uptake and delivery resulting from the expression of two different α globin chains.

The third example, the hemoglobin of Ruppell's griffon (*Gyps rueppellii*), is even more remarkable. The fact that a Ruppell's griffon was hit by an aircraft at an altitude of 11,300 m over Abijan, Ivory Coast in west Africa shows that these vultures can ascend to extremely high altitudes. The blood of Ruppell's griffon is extraordinary in that it contains four hemoglobins instead of the usual two components in other birds. The four hemoglobins (Hb A, Hb, A', Hb D, and Hb D') have the same β chain but differ in the α chain, implying the existence of four α loci, α^A, $\alpha^{A'}$, α^D, and $\alpha^{D'}$. Since the β chain of the four hemoglobins is identical to those of the European black vulture and the white-headed vulture, which ascend to altitudes of only about 4500 m, the α chains must account for the differences in the tolerance of hypoxic stress among these birds.

The α chain sequences of these birds show differences in the *respiratory box*, α34 to α38, which is of great structural and functional importance because it contains three $\alpha_1\beta_1$ and two $\alpha_1\beta_2$ contacts. At α34 ($\alpha_1\beta_1$ contact) the substitution Ile→Thr has occurred in the Hb A of Ruppell's griffon. With α34Thr, a hydrogen bond can be formed to β125Asp so that the low-affinity T structure is stabilized. For this

Table 7.15. Amino acids from the α globins of Ruppell's griffon (RG), the European black vulture (BV) and the white-headed vulture (WV) at positions that show variation among sequences

	Respiratory box		*Other positions*			
α chain	*34*	*38*	*12*	*104*	*123*	*137*
RG α^A	Thr (–)	Pro (–)	Asn	Cys	Ala	Thr
RG $\alpha^{A'}$	Ile (+)	Pro (–)	Asn	Cys	Ala	Thr
BV α^A	Ile (+)	Pro (–)	Thr	Cys	Ala	Thr
WV α^A	Ile (+)	Pro (–)	Thr	Cys	Ala	Ser
RG α^D	Ile (+)	Gln (+)	Thr	Cys	Ser	Ala
RG $\alpha^{D'}$	Ile (+)	Gln (+)	Thr	Asn	Ser	Ala
BV and WV α^D	Ile (+)	Pro (–)	Ala	Cys	Ala	Ala

reason, the oxygen affinity of Hb A is lower than those of Hb A', Hb D, and Hb D'. A similar explanation holds at α38 ($\alpha_1\beta_2$ contact): Ruppell's griffon Hb D and D' have the substitution Pro→Gln, which increases the oxygen affinity because the equilibrium between the T and R structures is shifted towards the R structure. A few differences are observed outside the respiratory box. A change at position 12 is unlikely to exert much influence on function because this position is on the surface and is highly variable among species. Whether or not the changes Cys→Asn at position 104 affects the oxygen affinity of $\alpha^{D'}$ is not clear because while α104 is not involved in any contact, Cys at α104 has been well conserved and Asn has been found only in the α chain of the opossum. Position 123 at the $\alpha_1\beta_1$ contact and the substitution Ala→Ser in α^D and $\alpha^{D'}$ weakens the hydrophobic interaction between Ala and β34Val by the hydrophilic Ser. However, this substitution was suggested to be of minor importance. The changes at position 137 are not yet understood; however, since all αA chains of birds, except for the white headed vulture, have Thr at this position, whereas all αD chains have Ala, the exchange Thr→Ala is unlikely to have special influence. In summary, the major changes occur only in the respiratory box, and for Ruppell's griffon three levels of affinity are predicted: Hb A < Hb A' < Hb D/D'. The advantages of a low-affinity hemoglobin (greater oxygenation of the tissues) and a high-affinity component (improved arterial oxygenation) are combined. At extreme high altitudes the high-affinity Hb D/D' serve to take up oxygen while at low altitudes the intermediate and low affinity Hb A' and Hb A ensure sufficient oxygenation. Hiebl et al. (1988) suggested that this three-stage cascade system constitutes the molecular basis for flight at altitudes of 11,300 m, at which the oxygen tension is only about 25% of that at sea level.

It is clear from the above discussion that a good knowledge of the structure and function of hemoglobin greatly facilitates the identification of changes that are important for the adaptation of life to high altitudes. In all the above examples it seems that the adaptation can be attributed to one or at most a few changes. The authors have therefore suggested that most of the other changes are functionally neutral. The same suggestion has been made by Perutz (1983) in an extensive review of adaptation of hemoglobin in different species. Of course, one should be wary that small functional differences are experimentally difficult to detect but may be of significance in the long history of evolution.

8

DNA Polymorphism in Evolution

The extent and the mechanism of maintenance of genetic variability in natural populations have always been a subject of great interest to population geneticists and evolutionists. As the debate over the evolutionary forces governing protein polymorphism was inconclusive, it has moved to the DNA level. DNA sequence data represent the highest level of genetic resolution and allow the development of more powerful statistical approaches than do protein polymorphism data. For example, the theory of gene genealogy and coalescence has been developed mainly for analyzing DNA polymorphism data. Although DNA technologies have not yet been extensively used in population genetics studies, available data have already provided considerable insight into the maintenance of DNA variation. In this chapter we review both experimental data and theoretical developments.

Measures of DNA Polymorphism

We considered gene diversity as a measure of genetic variability in a population. The gene diversity at a locus (h) is the probability that two sequences randomly chosen from the population are different; for a randomly mating population, h is also the heterozygosity. Although this measure has been commonly used for studying protein polymorphism, particularly for electrophoretic data, it may not be suitable for DNA sequence data, because the extent of genetic variation at the DNA level in nature is quite extensive. Indeed, when long sequences are considered, each sequence in the sample may be different from the other sequences, so that h is close to 1. When this is the case, h cannot discriminate among different loci or populations and is therefore no longer an informative measure of polymorphism.

For a pair of DNA sequences, it is more informative to consider the number of nucleotide differences between the two sequences rather than just whether they are different. Thus, for DNA sequence data, a more appropriate measure of polymorphism in a population is Π, the average number of nucleotide differences between two sequences randomly chosen from the population. When n sequences are taken randomly from the population, Π is estimated by

$$\Pi = \frac{1}{[n(n-1)/2T]} \sum_{i<j} \Pi_{ij} \qquad \text{...(1)}$$

where Π_{ij} is the number of nucleotide differences between the ith and jth sequences and n(n–1)/2 is the number of possible pairs. Alternatively, we may first compute the frequencies of different alleles and estimate Π by

$$\Pi = \frac{n}{n-1} \sum_{i,j} x_i x_j \Pi_{ij} \qquad \text{...(2)}$$

where x_i and x_j are the frequencies of the ith and jth alleles in the sample, respectively, and (n/(n–1) is the correction factor for sampling bias. It can be shown that Equation 1 and 2 given the same result.

A mutation model that is commonly used to study Π is the *infinite-site* model, which assumes that the number of nucleotide sites on the sequence is so large that each new mutation occurs at a site that has not been mutated before. Under this model and under the assumption of random mating, Watterson (1975) showed that the mean of Π is given by

$$E(\Pi) = \theta \qquad \text{...(3)}$$

and Tajima (1983) showed that the variance of Π is given by

$$V(\Pi) = \frac{n+1}{3(n-1)}\theta + \frac{2(n^2+n+3)}{9n(n-1)}\theta^2 \qquad \text{...(4)}$$

where $\theta = 4N_e u$, N_e is the effective size of a diploid population, and u is the mutation rate per sequence per generation; Equation 4 assumes no recombination between sequences. Note that the mean is independent of the sample size n, whereas n increases to infinity, the variance decreases to $\theta/3 + 2q^2/9$, which is the component of the variance that is due to random drift.

The Π value is likely to be larger for a long sequence than for a short one. In fact, Π increases with L because $u=L\upsilon$ (and hence $\theta=4N_e u$) increases with L, where υ is the rate of mutation per site per generation and L is the length of the sequence (i.e., the number of nucleotide sites on the sequence). For this reason, Π is not convenient for making comparisons of DNA sequence variation among loci, because L is likely to vary among loci. This drawback, however, can be easily removed by standardizing Π by L. That is, we consider

$$\pi = \Pi / L \qquad ...(5)$$

Nei and Li (1979) called π the *nucleotide diversity*, which is the number of nucleotide differences per site between two randomly chosen sequences. From Equation 5, one can easily shown that $V(\pi)=V(\Pi)/L^2$.

Another commonly used measure of DNA polymorphism is the number of segregating nucleotide sites (K) in a sample of sequences; a *segregating site* is a site that shows variation among the sequences in the sample. Under the infinite-site model and under random mating, Watterson (1975) showed that the mean and variance of K are given by

$$E(K) = a\theta \qquad ...(6)$$

$$V(K) = a\theta + b\theta^2 \qquad ...(7)$$

where

$$a = 1 + \frac{1}{2} + \cdots + \frac{1}{n-1} \qquad ...(8)$$

$$b = 1 + \frac{1}{2^2} + \cdots + \frac{1}{(n-1)^2} \qquad ...(9)$$

Obviously, K depends on the sequence length L but this dependence can be removed by considering $k=K/L$. The advantage of k over π is that it has a smaller stochastic variance, but a drawback is that k depends on the same size n.

A third measure of DNA polymorphism is the number of alleles (i.e., different sequences) in the sample (n_a). The mutation model commonly used to study this quantity is the *infinite-allele* model, which assumes that each mutation creates a new allele not currently present in the population. Under this mutation model and under random mating, Ewens (1972) showed that the mean of n_a is given by

$$E(n_a) = \theta\left(1 + \frac{1}{\theta+1} + \frac{1}{\theta+2} + \cdots + \frac{1}{\theta+n+1}\right) \qquad ...(10)$$

Since n_a is dependent on both the sample size and the sequence length, it is less convenient that π for making comparisons among loci or populations.

Note that all the above formulas for the means and variances are derived under the assumptions that mutations are neutral and that the population is at equilibrium.

Let us now consider how to compute π, K, and n_a from sequence data by using the 11 sequences of the alcohol dehydrogenase (Adh) gene in *Drosophila melanogaster* obtained by Kreitman (1983). The aligned sequences were 2379 nucleotides long, i.e., L=2379. Since there were 43 such sites, K=43; deletions and insertions were disregarded. It is easy to see that there are nine different alleles, i.e., $n_a=9$. One of the alleles is represented by three sequences, while the rest are each represented by one sequence. Thus, the frequencies x_1, ..., x_8 are each 1/11 and the frequency x_9 is 3/11. We now calculate the proportion of different nucleotides for each pair of alleles. For example, alleles 1-S and 2-S differ from each other by 3 nucleotides out of 2379 or $\pi_{12}=3/2379=0.13\%$. The π_{ij} values for all the pairs in the sample in Table 8.1. By using Equation 2 with Π_{ij} replaced by π_{ij}, the nucleotide diversity is estimated to be $\pi=0.0065$, so that we expect two randomly chosen sequences of 1000 base pairs to differ, on average, at 6.5 sites. Six of the alleles studied were slow-migrating electrophoretic variants (S), and five were the fast (F) type; S and F differ by only one amino acid. When the nucleotide diversity for each of the two electrophoretic classes is calculated separately, we obtain $\pi=0.0056$ for the S sequences and $\pi=0.0029$ for the F sequences. The twofold difference in π suggests that the S electrophoretic variant is considerably older than the F variant, so that the S-type sequences have no average diverged more at the DNA level.

At the present time, DNA polymorphism data remain limited, even for humans and *Drosophila*; *Drosophila* data will be discussed later in this chapter. The data from humans analyzed by Li and Sadler (1991), who used published (genomic or cDNA) sequences pairs (for 49 genes) that have been verified for sequencing accuracy. The data were mainly from white Americans and so the nucleotide diversity may be considered as the average value for the white American population. The average level of nucleotide diversity (π) is low, on average lower than 1 nucleotide difference per 1000 sites between two random sequences. The π value for Y-linked sequences might be even lower, because a world-wide sample of 38 sequences from the last intron (~739 bp) of the ZFY gene revealed no variation.

Table 8.1. Pairwise percent nucleotide differences (per 100 sites) among 11 sequences of the alcchol dehydrogenase locus in *Drosophila melanogaster*

	Sequence									
Sequence	1-S	2-S	3-S	4-S	5-S	6-S	7-F	8-F	9-F	10-F
1-S										
2-S	0.13									
3-S	0.59	0.55								
4-S	0.67	0.63	0.25							
5-S	0.80	0.84	0.55	0.46						
6-S	0.80	0.67	0.38	0.46	0.59					
7-F	0.84	0.71	0.50	0.59	0.63	0.21				
8-F	1.13	1.10	0.88	0.97	0.59	0.59	0.38			
9-F	1.13	1.10	0.88	0.97	0.59	0.59	0.38	0.00		
10-F	1.13	1.10	0.88	0.97	0.59	0.59	0.38	0.00	0.00	
11-F	1.22	1.18	0.97	1.05	0.84	0.67	0.46	0.42	0.42	0.42

For the regions, the highest one is observed at the fourfold degenerate site and is only 0.11%. The second highest level is observed at the twofold degenerate sites; the synonymous and nonsynonymous components are 0.05% and 0.01%, respectively, and the total diversity is 0.06%. The third highest level, π=0.04%, is in the 3' untranslated (UT) region, though it is not significantly higher than the values in the 5'UT regions and at the nondegenerate sites, both being 0.03%. These values are well correlated with the selective constraints in different regions of the gene; e.g., the nondegenerate sites and the fourfold degenerate sites are, respectively, subject to the most and least stringent constraints and have, respectively, the lowest and the highest level of nucleotide diversity.

Gene Genealogy and the Coalescent Theory

This section reviews basic concepts of gene genealogy and the coalescent theory, which has proved to be a powerful approach to the statistical analysis of DNA polymorphism data.

Consider a sample of n sequences of a DNA region from a population. If the DNA region is completely linked so that no recombination occurs between sequences, then the n sequences are connected by a single phylogenetic tree, i.e., a genealogy. In the genealogy, the n sequences can be traced back first to n – 1 ancestral

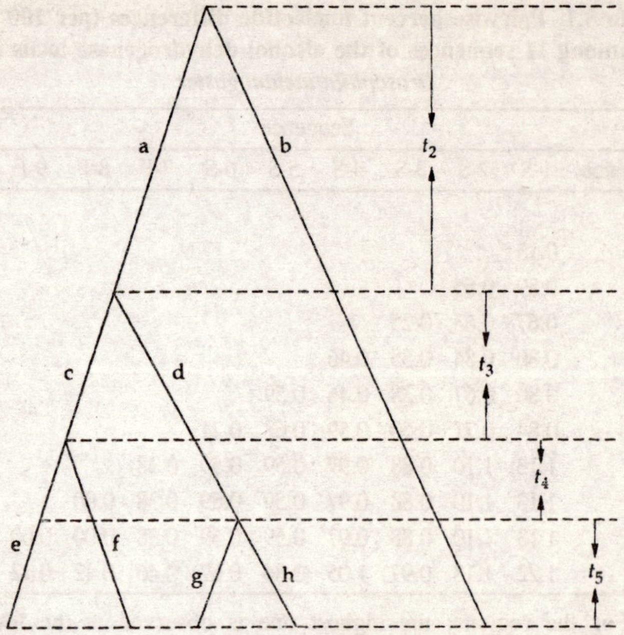

Fig. 8.1. An example of genealogy of five sequences with the root of the genealogy at the top.

sequences, next to n – 2 ancestral sequences, and so on until finally to a single common ancestral sequence. Using a different terminology, we can say that the n sequences coalesce first to n – 1 ancestral sequences, then to n – 2 ancestral sequences, and so on until finally to a single common ancestral sequence. The mathematical theory that treats the genealogical branching process backward in time as a *coalescence process* is called the *coalescent theory*. This powerful theory for studying various quantities pertaining to gene genealogies was initiated by Kingman (1982), Hudson (1982), and Tajima (1983). In the following we shall consider a diploid random mating population in the absence of natural selection and migration and for simplicity, we assume that the effective size N_e is equal to the actual population size N. For more complex situations involving selection, recombination or migration, readers may refer to Kaplan et al (1988, 1991) Takahata and Nei (1990), and Hudson (1993).

An essential quantity in the coalecent theory is the time duration (t_m) required for the coalescence from m sequences to m – 1 sequences. Let us start with the simplest case: two sequences randomly drawn from the population, i.e., m = 2. The probability that the two sequences are derived from a common ancestral sequence in the preceding generation is 1/(2N) because there are 2N sequences in the population and because sampling is with replacement, i.e., the same gene can be repeatedly sampled. Obviously, the probability that they are derived from two different sequences in the preceding generations is 1–1/(2N). Therefore, the probability that the two sequences are derived from a common ancestral sequence 2 generations ago is [1–1/(2N)][1/(2N)] and the probability that the two sequences are derived from a common ancestral sequence t + 1 generations ago is

$$[1-1/(2N)]^t[1/(2N)] \approx [1/(2N)]\exp[-t/(2N)] \quad ...(11)$$

From this distribution one can show that the mean and variance of t_2 are

$$E(t_2) = 2N \quad ...(12)$$

$$V(t_2) = 4N^2 \quad ...(13)$$

In general, the time t_m for the coalescence from m to m – 1 sequences follows the exponential distribution with parameter m(m – 1)/M:

$$g(t_m) = \frac{m(m-1)}{M}\exp\left[\frac{-m(m-1)}{M}t_m\right] \quad ...(14)$$

where $M = 4N_e$. The mean and variance of t_m are

$$E(t_m) = \frac{M}{m(m-1)} \quad ...(15)$$

$$V(t_m) = \left[\frac{M}{m(m-1)}\right]^2 \quad ...(16)$$

The genealogy of n sequences has 2(n – 1) branches. A branch is said to be *external* if it directly connects to an external node (i.e., a node at the tip of the tree), otherwise it is said to be *internal*. Since there are n external node (sequences), n of the 2(n – 1) branches are external and n – 2 branches are internal. In Fig. 8.1, b, e, f, g, and h are external and a, c, and d are internal. Note that we consider a rooted tree, so that branch a is internal. The motivation of distinguishing between internal and external branches is that this will be useful later when we consider detection of natural selection from sequence data.

Let J, I, and L be, respectively, the total time length of all branches, the total time length of internal branches and the total time length of external branches. Fu and Li (1993) showed that

$$E(L) = M \quad ...(17)$$

which is independent of the sample size n. Thus, regardless of the number of sequences sampled, the expected total time length of external branches is always $M=4N_e$ generations. The expectations of I and J are

$$E(I) = (a-1)M \quad ...(18)$$

$$E(J) = E(I) + E(L) = aM \quad ...(19)$$

where a is given by Equation 8.

Let us now consider mutations. We assume that the rate of mutation per sequence per generation is u and that the number of mutations in a sequence in a time period t follows the Poisson distribution with parameter ut. Let η_e and η_i be the total numbers of mutations in the external and internal branches, respectively, and let $\eta=\eta_i+\eta_e$ be the total number of mutations that occurred in the entire genealogy of n sequences. Fu and Li (1993a) showed that

$$E(\eta_e) = \theta \quad ...(20)$$

$$V(\eta_e) = \theta + c\theta^2 \quad ...(21)$$

$$E(\eta_i) = (a-1)\theta \quad ...(22)$$

$$V(\eta_i) = (a-1)\theta + \left[b - \frac{2a}{n-1} + c\right]\theta^2 \quad ...(23)$$

$$E(\eta) = a\theta \quad ...(24)$$

$$V(\eta) = a\theta + b\theta^2 \quad ...(25)$$

where b is given by Equation 9 and

$$c=1 \qquad \text{if } n = 2, \text{ but}$$

$$= \frac{2[na-2(n-1)]}{(n-1)(n-2)}, \qquad \text{if } n > 2$$

Under the infinite-site model, which assumes that every mutation occurs at a new site of the sequences, the number of segregating (polymorphic) sites (K) among the sequences in the sample is equal to the total number of mutations η. Therefore, the mean and variances of η should be the same as those of K and indeed, Equations 24 and 25 are identical to Equations 6 and 7.

It is interesting to know the relationships among these quantities and he quantity Π defined by Equaticn 1. For example, Fig 8.2 shows that η_e and η_i become almost independent of each other (i.e., have a low correlation coefficient) when the sample size n becomes larger than 10, whereas the correlation between η and Π is strong even when the sample size is quite large.

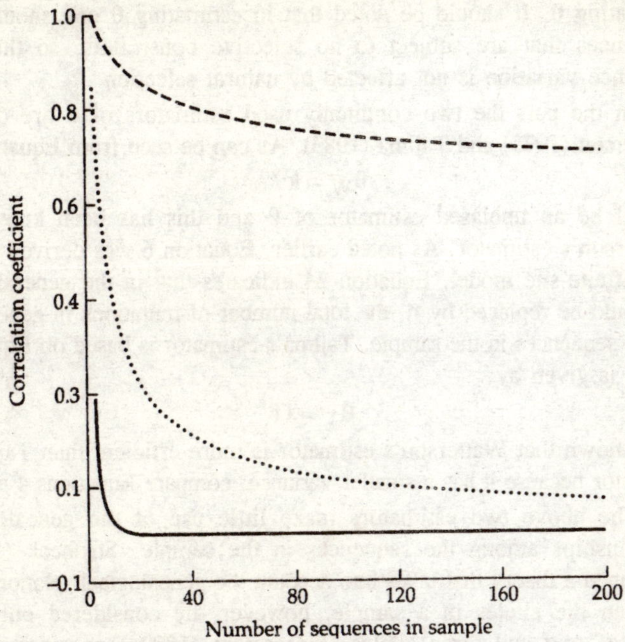

Fig. 8.2. Correlation coefficients. The solid, dotted, and dashed curves show the correlation coefficients between η_e and η_i, between η_e and Π, and between η and Π respectively.

Estimation of Population Parameters

A great difficulty in population genetics is our inability to measure any of the basic parameters that are involved in the theory. For example, we do not know the long-term effective population size of any species or the mutation rate u for any gene or sequence. Fortunately, the situation is improving as a result of the introduction of various molecular techniques into population genetics. In particular, the feasibility of obtaining a fairly large number of DNA sequences from a population has enabled us to obtain better estimates of certain parameters such as θ. As can be seen from the preceding sections, θ is a parameter that appears in many formulas. Indeed, θ plays a prominent role in the stochastic theory of population genetics and there has been much interest in the technical aspects of estimating θ. Because the two parameters N_e and u usually do not appear separately in a formula, we are not able to estimate them separately but can estimate only their product through $\theta=4N_eu$. Of course, if we know N_e or u, we can compute the other form θ. We discuss below methods for estimating θ. It should be noted that in estimating θ one should use sequences that are subject to no selective constraints, so that the sequence variation is not affected by natural selection.

In the part the two commonly used estimators of θ are due to Watterson (1975) and Tajima (1983). As can be seen from Equation 6.

$$\theta_W = k / a \qquad \text{...(26)}$$

should be an unbiased estimator of θ and this has been known as Watterson's estimator. As noted earlier, Equation 6 was derived under the infinite-site model. Equation 24 indicates that in the general case K should be replaced by η, the total number of mutations in genealogy of the sequences in the sample. Tajima's estimator is based on Equation 3 and is given by

$$\theta_T = \Pi \qquad \text{...(27)}$$

It is known that Watterson's estimator is more efficient than Tajima's estimator because it has a smaller variance; compare Equations 4 and 7.

The above two estimators make little use of the genealogical relationships among the sequences in the sample. Strobeck (1983) developed a theory that takes into account the genealogical relationships between the alleles in a sample, however, he considered only the cases of two and three alleles. Felsenstein (1992a) considered any number of sequences and suggested that much improvement in efficiency over Watterson's estimator can be achieved by incorporating the genealogical relationships among sequences. Fu and Li (1993b) investigated the maximum amount of improvement that might be achieved in practice.

The tree is divided into segments by the horizontal lines through each of the branching nodes (or tips). Let η_j be the total number of mutations that occurred between the (j–1)-th node and the jth node, j=2,3,..., n. Suppose that in an ideal situation, we can observe or infer without error the η_j values. Then, using the coalescent theory and the maximum likelihood method, Fu and Li (1993b) showed that the optimal estimator θ_m is the solution of the following equation for θ

$$\sum_{m=2}^{n} \frac{\eta_m + 1}{\theta + m - 1} = \frac{\eta}{\theta} \qquad \text{...(28)}$$

and that the variance of this estimator is

$$V_{min} = \theta \left[\sum_{m=2}^{n} \frac{1}{\theta + m - 1} \right]^{-1} \qquad \text{...(29)}$$

Since θ_m is based on $\eta_2, \ldots, \eta_n$ whereas θ_W is based only on η (i.e., the sum of $\eta_2, \ldots, \eta_n$), θ_m uses more information and should be more efficient than θ_W. Indeed, Fu and Li (1993b) showed that

$$V(\theta_m) \leq V(\theta_W)$$

The ratio $V(\theta_m)/V(\theta_W)$ indicates the relative efficiency of θ_W to θ_m. Thus, θ_m can be much more efficient than θ_W and there is considerable room for improvement. For this reason, there is currently much effort to improve the estimation of θ. Felsenstein (1992b), Griffiths and Tavare (1994), and Kuhner et al. (1995) use the maximum likelihood approach, while Fu (1994a) proposes an approximate approach that is computationally less demanding than the likelihood approach.

Fu's estimator is based on the following observation. Since in practice the η_j's cannot be estimated without error or, in other words, we have less information that knowing precisely $\eta_2, \ldots, \eta_n$, we cannot have a better estimator than θ_m, i.e., a variance smaller than V_{min}. Thus, if one can obtain an estimator that has a variance very close to V_{min}, then that estimator should be close to the best estimator. Fu

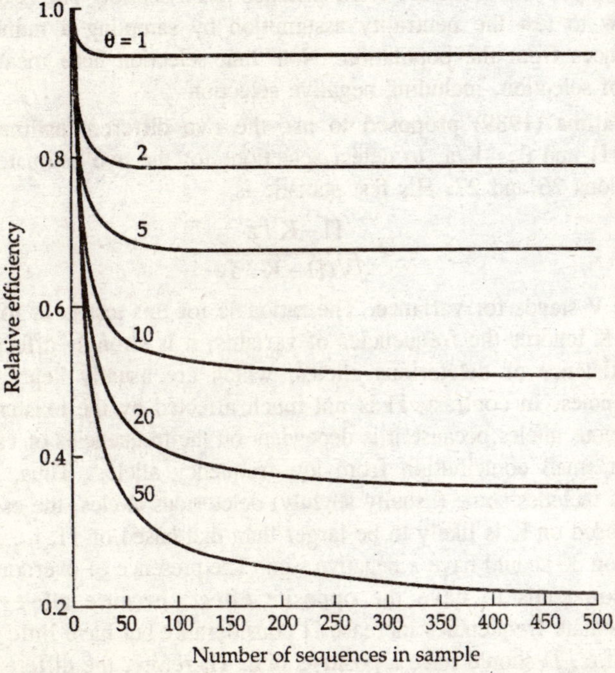

Fig. 8.3. Efficiency of Watterson's $\hat{\theta}_W$ estimate of θ relative to $\hat{\theta}_m$.

(1994a) showed that by using the simple UPGMA method to reconstruct a genealogy and to infer the number of mutations on each branch of the genealogy from the sequence data one can in fact obtain an estimate of θ with a variance very close to V_{min}. A computer program is available from Dr. Y.-X Fu.

Note that we have so far assumed no recombination. If the rate of recombination per sequence per generation (r) is low, so that $4N_er$ is of the order of 1, the effect of recombination is not serious and the above approach by Fu is still applicable. However, if the recombination rate is higher, the effect needs to be taken into account; Fu (1994b) has developed a new estimator for this purpose.

Detection of Natural Selection

In molecular population genetics or evolutionary studies, one often assumes that a DNA region is subject to no natural selection, so that all mutations in the region are selectively neutral. For example, noncoding regions such as intergenic regions or introns are usually assumed to be subject to no selective constraints, and in coding regions synonymous mutations are often assumed to be neutral. The issue here is how to test the neutrality assumption by sampling a number of sequences from the population. Note that selection here means any type of selection, including negative selection.

Tajima (1989) proposed to use the two different estimates of $\theta, \theta_T = \Pi$ and $\theta_W = K/a$, to detect selection; for the two estimates, see Equations 26 and 27. His test statistic is

$$D \frac{\Pi - K/a}{\sqrt{V(\Pi - K/a)}} \qquad ...(30)$$

where V stands for variance. The rationale for this test is as follows. Since K ignores the frequencies of variants, it is strongly affected by the existence of deleterious alleles, which are usually kept at low frequencies. In contrast, Π is not much affected by the existence of deleterious alleles because it is dependent on the frequencies of variants with a small contribution from low-frequency alleles. Thus, if the sample includes some (usually slightly) deleterious alleles, the estimate of θ based on K is likely to be larger than that based on Π, i.e., D in Equation 30 should have a negative sign. The presence of overdominant selection tends to have the opposite effect, because alleles with intermediate frequencies increase Π considerably but have little effect on K, i.e., D should have a positive sign. Therefore, the difference $\Pi - K/a$ can be used to detect the presence of selection. The denominator is intended to normalize the statistic. Tajima (1989) suggested using a beta distribution as an approximation of the distribution of D so that critical values of the test can be obtained, but a more accurate approach is to use Monte Carlo simulation to obtain the critical values of the statistic.

One can construct a test similar to the above one by using any pair of estimates of θ as long as the variance of their difference can be calculated. However, in order for such a test to be useful, the two estimates must be sufficiently different when selection is present. Because the expectations of η_e and $\eta_i/(a - 1)$ are both equal to θ under neutral mutation and because they are likely to be different when selection is present, Fu and Li (1993a) proposed the following test:

$$G = \frac{\eta_e - \eta_i / (a-1)}{\sqrt{V[\eta_e - \eta_i / (a-1)]}} \quad ...(31)$$

As is well known, deleterious mutations are usually quickly eliminated from the population, and those that are present in the sample are most likely to have arisen recently. Recent mutations are close to the tips (external nodes) in the genealogy and therefore are mostly included in the value of η_e. In contrast, mutations in the internal branches are most likely to be neutral and $\eta_i/(a - 1)$ is not strongly affected by the presence of selection. Therefore, these two estimates of θ should be useful for testing for the presence of selection. An advantage of G over D is that η_e and η_i are weakly correlated, whereas Π and $\theta_w = K/a$ are strongly correlated.

Since Π is less affected than η_e by the presence of selection, an obvious variant of test G is

$$F = \frac{\Pi - \eta_e}{\sqrt{V(\Pi - \eta_e)}} \quad ...(32)$$

The critical values of the G and F statistics have been provided by Fu and Li (1933a), who also gave an example of the application of these tests to DNA sequence data.

Simonsen et al. (1995) simulation study of the relative performances of different tests under a selective sweep suggested that Tajima's test given by Equation 30 tends to be more powerful than the others. However, Charlesworth et al (1995) result from limited simulations under background selection suggested the superiority of Fu and Li's (1993a) tests over Tajima's test. Readers may also refer to the maximum likelihood approach by Golding and Felsenstein (1990), which is a between-species method but can also be applied to within-species polymorphism data.

TESTING THE NEUTRAL MUTATION HYPOTHESIS

The tests represented in the preceding section are for testing the hypothesis that all mutations in a DNA region are selectively neutral, but not for testing the neutral mutation hypothesis. The latter hypothesis postulates that the majority of mutations that have contributed significantly to the genetic variation in natural populations are neutral or nearly neutral. This assumption is much weaker than the assumption that all mutations are neutral. Indeed, the neutral mutation hypothesis assumes that the most prevalent type of selection is purifying (negative) selection. Thus, for example, even if all the rare variants at a locus are deleterious, the neutral mutation hypothesis still holds as long as the majority of the more common variants are selectively neutral or nearly so; note that in this case the hypothesis that all mutations are neutral does not hold. Clearly, we need different tests for testing the neutral mutation hypothesis.

According to the neutral mutation hypothesis, both the variation within populations and the differences between populations are mainly due to neutral or nearly neutral mutations. In other words, polymorphism is a transient phase of molecular evolution, and the rate of molecular evolution is positively correlated with the level of within-population variation. This prediction had previously been used to examine protein polymorphism and evolution. Recently, tests based on using within- and between-population DNA sequence comparisons have been developed. We discuss two such tests below.

Hudson-Kreitman-Aguade Test

This test by Hudson, Kreitman, and Aguade (1987) is known as the *HKA test*. It is to test whether the levels of within- and between-population DNA variation are positively correlated, as predicted by the neutral mutation hypothesis.

Consider m regions (loci) in species 1 and 2. Suppose that at each of the m regions n_1 and n_2 random sequences have been obtained from species 1 and 2, respectively. Let K_1 be the number of segregating sites at locus i in the sample of n_1 sequences from species 1; K_2, is similarly defined for species 2. Let D_i be the number of nucleotide differences at locus i between a random sequence from the sample from species 1, and a random sequence from the sample from species 2. The observations K_{1i} and K_{2i} (i=1, ..., m) are measures of within-species variation and the D_i are measures of between statistically consistent with the former.

The HKA model makes the same assumptions (e.g., neutrality, equilibrium, and no recombination) as those in the derivation of Equations 6 and 7. In addition, it assumes that all loci are unlinked (or evolve independently), that species 1 and 2 are at equilibrium at the time of sampling with effective population sizes N and Nf, respectively, and that the two species were derived t generations ago from a single ancestral population that was at equilibrium with an effective size of N(1+f)/2, i.e., the average of the population sizes of species 1 and 2. The test statistic is

$$X^2 = \sum_{i=1}^{m}\left[K_{1i} - \hat{E}(K_{1i})\right]^2 / V(K_{1i}) + \sum_{i=1}^{m}\left[K_{2i} - \hat{E}(K_{2i})\right]^2 / \hat{V}(K_{2i})$$

$$+\sum_{i=1}^{m}\left[D_i - \hat{E}(D_i)\right]^2 / \hat{V}(D_i) \qquad ...(33)$$

where $\hat{E}$ and $\hat{V}$ denote estimates of expectation and variance, respectively. The estimated expectation and variance of K_{1i} can be obtained by Equation 6 and 7, i.e.,

$$\hat{E}(K_{1i}) = a_1\theta_i \qquad ...(34)$$

$$V(K_{1i}) = a_1\theta_i + b_1\theta_i^2 \qquad ...(35)$$

where a_1 and b_1 are given by Equation 8 and 9 with n replaced by n_1. The equations for the expectation and variance of K_{2i} are the same as above except that θ_i and n_1 are replaced by $f\theta_i$ and n_2 respectively. The estimated expectation and Variance of D_i are given by

$$E(D_i) = [t + (1+f)/2]\theta_i \qquad ...(36)$$

$$V(D_i) = E(D_i) + [\theta_i(1+f)/2]^2 \qquad ...(37)$$

which were obtained by Li (1977) and Gillespie and Langley (1979). The above equation involve θ_i, f, and t, which can be estimated from the following system of m+2 equations:

$$\sum_{i=1}^{m} K_{1i} = a_1\sum_{i=1}^{m}\hat{\theta}_i \qquad ...(38)$$

$$\sum_{i=1}^{m} K_{2i} = \hat{f}a_2\sum_{i=1}^{m}\hat{\theta}_i \qquad ...(39)$$

$$\sum_{i=1}^{m} D_i = \left[\hat{t} + (1+\hat{f})/2\right] \sum_{i=1}^{m} \hat{\theta}_i \qquad \text{...(40)}$$

$$K_{1i} + K_{2i} + D_i = \hat{\theta}\left[\hat{t} + (1+\hat{f})/2 + a_1 + \hat{f}a_2\right], \quad i = 1, \ldots, m-1 \quad \text{...(41)}$$

This system of equations is obtained from Equations 34-37 and the analogous equations for $E(K_{2i})$, by summing different combinations of the equations and replacing the expectations of the random variables with the observed values. From computer simulations, Hudson et al. (1987) found that if n_1, n_2 and t are sufficiently large, X^2 is approximately χ^2 distributed with 2m − 2 degrees of freedom; the degree of freedom is 2m − 2 because m + 2 parameters are estimated from the above system of equations and should be deducted from the 3m degrees of freedom in Equation 33. In essence, the HKA test is goodness-of-fit test, i.e., one first uses Equations 38-41 and the observations K_{1i}, K_{2i}, and D_i to estimate the parameters θ_i's, f, and t under the assumption of neutrality, and then puts the estimated parameters into Equation 33 to see if there is any significant deviation from the prediction of the neutral mutation hypothesis. The HKA test is conservative for rejecting selective neutrality because of the assumptions of complete linkage within loci and no linkage between loci.

The HKA test has been applied to sequence data from three contiguous regions: the 5' flanking region of *Adh*, the *Adh* coding region and the *Adh-dup* coding regions in *Drosophila melanogaster* and *D. simulans*. The *Adh-dup* gene is 3' to the *Adh* gene and is related to *Adh* by an ancient tandem duplication. The data consist of 11 sequences across the above three regions from *D. melanogaster* and 1 sequence from *D. simulans*. Since no within-species variation can be computed from the single sequence available from *D. simulans*, the HKA test is modified to

$$X^2 = \sum_{i=1}^{m}\left[K_i - \hat{E}(K_i)\right]^2 / \hat{V}(K_i) + \sum_{i=1}^{m}\left[D_i - \hat{E}(D_i)\right]^2 / \hat{V}(D_i)$$

where K_i refers to the number of segregating sites in region i in the sample from *D. melanogaster*. For this reason, the degree of freedom becomes m − 1. The above test is based on the number of segregating sites and is denoted as "Seg". An alternative test is to substitute the average pairwise difference (Pwd) for the number of segregating sites; Pwd is Π as defined in Equation 1. This test may be more sensitive to among-region differences when the observed polymorphism frequency distributions differ across regions. However, since the expected variance of Pwd exceeds the expected variance of the number of segregating sites (K), Pwd has less power to reject the null hypothesis when the frequency distributions are the same among regions.

Kreitman and Hudson (1991) considered silent sites, which are defined as positions at which mutations do not change the polypeptide sequence encoded by the gene. In the 5' flanking region (1243 bp), every site is silent. In a coding region a site can be only partially silent (synonymous), and the number of silent sites is equal to the number of possible silent nucleotide changes in the region divided by 3; the number of silent sites is 319.2 for *Adh* (903 bp) and 599.8 for *Adh-dup* (1293 bp). The 5' flanking region vs. *Adh* test and the *Adh* vs. *Adh-dup* test are both significant at the 5% level, whereas the 5' flanking region vs. *Adh-dup* test is not. These results implicate the *Adh* coding region as the cause of departure from neutrality. The departure occurred because the within-species polymorphism in the *Adh* coding region was higher than expected. This pattern fits the suggestion that the *Adh* fast (Adh^f) and slow (Adh^s) electrophoretic alleles have been maintained by balancing selection, because levels of neutral polymorphism are expected to be elevated at sites sufficiently

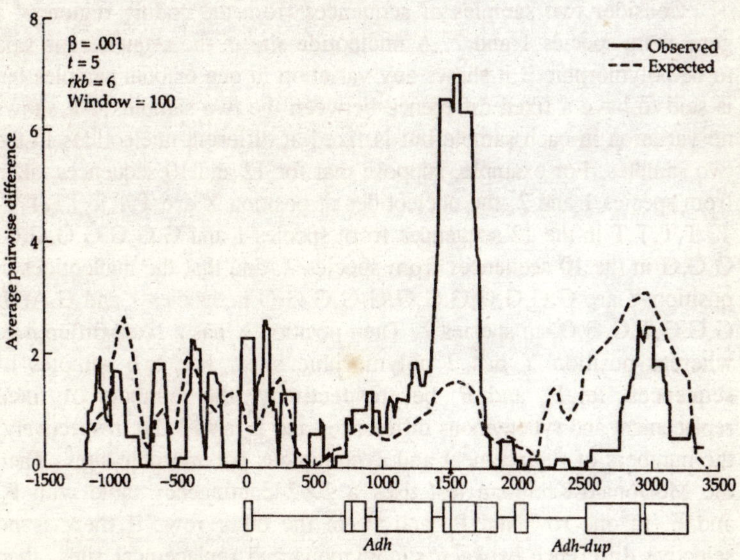

Fig. 8.4. Sliding window of observed and expected polymorphism levels in a sample of 11 Drosophila melanogaster alleles.

tightly linked to a balanced polymorphism. Indeed the level of polymorphism is highest at position 1490, the site where the threonine-lysine difference distinguishing the *Adh*^{f} and *Adh*^{s} alleles, and that the polymorphism level is also very high at sites closely linked to the balanced polymorphism. Thus, the high polymorphism at silent sites in the *Adh* coding region is apparently not due to selection at these sites *per se* but is maintained because of their linkage to a balanced polymorphism.

McDonald-Kreitman Test

McDonald and Kreitman (1991) proposed a simple method to contrast the patterns of within-species polymorphism and between-species divergence at synonymous (silent) and nonsynonymous (replacement) sites in the coding region of a gene. Obviously, in a coding region the synonymous sites and the replacement sites should have the same evolutionary history because they are tightly linked. Therefore, if polymorphism and divergence at both types of site are due to neutral mutations, the ratio of replacement to synonymous differences between species should be the same as the ratio of replacement to synonymous polymorphisms within species. A significant difference between the two ratios can therefore be used to reject the neutral mutation hypothesis.

Consider two samples of sequences from the coding region of a gene from species 1 and 2. A nucleotide site in the sequences is said to be polymorphic if it shows any variation in one or both samples but is said to have a fixed difference between the two samples if it shows no variation in each sample but is fixed at different nucleotides in the two samples. For example, suppose that for 12 and 10 sequences taken from species 1 and 2, the nucleotides at position X are T,T,T,TT,T,T, T, T,T,T,T in the 12 sequences from species 1 and G,G,G,G,G,G,G, G,G,G in the 10 sequences from species 2, and that the nucleotides at position Y are G,G,G,G,G,G,G,G,G,G,G,G in species 1 and G,A,G, G,G,G,T,G,G,G, in species 2. Then position X has a fixed difference, whereas position Y has 2 polymorphic sites. For two samples of sequences, let F_R and F_S be, respectively, the numbers of fixed replacement and synonymous differences and P_R and P_S be, respectively, the numbers of replacement and synonymous polymorphic sites. Then the McDonald-Kreitman test uses a 2×2 contingency table with F_R and F_S in one row and P_R and P_S in the other row. If there is no selective difference between synonymous and replacement sites, then the test should be nonsignificant.

McDonald and Kreitman (1991) applied the above test to DNA sequences of the coding region of the alcohol dehydrogenase (*Adh*) gene from *Drosophila melanogaster*, *D. simulans*, and *D. yakuba*. The ratio of replacement changes to synonymous changes that are fixed between species is significantly greater than the ratio of replacement to synonymous polymorphisms. Indeed, the former is 7/17=0.41, whereas the latter is only 2/42=0.05; Fisher' test and the G-test both give a probability of $P<0.006$. A major problem in this study is the rule that the between-species variation at a site is neglected if that site is polymorphic in any of the three species. This rule tends to underestimate between-species variation. For example, at nucleotide position 1590, all the alleles in *D. melanogaster* have T, half the alleles in *D. simulans* have T, and the other half have C and all the alleles in *D. yakuba* have C. At this position, McDonald and Kreitman assigned a single synonymous polymorphism within *D. simulans* but no difference between species. However, under their assumption of monomorphism in the common ancestral species, a substitution must have occurred between *D. melanogaster* and *D. yakuba*. Of course, recounting a few such cases does not change the conclusion because the P value is rather small. Another caveat is that only 6 sequences were sampled from *D. simulans* and so a fixed difference may become a polymorphic difference if the sample size is increased. Nevertheless, the high ratio of fixed replacement to fixed synonymous changes is difficult to explain under the neutrality assumption. Rather, it would be easier to explain the observation under the assumption that most of the observed fixed replacement differences were due to adaptive fixation of selectively advantageous mutations. A mutation that became fixed by selection would have been a polymorphism for a shorter time period than would a mutation that became fixed by random drift. For this reason, an adaptive substitution is less likely to appear as a polymorphism than is a neutral substitution. Interestingly, Sawyer and Hartl (1992) estimated that a small selective advantage ($\sim 1.5 \times 10^{-6}$) would be sufficient to explain the observed data.

An even more extreme case was found by Eanes et al. (1993) for the glucose-6-phosphate dehydrogenase (G6PD) gene in *D. melanogaster* and *D. simulans*. The ratio of fixed replacement differences to fixed synonymous differences is 21/26=0.81, whereas the ratio of replacement polymorphisms to synonymous polymorphisms is only 2/36=0.06; the G-test gives $P<<0.001$. This implies a tenfold excess of replacement changes over that expected if the G6PD in *D. melanogaster* is well established as the initial enzymatic step in the pentose shunt pathway, and the rapid amino acid change in G6PD has been suggested to be due to selective response to varying pentose shunt function in changing environments.

Another case of deviation from the neutrality expectation is the *jingwei* gene in *D. melanogaster* and *D. simulans*. In addition, at the *zeste* locus, there is a higher proportion of replacement differences among fixed sites than among polymorphic sites, though a statistical test would lack power because no polymorphic replacement difference was found. On the other hand, the patterns at PGI, 6PGD, GAPDH, and GPDH and at yp2 and per reflect molecules that are either very constrained or appear to be changing in a neutral fashion when the McDonald-Kreitman test is applied. As the number of loci studied is still limited, it is too early to draw a general conclusion on the proportion of genes whose polymorphism-divergence pattern cannot be explained by the neutralist view.

Note that all the above tests assume that synonymous mutations are neutral, but this is apparently not the case for *Drosophila* genes. However, Charlesworth (1994) suggested that the McDonald-Kreitman test is conservative, because nonsynonymous mutations are subject to stronger negative selection than are synonymous mutations. Akashi (1995) discussed how codon usage bias may be taken into account when conducting the McDonald-Kreitman test.

Joint Effects of Linkage and Selection

The neutralist prediction of a positive correlation between intraspecific variation and interspecific divergence can be violated by the presence of positive Darwinian (directional) selection. When directional selection drives an advantageous mutation through a population to fixation, much of the neutral variation at linked sites is eliminated during the process. This phenomenon, known as the "hitchhiking effect" or "selective sweep", was first analyzed by Maynard Smith and Haigh (1974) and later by Kaplan et al. (1989), Stephan et al. (1992), and others. This effect is expected to increase with the strength of linkage. Thus, if advantageous substitutions occur at an appreciable rate, the level of standing variation should be low in regions with a severely restricted rate of recombination. In general, a positive correlation should exist between the level of polymorphism and the rate of recombination in a region. The hitchhiking process, however, is not expected to affect the rate of divergence between species at linked neutral sites, though it affects the rate at linked sites under selection. Thus, in contrast to the neutralist prediction, positive selection can lead to uncoupling of levels of polymorphism and divergence. The uncoupling can also occur by the presence of balancing selection, which can increase the level of polymorphism at linked loci.

Aguade et al. (1989) seem to have been the first to use the hitchhiking effect as an indirect means to detect the action of directional selection at the DNA level. The *yellow* gene and the *achaete-scute* (ASC) region are located near the tip of the X chromosome in *D. melanogaster*, which has a very low rate of recombination. A survey of restriction-map variation at the *yellow-ASC* (y-ASC) region in 64 lines of *D. melanogaster* from nature showed reduced nucleotide variation, although no apparent reduction was observed in other populations. In fact, in that survey only 9 polymorphisms were detected out of 176 restriction sites (approximately 2112 nucleotide-site equivalents) and 6 out of the 9 polymorphisms were present only once in the sample; given that the average π for the genome is 0.05, the probability of observing such a sample is only 2×10^{-6}. A population-bottleneck effect was considered unlikely because the same sample was analyzed for other regions of the X chromosome with no evidence of reduction in levels of variation. An even greater reduction in polymorphism in the *achaete* region was found in the sibling species *D. simulans*: no nucleotide variation was detected in a sample of 103 lines when 390 site equivalents were studied. The reduction in variation in both species is obviously not due to purifying selection (selective constraints), because the divergence in this region between the two species is comparable to those in other regions: 6% compared to 7% and 6% silent divergences at the 5' *Adh* and *hsp*82 regions, respectively. From these lines of evidence, it is reasonable to conclude that the reduction in nucleotide variation in the y-ASC region is due to a hitchhiking effect. The same conclusion was drawn for the *cubitus interruptus Dominant* locus located on the nonrecombining fourth chromosome: sequencing of a 1.1-kb region of the gene in 10 natural lines of *D. melanogaster* and 9 of *D. simulans* had uncovered only one polymorphism. In addition, no variation was observed at the *su-f* region in *D. melanogaster* and *D. simulans* and in the *fw* locus of *D. ananassae*.

Much interest in the issue of polymorphism vs. linkage was generated by Begun and Aquadro (1992), who showed that for 20 gene regions from across the genome of *D. melanogaster* the level of nucleotide diversity is positively correlated with the regional rate of recombination. This could not be explained by variation in mutation rates or in functional constraints because there was no correlation between recombination rate and DNA sequence divergence between *D. melanogaster* and *D. simulans*. They concluded that the correlation resulted from hitchhiking associated with the fixation of advantageous mutants. The positive correlation is supported by further data from the X and third chromosomes. The amount of DNA variation is positively correlated with the rate of recombination in each of these two chromosomes.

Charlesworth et al. (1993) pointed out that, in addition to the hitchhiking effect, selection against deleterious mutations can also reduce nucleotide variation at linked sites. This type of selection, called *background selection*, and hitchhiking essentially involve the same process, namely, the elimination of neutral or nearly neutral variants as a result of selection at linked sites. Background selection, like positive selection, reduces the effective population size at nearby sites. For an autosomal region with no recombination and with a total mutation rate of U per generation and selection s against mutant homozygotes, and hs against heterozygotes (h denotes degree of dominance), the effective size in a random mating population is reduced from N_e to f_0N_e, where

$$f_0 \approx \exp[-U/(2hs)] \qquad \text{...(42)}$$

For an X-linked region with the same parameters and with no recombination,

$$f_0 \approx \exp[-1.5U/(2hs+s)] \qquad \text{...(43)}$$

Therefore, the expected nucleotide diversity (π) at a neutral site is reduced from $4N_e\upsilon$ to $4f_0N_e\upsilon$, where υ is the mutation rate per site. The background effect decreases rapidly with increasing recombination frequency. Based on their simulation study and data on U, s, hs, and recombination rates in *D. melanogaster*, Charlesworth et al. (1993) estimated that the reduction in π in this species can be very high. Background selection may be sufficient for explaining relatively mild reductions in π in certain regions, though it may not account for drastic reductions such as that observed in the *su-f* region.

It is difficult to distinguish between hitchhiking and background selection effects because they lead to many similar observable consequences for the pattern of neutral variability. Aquadro et al (1994) suggest that one way to evaluate the relative contributions of hitchhiking and background selection is to compare the levels of DNA variation in regions of equivalent recombination on the X chromosome and autosomes. Because X-linked genes are hemizygous in males, recessive

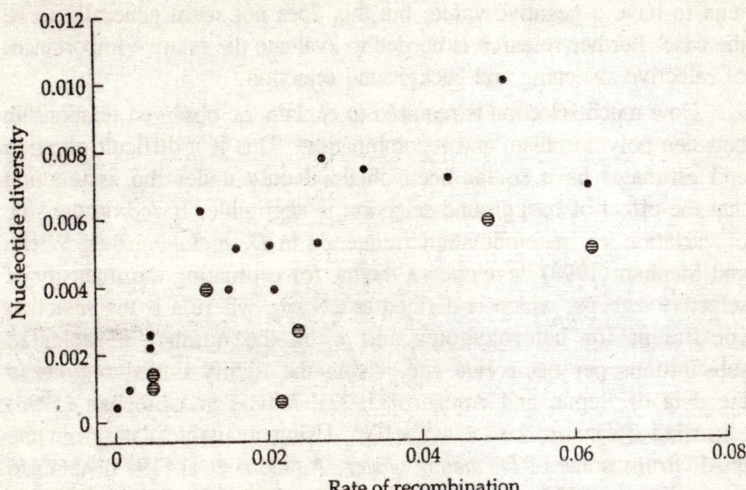

Fig. 8.5. Scatterplot of nucleotide diversity versus rate of recombination for 15 third-chromosome gene regions and seven X-linked gene regions.

advantageous mutations are fixed more rapidly on the X chromosome than on an autosome, the hitchhiking effect on nucleotide variation is likely to be stronger on X-linked genes than no autosomal genes. So, X-linked loci should tend to have lower levels of variation than autosomal loci. In contrast, background selection is a mutation-selection equilibrium model in which recessive deleterious mutations are eliminated more efficiently from the X chromosome than the autosome, so that the X chromosome has a larger proportion of mutation-free gametes compared to the autosomes. [This prediction follows from Equations 42 and 43 because $1/(2hs) > 1.5/(2hs+s)$ as long as $h<1$.] For this reason, X-linked genes are expected to have higher levels of variation (all else being equal), after correcting for different effective population size for X-linked and autosomal genes. Limited data suggest that, for gene experiencing similar recombination rates, sequence variation is generally lower for X-linked genes than for autosomal genes, supporting the notion that the hitchhiking effect is an important cause of the correlation between recombination and variation. However, the data is too limited to draw a firm conclusion.

The issue, therefore, remains unsettled. Gillespie (1994) argued that directional selection is usually too weak for the hitchhiking effect to be effective. Aguade and Langley (1994) and Braverman et al. (1995) noted that, under the selective-sweep hypothesis, tightly linked loci should tend to have low frequency alleles so that Tajima's D should tend to have a negative value, but this does not seem generally to be the case. Further research is needed to evaluate the relative importances of selective sweeping and background selection.

How much selection is required to explain the observed relationship between polymorphism and recombination? This is a difficult question and estimates have so far been obtained only under the assumption that the effect of background selection is negligible. Based on the sloe of variation vs. recombination frequency in *D. melanogaster*, Wiehe and Stephan (1993) developed a theory for estimating the intensity of selective sweeps, which is defined as $2N_es\upsilon_a$, where s is the selection coefficient for heterozygotes and υ_a is the number of selected substitutions per nucleotide site. Using the tightly linked regions in the data of Begun and Aquadro (1992), Wiehe and Stephan (1993) estimated $2N_es\upsilon_a$ to be $> 1.3\times10^{-8}$. Using available data from the third chromosome of *D. melanogaster*, Aquadro et al (1994) obtained an estimate of $2N_es\upsilon_a=5.4\times10^{-8}$. If $N_e=2\times10^6$ and $s=0.01$, then $\upsilon_a=1.4\times10^{-12}$. If the mutation rate per site is $\upsilon=10^{-9}$, then about 0.14% of the mutations are selectively advantageous. This number is very small but becomes larger if s is smaller, which is likely to be the case.

MHC Polymorphism

In this and the next section we shall consider some examples where the presence of natural selection was detected by methods not described in the preceding section.

The *major histocompatibility complex* (MHC) is a large multigene family whose products control the immune system's capacity to recognize foreign proteins. MHC molecules bind foreign antigens (peptides) and display them on the cell surface. Such an assembly can be recognized by T cell receptors—the recognition activates the T lymphocytes, and the activated lymphocytes initiate appropriate immune responses to destroy the invaded cell or the invader itself. In the case of grafted tissue, the T cells recognize as foreign the MHC molecules of the donor, if they are different from those of the host, and the immune attack is then directed against the transplant. However, if the MHC molecules are recognized as "self," no immune response is initiated. This is the origin of the term "histocompatibility."

The human MHC, usually referred to as the *human leukocyte antigen* (HLA) system, is located on chromosome 6 in a region that is more than 4 million bp in length and contains more than 100 genes. Only some of these genes encode molecules involved in the presentation of peptides to T lymphocytes. Others include genes that control the degradation of proteins into peptides and transport of the peptides across membranes; genes that specify other components of the immune system; genes that subserve functions that are either unknown or unrelated to the immune response; and nonfunctional genes. The MHC genes proper are classified into two classes, I and II; the two classes are evolutionarily related, though distantly.

It has long been realized that some MHC loci are highly polymorphic because tissue grafts are invariably rejected by the host whenever they are contributed by an unrelated donor. This has indeed been borne out by biochemical and molecular data. For example, heterozygosity at the human *HLA-A* and *-B* loci is about 90% in many populations, and the number of alleles detected in a sample of 200 individuals ranges approximately from 15 to 30. Similarly, the number of major alleles at the mouse H2-K and -D (or H-2K and -2D) loci are 54 and 55, respectively. How such high polymorphism is maintained has been a subject of debate for decades. Recently there has been significant progress towards the resolution of this controversy.

One simple analysis that can provide insight into the mechanism of maintenance of polymorphism is Watterson's (1978) homozygosity test, which compares the observed homozygosity in a sample of size n containing k alleles to the expected homozygosity conditional on n and k under neutrality. If purifying selection is operating, the observed value is likely to be higher than the expected value (owing to the presence of low frequency alleles), whereas if balancing selection (e.g., overdominant selection, frequency-dependent selection) is operating, the reverse is expected to be true. Hedrick and Thomson (1983) compared the observed homozygosity in 22 samples at both the HLA-A and -B loci with the neutral expectation. In all cases the observed homozygosity is lower than the expected value, and they concluded that some form of balancing selection was responsible for the high polymorphism.

Hughes and Nei (1988) reasoned that if there is any positive Darwinian selection at MHC loci, it must occur at the *antigen recognition site* (RAS), also known as the peptide binding region. A class I MHC molecule has three extracellular domains (α_1, α_2 and α_3), a transmembrane portion, and a cytoplasmic tail. The π_3 domain associates noncovalently with β_2-microglobulin, a polypeptide encoded by a gene that lies outside the MHC but is distantly related to the MHC genes. The ARS of the class I MHC molecule is located in the α_1 and α_2 domains and consists of 57 amino acid residues.

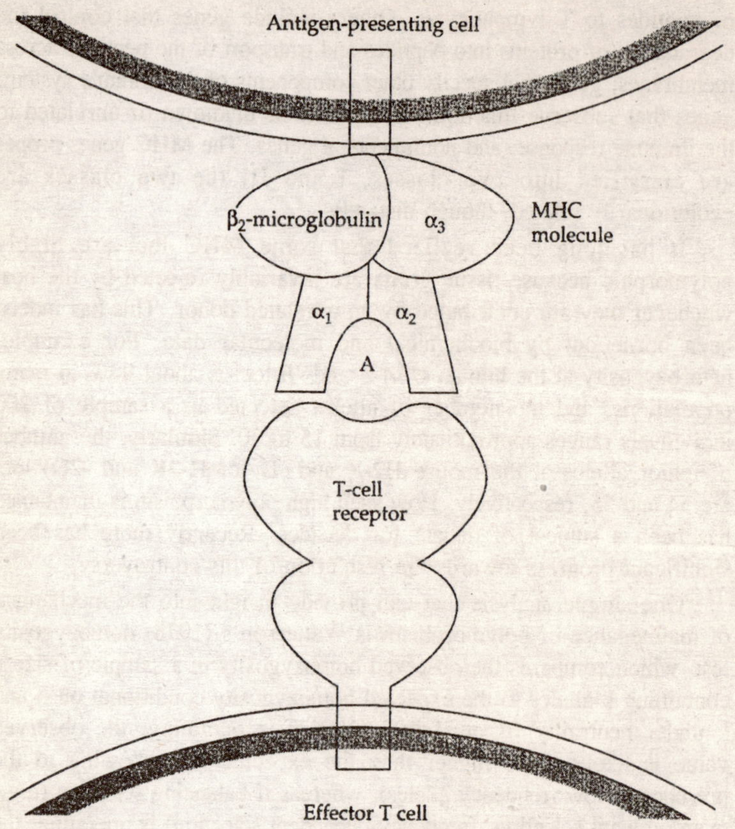

Fig. 8.6. A class I MHC molecule presenting an antigen to a T cell.

To see if natural selection occurs at the ARS, Hughes and Nei (1988) compared the pattern of nucleotide substitution at the ARS with the pattern in regions outside the ARS. In each region, they computed the number of substitutions per synonymous site (K_S) and the number of substitutions per nonsynonymous site (K_A) between alleles. In the absence of selection on a DNA region, K_S and K_A should be more or less the same. If there is purifying selection, K_S is expected to be higher than K_A, because selection would occur at the amino acid level. By contrast, if there is positive Darwinian selection, K_A is expected to be higher than K_S. In the case of the human HLA-A locus, K_A is much higher than K_S in the ARS, the difference being significant at the 0.1% level when the standard errors computed by Nei and Jin's (1989) method are used. In the other regions, K_S is higher than K_A. The same pattern of nucleotide substitution is also observed for the

HLA-B and HLA-C loci. It is interesting to see that K_A in the ARS is significantly higher than K_S in any region. This clearly indicates that amino acid substitutions in the ARS are elevated by positive Darwinian selection. By contrast, K_A is much smaller than K_S in the non-ARS regions, suggesting that amino acid substitutions are generally subject to purifying selection in these regions.

The K_A in the ARS is again substantially higher than the K_S in any region. (The unusually low value of K_S for α_3 may have been caused by interlocus genetic exchange). Therefore, the pattern of nucleotide substitution for the mouse MHC loci is essentially the same as that for the human MHC loci. The same conclusion holds for class II loci.

Another extraordinary feature of the MHC polymorphism is the high nucleotide diversity among MHC alleles. In other genetic systems, alleles at a given locus usually differ by a few nucleotide substitutions at most. In the MHC, some alleles differ by 100 or more substitutions. This observation means that these alleles have diverged from each other a long time ago. In fact, nucleotide sequencing work has revealed alleles that are shared by species that diverged millions of years ago. For example, Figueroa et al. (1988) found that two allelic lineages marked with deletions/insertions at the Aβ1 locus are shared by mice and rats, which diverged at least 10 million years ago. Such polymorphisms are obviously more ancient than the species themselves and are called ***trans-species polymorphisms***. In such a case, the

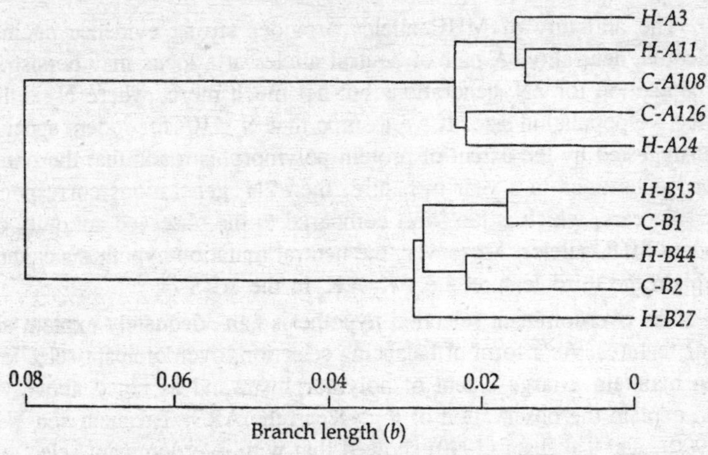

Fig. 8.7. Phylogenetic tree for alleles from the class I MHC loci A and B in the human (HLA-A and -B) and the chimpanzee (CHA-A and -B).

divergence of the allelic lineages predates the divergence of the species. Fig. 8.7 shows such an example from the human HLA-A and -B loci and the chimpanzee C-A and -B loci. The human allele H-A11 is closer to the chimpanzee allele CA108 than to any other human alleles at the A locus, whereas the human allele H-A24 is closer to the chimpanzee allele C-A126 than to any other human alleles. This comparison implies that the two human alleles H-A11 and H-A24 diverged before the human-chimpanzee split and so did their counterparts in the chimpanzee, i.e., alleles C-A108 and C-A126. Thus, a pair of polymorphic allelic lineages is shared by human and chimpanzees.

Many other examples of trans-species polymorphism at MHC loci have been found in primates and rodents. A most surprising case is that certain human MHC alleles appear to have diverged before the separation of prosimian and anthropoid primates more than 65 million years ago. The antiquity and the large number of MHC alleles in humans strongly suggest that the lineage leading to modern humans has never gone through any severe population bottle-neck.

In summary, there are four extraordinary features of MHC polymorphism: (1) the extent of polymorphism is extremely high, (2) the number of nucleotide differences between alleles is unusually large, (3) in the ARS the rate of nonsynonymous substitution is higher than the rate of synonymous substitution, and (4) allelic lineages may persist in a population for tens of millions of years. Any hypothesis for the origin and maintenance of MHC polymorphism must be able to explain all these features.

The antiquity of MHC alleles provides strong evidence against selective neutrality. A pair of neutral alleles at a locus may persist in a population for $2N_e$ generations but not much more, where N_e is the effective population size. If we assume that $N_e = 10^5$ for rodent species as suggested by the extent of protein polymorphism and that there are two generations in a year in nature, then $2N_e$ generations correspond to 10^5 years, which is too short compared to the observed antiquity of rodent MHC alleles. Moreover, the neutral mutation hypothesis cannot explain the third feature, i.e., $K_A > K_S$ in the ARS.

The overdominant selection hypothesis can adequately explain all four features. As a form of balancing selection, overdominant selection can maintain a large extent of polymorphism and as noted above, it can explain the observation of $K_A > K_S$ in the ARS. Takahata and Nei (1990) and Takahata (1990) showed that with overdominant selection the coalescence time for alleles in a population can be tens of millions of years under reasonable selection, population, and mutation parameters. As a consequence, large divergences between alleles can occur. The overdominance hypothesis was originally proposed by Doherty and Zinkernagel (1975), who showed that different MHC molecules bind different foreign antigens and argued that a heterozygote having two different MHC molecules would have a better chance to third foreign antigens and thus would be more resistant to infectious diseases than a homozygote with only one type of MHC molecule.

Frequency-dependent selection was one of the most popular early hypotheses for the maintenance of MHC polymorphism. The most popular model of frequency-dependent selection assumes that host individuals carrying a recently arisen mutant allele have a selective advantage because pathogens will not have had time to evolve the ability to infect host cells carrying a new mutant antigen. Hughes and Nei (1988) noted that this model will result in a constant turnover of alleles in the population because old alleles lose resistance to pathogens and will become lost by random drift. While this model is expected to generate a higher value of K_A than K_S, it can explain neither the high degree of polymorphism nor the long persistence of MHC alleles; this indeed has been shown theoretically. Another form of frequency-dependent selection assumes minority advantage. Since this form of selection can enhance the retention of alleles as does overdominant selection, it can also explain the above four features of MHC polymorphism. Takahata and Nei (1990), however, argued that while it is possible that a new mutant allele has a selective advantage over old ones because there are no pathogens adapted to it, it is difficult to imagine why an allele should regain a selective advantage when it become old and again rare in the population.

Hill et al. (1991) have provided evidence that in West Africa specific class I and class II alleles are associated with resistance of *Plasmodium falciparum* malaria and suggested that their data favour the hypothesis of frequency-dependent selection rather than overdominant selection. Hughes and Nei (1992) argued that the study of Hill et al. was not a test of the overdominance hypothesis because this hypothesis depends on the exposure of the population to two or more pathogens, but Hill et. al did not consider two or more stains of *P. falciparum* and did not study other pathogens. Moreover, among individuals having the protective DRw13.02 haplotype, there is a higher proportion of heterozygotes among those with severe malarial anaemia than with mild controls. This observation suggests that the fitness of DRw13.02

homozygotes is higher than that of DRw13.02 heterozygotes, so that the election is directional (i.e., not frequency dependent) and thus cannot maintain polymorphism. In response, Hill et al. (1992) argued that their study might be a test of the overdominance hypothesis because *P. falciparum* shows striking polymorphism in many of its immunodominant antigens and because most Gambian children with malaria harbor more than one antigenically distinct parasite strain. They argued further that the finding that alleles with frequencies of 0.13-0.15 are protective is not incompatible with minority-advantage models, for these alleles might have been even more protective when they were less common. As the issue remains controversial, further studies are needed to distinguish between the two types of natural selection for maintaining MHC polymorphism.

Other hypotheses such as a high mutation rate in the MHC complex, gene conversion, mating preference, and maternal-fetal incompatibility either lack solid evidence or may not explain the above four observations adequately. For example, neither a high mutation rate nor gene conversion can explain why the rate of nonsynonymous substitution is higher than that of synonymous substitution in ARS (but not in other regions).

The genetic systems governing self-incompatibility in plants exhibit many features similar to those of the MHC system. In many plant species, self-incompatibility is controlled by a single locus, the S locus. In gametophytic self-incompatibility, pollen deposited on a stigma fails to germinate and grow into the style of the pistil if the S allele expressed by the pollen matches one of the two alleles in the pistil. The S system shares three striking features with the MHC system. First, it carries a very large number of alleles e.g., a rare *Oenothera* species with a population size of only ~500 individuals contained 45 distinct S alleles. As Wright (1939) showed theoretically, the self-incompatibility system favours the introduction of new alleles because the probability of avoiding a pistil bearing the same allele as the pollen is greatest for rare alleles. This reduces the rate of loss of new alleles from a population, as compared to the case of neutral alleles, and for this reason the S locus can maintain many alleles. Second, it exhibits trans-species polymorphisms. Self-incompatibility is a type of balancing selection and so an allele can persist for a very long time. According to Ioerger et al.'s (1990) estimates from nucleotide sequence data, some allelic lineages predated the divergence of three species in Solanacea (an ornamental tobacco plant, a wild petunia, and a wild potato) some 27 to 36 million years ago. Third, it shows extremely high within-species allelic diversity. Indeed, the amino acid sequence similarity between two alleles in a species can be lower than 40%; the high allelic divergence follows from the antiquity of alleles. The sporophytic self-incompatibility system such as that in *Brassica* also shows similar features as the gametophytic system, though it appears to be controlled by multiple loci.

Protein and DNA Polymorphisms in American Oysters: Contrasting Pictures of Geographic Differentiation

Population geneticists have always been interested in geographically structured populations, because the pattern of genetic differentiation among different parts of such a population may provide insight into the mechanism of maintenance of genetic polymorphism and insight for understanding how genetic differentiation may occur because of environmental variation or isolation by distance. In the 1970s and early 1980s, starch gel electrophoresis of soluble proteins was a popular tool for conducting such studies. One example was the study of electrophoretic variation in proteins encoded by nuclear genes in American oysters (*Crassostrea virginica*) sampled from Massachusetts to Texas. Surprisingly, the study revealed a near uniformity of allele frequencies throughout this range, despite the fact that it stretches over 4000 miles. The result was attributed to high gene flow due to "the rather long planktonic stage of larval development, since this species has the ability to disperse zygotes over great distances when facilitated by tidal cycles and oceanic currents".

A very different pattern, however, was observed in a later study using restriction analysis of mtDNA, which showed a major genetic "break" distinguishing all sampled populations north versus south of the mid-Atlantic coast of Florida. The latter finding was in striking concordance with a previous restriction analysis of mtDNA in the horseshoe crab. In both mtDNA studies, numerous restriction site changes were involved, and the estimated mean nucleotide sequence divergence between the north and south regions was high, about 2.6% and 2.0% respectively.

There are several possible explanations for the apparent inconsistency between nuclear and cytoplasmic structures in oysters. The discrepancy could be due to biological or demographic factors, such as a higher rate of interpopulation gene flow mediated by sperm rather than by eggs, or directional selection favouring different mtDNA haplotypes in the two regions. Alternatively, it could be due to a

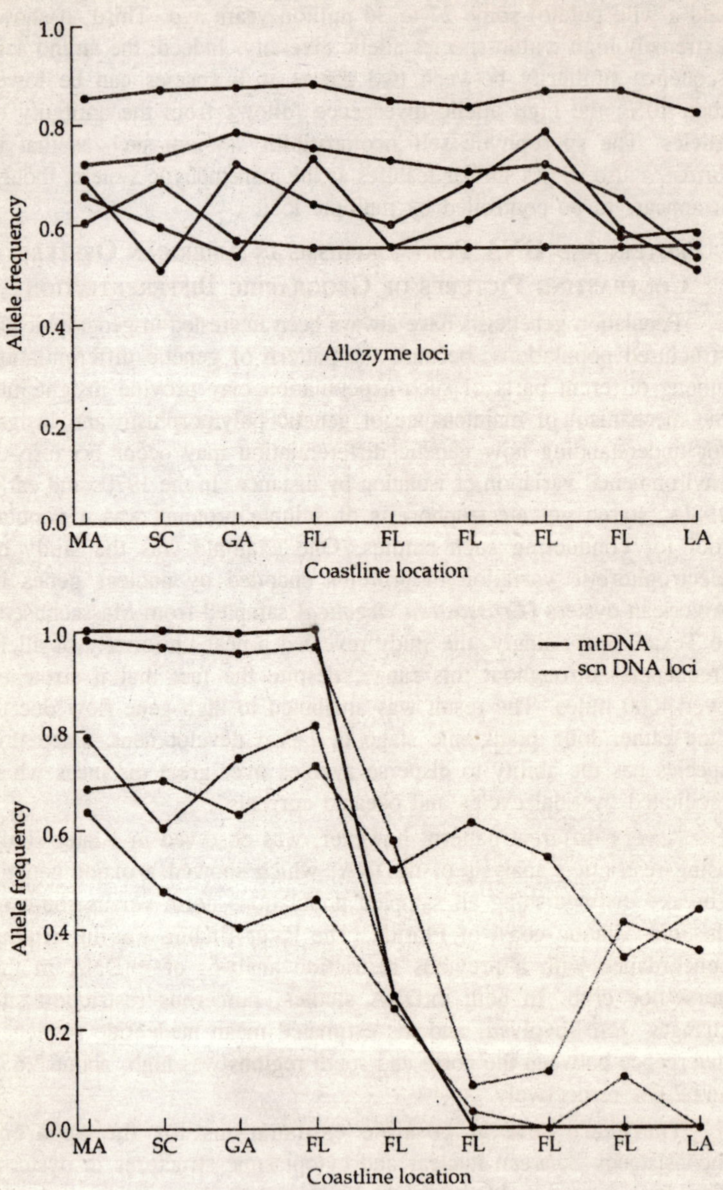

Fig. 8.8. Allele frequencies in American oyster populations. (Top) Frequencies of the most common alleles at five polymorphic allozyme loci; (Bottom) five DNA loci along a coastline transect running from Massachusetts (MA) through South Carolina (SC), Georgia (GA), Florida (FL), and Louisiana (LA).

slower rate of evolutionary change in allozyme frequencies, or balancing selection at multiple allozyme loci.

To distinguish between the above two classes of competing hypotheses Karl and Avise (1992) conducted an analysis of restriction fragment length polymorphism (RFLP) in single-copy nuclear (scn) DNA. They designed PCR primers suitable for amplifying each of four anonymous nuclear loci (a locus is anonymous if its function and genomic location are unknown). Nuclear DNA was isolated from each of 277 oysters collected at 9 locations between Massachusetts and Louisiana. Restriction site polymorphisms from all four loci revealed a pronounced genetic break along the eastern coast of Florida, as was true for the mtDNA study.

The similar pattern of geographic differentiation registered in mtDNA and scnDNA indicates that biological factors differentially affecting mitochondrial and nuclear genomes cannot account for the original discrepancy between the mtDNA and allozyme data sets. Thus, the geographic uniformity in allozymes is probably not due to greater dispersal of male gametes or to the slow rate of evolutionary sorting of nuclear lineages from an ancestral gene pool, because effects of such processes should also be reflected in the distributions of scnDNA alleles. Neither is the population genetic break in mtDNA attributable to directional selection on mtDNA, unless the kinds of selection favouring different mtDNA haplotypes in the Atlantic versus the Gulf extend to scnDNA alleles as well. Indeed, it is difficult to imagine a selective force that could have molded in such a consistent fashion the mtDNA break across species as different as the oyster, horseshoe crab, fishes, reptile, and bird that have been assayed. One remaining possibility is that several of the allozyme loci have been under uniform balancing selection, and thus, do not record the population subdivision suggested by the mtDNA and scnDNA data. This explanation, of course, requires that the same type of balancing selection operates in both the Atlantic and the Gulf populations. At any rate, either balancing selection must have acted on some allozyme loci, or diversifying selection must have acted on mtDNA and some scnDNA loci. This deduction underlines the need for caution in inferring population structure and gene flow from any single gene or perhaps even class of marker loci.

9

Transposition and Horizontal Transfer

"Jumping Genes" (transposable genetic elements) were first discovered in maize by Barbara McClintock in the late 1940s. She found that genes associated with the development of colour pigments in kernels could be turned on or off at abnormal times by the action of certain genetic elements, which she called "controlling elements," that apparently could move from site to site on different maize chromosome. However, the idea of jumping genes ran contrary to the traditional view that genomes are stable, static entities in which genes can be assigned to specific loci and retain their precise chromosomal position over long periods of evolution. Since the traditional concept was not casually arrived at but came from a large body of genetic data. McClintock's observation was thought to be rare phenomenon and not a general interest.

With the great advances in molecular biology techniques in the 1970s and 1980s, transposition of genetic elements was found to be a widespread phenomenon not only in maize but also in other organisms. DNA sequences that possess an intrinsic capability to change their genomic location are called *mobile elements* or *transposable elements* (TEs). The new data stimulated vigorous debates on the biological significance of TEs. Since TEs can "jump about" in a genome and can move genetic materials from one genomic location to another, they can produce novel gene mutations and chromosomal rearrangements that cannot be produced by other known mechanisms. For this reason, many authors argued that TEs play an important role in evolution, especially in speciation events. On the other extreme, Doolittle and Sapienza (1980), Orgel and Crick (1980), Temin (1980) and others regarded TEs merely as genetic parasites. They noted that movement of TEs in a genome is more likely to produce deleterious rather than beneficial effects and argued that TEs persist in nature not because they confer advantages to their hosts but because they can multiply faster than other parts of the host genome. This great divergence in opinion notwithstanding, it is now generally recognized that the structural organization of genomes is much more fluid and prone to evolutionary changes than previously thought. Although the evolutionary implications of DNA transposition remain not fully understood, progress in both empirical and theoretical population studies of TEs has considerably increased our understanding of the spread and maintenance of TEs in populations. Moreover, molecular studies have greatly increased our understanding of the mechanism and the effects of DNA transposition.

DNA can also move between organisms. In fact, it has long been known that a bacterium can absorb DNA from its surroundings and functionally integrate the exogenous DNA into its genome, a phenomenon known as *transformation*. There are also other mechanisms by which genes can be transferred across bacterial species boundaries. Thus, cross-species (horizontal) gene transfer might have played a significant role in bacterial evolution; evidence for such a role has been increasing. Furthermore, recent evidence suggests that horizontal gene transfer can occur from prokaryotes to eukaryotes and vice versa. There is also strong evidence for the transfer of TEs among eukaryotes as well as among prokaryotes. Therefore, although the possibility of horizontal gene transfer was originally viewed by biologists with much skepticism, it may indeed by a widespread phenomenon.

TRANSPOSITION AND RETROPOSITION

Transposition is defined as the movement of genetic material from one chromosomal location to another. There are two types of transposition, distinguished by whether the TE is replicated or not. In the *nonreplicative* type of transposition, the element itself moves from one site to another. In *replicative* (or *duplicative*) transposition, the TE is copied, and one copy remains at the original site, while the other inserts at a new site. Thus, replicative transposition is characterized by an increase in the number of copies of the TE. Some TEs use only one type of transposition; others use both the replicative and nonreplicative pathways.

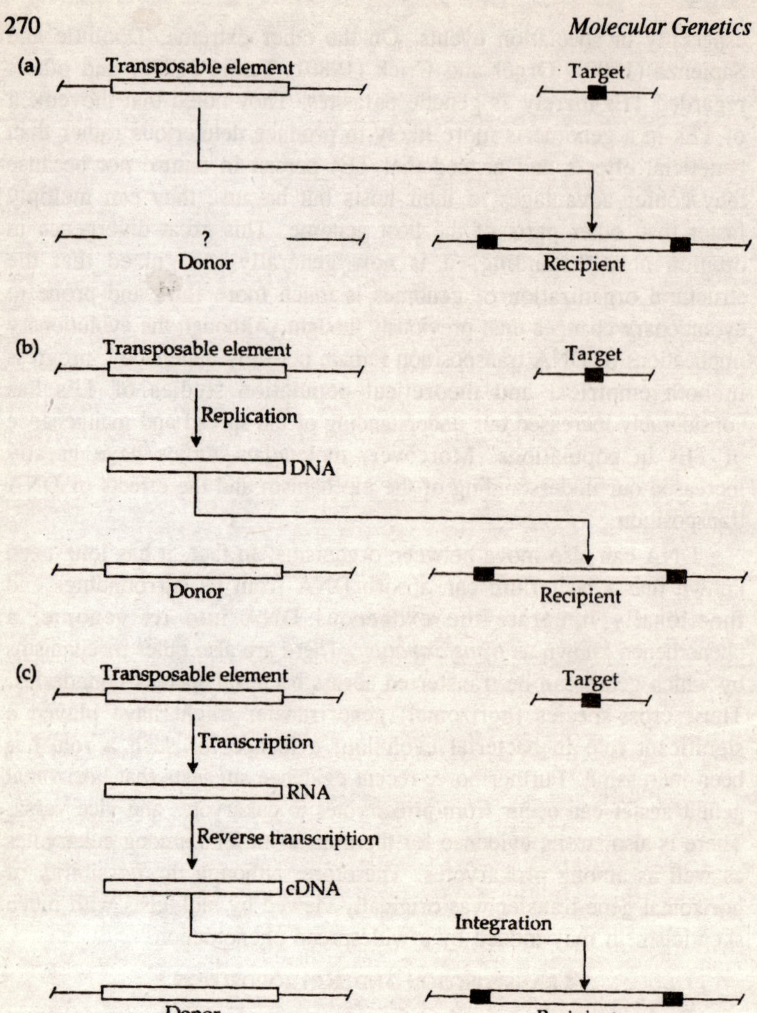

Fig. 9.1. (a) Nonreplicative transposition. (b) Replicative transposition.

In the above types of transposition, the genetic information is carried by DNA. It is known that genetic information can also be transposed through RNA. In this mode, the DNA is transcribed into RNA, which is then reverse-transcribed into cDNA. In order to distinguish between the two modes, the RNA-mediated mode has been termed *retroposition*. In contrast with DNA-mediated transposition, retroposition is always of the duplicative form, because it is a reverse-transcribed copy of the element, not the element itself, that is transposed. Both transposition and retroposition are found in eukaryotic and prokaryotic organisms. When a TE is inserted into a host genome, a small segment of the host DNA at the insertion site (usually 4-12 bp) is duplicated. The duplicated repeats are in the same orientation and are called *direct repeats*. This characteristic is hallmark of transposition and retroposition.

Some TEs can transpose themselves in all cells; others are highly specific. For example, P elements in *Drosophila melanogaster* are usually mobile only in germ cells. The genomic locations of the recipient sites for transposition also show variation among different TEs. Some elements show an exclusive preference for a specific genomic location. For example, IS4 incorporates itself exactly and always at the same point in the galactosidase operon of *Escherichia coli*. Others, such as bacteriophage Mu, can transpose themselves at random to almost any genomic location. Many TEs show intermediate degrees of genomic preference. For example, 40% of all Tn10 transposons in *E. coli* are found in the lacZ gene, while Drosophila P elements have an affinity for the X chromosome and also prefer to insert into sequences 5' to the coding region of genes, rather than into the coding region itself. Some TEs exhibit higher affinities for a particular type of nucleotide composition. For example, IS1 favours AT-rich insertion sites. However, it should be noted that the distribution pattern of elements in a genome is affected by both natural selection and the rate of recombination.

Transposable Elements

According to their mode of transposition and the types of genes they contain, which may not be a reflection of common ancestry, transposable elements have been classified into three types: insertion sequences, transposons, and retroelements. In addition, there are genomic sequences that have been produced by retroposition of RNA but lack the intrinsic capability to transpose themselves. Such sequences are called retrosequences and will be discussed in later section on retrogenes and retropseudogenes.

Insertion Sequences

Bacterial insertion sequences were discovered in the late 1960s and early 1970s during investigations of expression of genes in E. coli and bacteriophage lambda. They were originally found as unstable mutations but were subsequently shown by hybridization and heteroduplex analysis to be insertions of the same few DNA segments in different positions and orientations. The similarity of bacterial insertion sequences to the genetic elements in maize discovered by McClintock (1952)

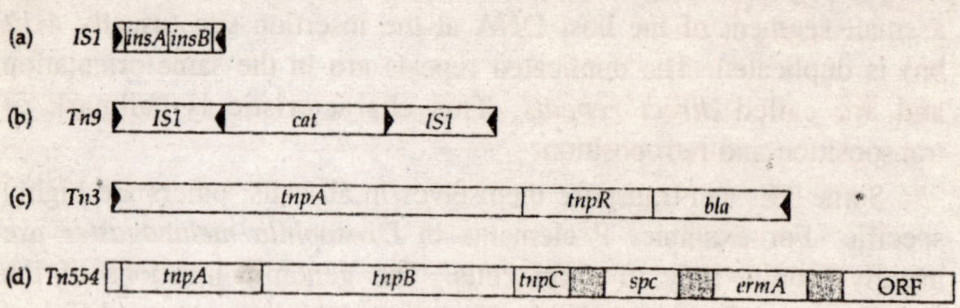

Fig. 9.2. Schematic representation of four transposable elements in bacteria.

became clear when it was recognized that they are residents of the *E. coli* genome and that the observed insertion mutations were examples of their movement of new genetic locations.

Insertion sequences (IS) are the simplest transposable elements. They carry no genetic information except that which is necessary for transposition. Insertion sequences are usually 700-2500 bp in length. They have been found in bacteria, bacteriophages, and plasmids. Moreover, the maize controlling elements are also insertion sequences (IS) by this definition. Bacterial insertion sequences are denoted by the prefix IS followed by a type number. The structure of an insertion sequence IS1 from *E. coli* and *Shigella dysinteria*, is approximately 770 bp in length, including two inverted nonidentical terminal repeats, 23 bp each. It contains two reading frames, InsA and InsB, which encode one or two forms of transposase, an enzyme that catalyzes the insertion of transposable elements into insertion sites. There are dozens of different types of insertion sequences in *E. coli*, and the genomes of most strains isolated from the wild contain variable numbers of each.

Transposons

In the mid 1960s it became evident that certain genes responsible for resistance to antibiotics in bacteria are capable of transferring from one DNA molecule to another. The first direct evidence that such transfer is by a process analogous to the insertion of Is elements was reported by Hedges and Jacob (1974), who postulated that the gene for ampicillin resistance is carried by a DNA element that can be transposed between molecules. They called such an element a *transposon* (Tn). Later, the term was used to denote any element that carries one or more genes that encode other functions in addition to those related to transposition; such genes are known as exogenous genes. Some authors call all transposable elements, including IS elements, transposons.

Since transposons also carry exogeneous genes they are usually larger than IS elements, about 2500-7000 bp long. In bacteria, some transposons are complex or compound transposons, so named because two complete independently transposable insertion sequences in either orientation flank one or more exogenous functional genes. Interestingly, in complex transposons, the entire transposon can transpose as a unit, and also one or both of the flanking insertion sequences can transpose independently.

Other bacterial transposons, as well as many eukaryotic transposons, are flanked only by short repeated sequences in various orientations and contain no insertion sequences. Not all transposons, however, are symmetrical in structure. Some have asymmetrical ends, lacking either inverted or direct terminal repeats. The coding regions of some transposons in animals, e.g., P elements in *Drosophila*, are interrupted by introns.

Transposons in bacteria often carry genes that confer antibiotic (e.g., Tn554) heavy-metal (e.g., Tn21), or heat (e.g., Tn1681) resistance to their carriers. Plasmids can carry such transposons from cell to cell, and as a consequence, resistance can quickly spread throughout populations of bacteria exposed to such environmental factors.

Fig. 9.3. Schematic structure of a complete P element in Drosophila melanogaster.

Several phages in bacteria are in fact transposons. For example, bacteriophage Mu is a very large transposon (~38 kb) that encodes not only the enzymes that regulate its transposition, but also a large number of structural proteins necessary to construct the packaging of its DNA.

Transposons of many types are quite widespread in the genomes of animals, plants, and fungi. *Drosophila melanogaster*, for instance, contains multiple copies of 50-100 different kinds of transposons.

Retroelements

As mentioned above, there are transposable elements that transpose by retroposition; all such elements contain a gene for the enzyme reverse transcriptase. Moreover, there are genetic entities that contain a gene for reverse transcriptase but do not have any intrinsic capability to transpose. Temin (1989) proposed to call the entire group of reverse transcriptase-bearing entities, whether transposable or not *retroelements*; the term retroids was coined earlier by Fuetterer and Hohn (1987).

Among the known retroelements, retroviruses are the most complex. Their genome contains at least three genes: gag, pol, and env. The gag gene encodes a polyprotein that is processed into several small structural proteins that are the major proteins of the nucleocapsid and matrix of the virion. The pol gene encodes a polyprotein that consists of a reverse transcriptase (RT); RNase H (RH), which destroys the RNA strand of the RNA; an integrase (IN), which is responsible for the integration of the double-stranded DNA intermediate made from the RNA transcript into the chromosomes of the host genome; and possibly also an aspartate proteinase (PR), which is responsible for processing the primary translation products of the gag and pol genes into the various structural and enzymatic components of the virion. The env gene encodes the envelope protein. Many retroviruses possess additional genes; for example, the AIDS virus (HIV) possesses at least six additional genes. The coding region of a retrovirus is flanked by long terminal repeats (LTRs), which contain important control sequences, e.g., a promoter, an enhancer, and a poly(A) additional signal.

The next most complex retroelements are *retrotransposons*, which are TEs with LTRs. With few exceptions, retrotransposons do not have the env gene for constructing virion particles and therefore cannot independently transport themselves between cells. Most retrotransposons have a gene organization similar to that of either the gypsy or copia elements. The gene content and organization of gypsy elements are very similar to those of retroviruses. The tree open-reading frames (ORFs) have sizes and arrangements similar to the gag, pol, and env genes of retroviruses. In fact, recent data suggest that gypsy is indeed an infectious retrovirus, and since gypsy is found in *Drosophila*, this finding changes the traditional view that retroviruses exist only in vertebrates. The other major subgroup of retrotransposons is represented by the element copia in *D. melanogaster*. Note that copia does not contain the env gene. Note also that the integrase domain in copia precedes the reverse transcriptase domain, rather than follows the RNase H domain, as in retroviruses and gypsy. Retrotransposons similar to copia have been found in many eukaryotes including the Ty1 elements in yeast (Ty stands for transposon yeast) and the Ta1 of *Arabidopsis thaliana*.

Retroposons, unlike retrotransposons, do not have LTRs. They are sometimes called non-LTR retrotransposons or nonviral retrotransposable elements. Because the best known elements within this group are the long interspersed nucleotide elements (LINEs) found in mammalian

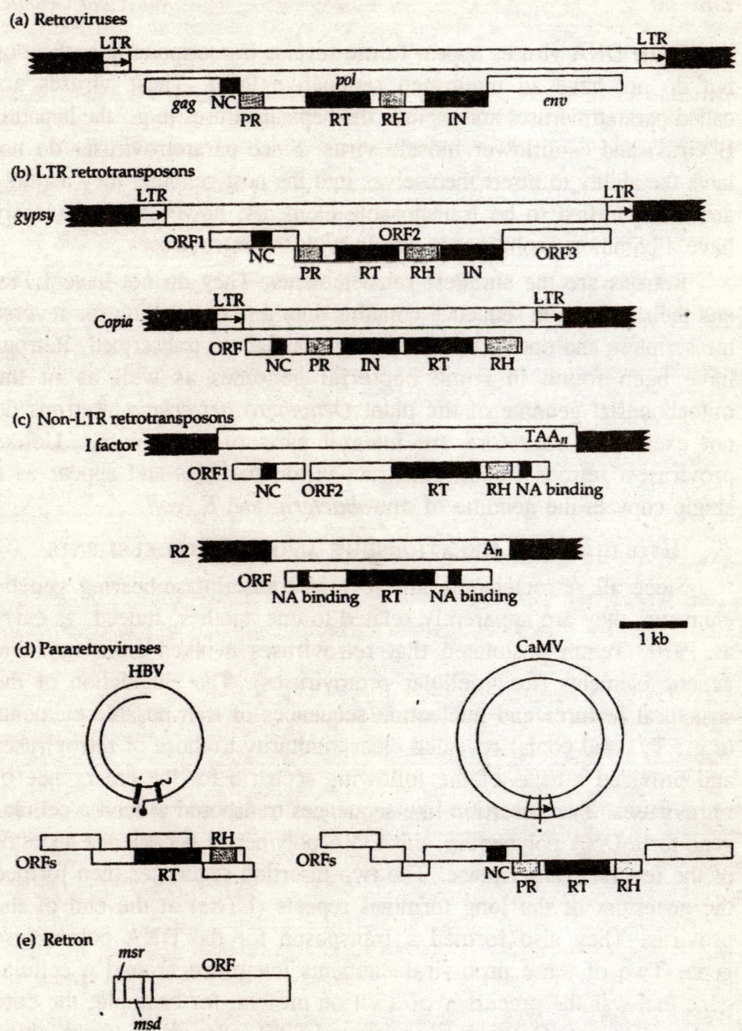

Fig. 9.4. Structural comparison of retroelements.

genomes, they have also been called LINE-like elements. Two examples of retroposons found in *D. melanogaster*. The retroposons are probably the most varied group of retroelements, and the I factor and R2 are not representative of any major subgroup. Most retroposons share at least a few of the features associated with the gag and pol genes of retroviruses. Most retroposons, like the I factor, contain two ORFs, while others, like R2, contain a single ORF.

Some DNA viruses appear to use reverse transcription in replication but do not have an integrated replicative form. These viruses are called pararetroviruses and include the hepadnaviruses (e.g., the hepatitis B virus) and cauliflower mosaic virus. Since pararetroviruses do not have the ability to insert themselves into the host genome, they actually are not qualified to be transposable elements; however, they clearly have a common evolutionary origin with the retroviruses.

Retrons are the simplest retroelements. They do not have LTRs and their nucleotide sequence contains only a region coding for reverse transcriptase and one or two other regions that are transcribed. Retrons have been found in some bacterial genomes as well as in the mitochondrial genome of the plant *Oenothera berteriana*. Retrons do not excise and therefore are integral parts of the genome. Unlike proviruses, retrons cannot construct virion particles and appear as a single copy in the genome of *myxobacteria* and *E. coli*.

Evolutionary Relationships among Retroelements

Since all retroelements are reverse transcriptase-bearing genetic elements, they are apparently related to one another. Indeed, as early as 1970, Temin postulated that retroviruses evolved from movable genetic elements (i.e., cellular protoviruses). The elucidation of the structural features and nucleotide sequences of transposable elements (e.g., Ty1 and copia) revealed clear similarity to those of retroviruses and provided a basis of the following scenario for the emergence of retroviruses. Two insertion-like sequences transposed around a cellular gene for a DNA polymerase; this DNA polymerase became the ancestor of the reverse transcriptase. The two insertion sequences then formed the ancestors of the long terminal repeats (LTRs) at the end of the provirus. They also formed a transposon for the DNA polymerase gene. Two of these protoviral elements integrated around a cellular gene that was the precursor of a virion protein, for example, the core protein precusor. Deletion of an internal LTR followed by transposition of the region containing the polymerase gene and the new gene gave an element with two structural genes. Repeating this process of integration of two elements around a cellular gene that was a precusor of another virion protein (for example, the envelope protein), followed by deletion of an internal LTR, gave an element with the structure of a provirus.

In the above scheme Temin (1980) proposed that the viral reverse transcriptase evolved from DNA polymerase because at that time no cellular reverse transcriptase genes had been found in either eukaryotes

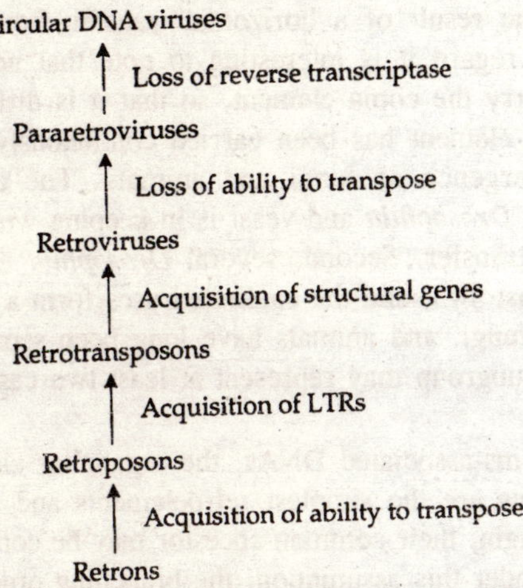

Fig. 9.5. Schematic representation of possible evolutionary pathways of retroelements.

or prokaryotes. Later, cellular genes for reverse transcriptase were found in myxobacteria and *E. coli*. This observation suggests that reverse transcriptase existed before retroviruses and not the reverse. Temin (1989) therefore revised his scheme as follows: The path of evolution went from retrons, which are the simplest reverse transcriptase-bearing entities, to retroposons by gaining the ability to transpose, next to retrotransposons by gaining LTRs, then to retroviruses by the acquisition of structural genes, and subsequently to pararetroviruses by loss of the ability to transpose. Of course, it is possible that some of the present-day retrotransposons have been derived from retroviruses, rather than the other way around.

Many authors have studied the evolutionary relationships among retroelements. Fig 9.6 shows a phylogeny of retroelements inferred from reverse transcriptase sequences. The tree can be divided into two major branches, as indicated by the arrow on the tree. The major branch on the left half of the tree contains all the non-LTR retroelements. Within this branch, the ms-associated DNAs, the organellar elements, and the group II introns form three separate monophyletic subgroups. In addition, with the exception of the three trypanosome elements CZAR, SLAC, and CRE1, all the other retroposons form a monophyletic group, which includes the I factor, R2 (R2 Dm and R2Bm) and the LINE (L1Hs and L1Md) elements. The major branch on the right half of the tree contains all the LTR retroelements. Two subgroups are worth noting. First, the copia and the Ty1 elements form a subgroup. Since the Ty1 and copia elements are only distantly related to other retrotransposons, it has been suggested that Ty1 is the result of a horizontal transfer from *Drosophila* to yeast. In this regard it is interesting to note that not all species of *Drosophila* carry the copia element, so that it is difficult to imagine that the copia element has been carried continuously in the genome since the divergence of fungi and animals. The close ecological dependence of *Drosophila* and yeast is in keeping with the possibility of horizontal transfer. Second, several *Drosophila* elements (gypsy, 412, etc.), yeast Ty3, and the caulimoviruses form a large subgroup. Since plants, fungi, and animals have long been separated from one another, this subgroup may represent at least two cases of horizontal transfer.

Since the ms-associated DNAs, the organellar elements, and the group II introns are the simplest retroelements and are probably of prokaryotic origin, their common ancestor may be considered the root of the tree. Under this assumption, the branching order of the tree is consistent with Temin's (1980) view that the evolution of retroelements went from retrons to retroposons, to retrotransposons, and then to retroviruses, though this tree does not support the view that

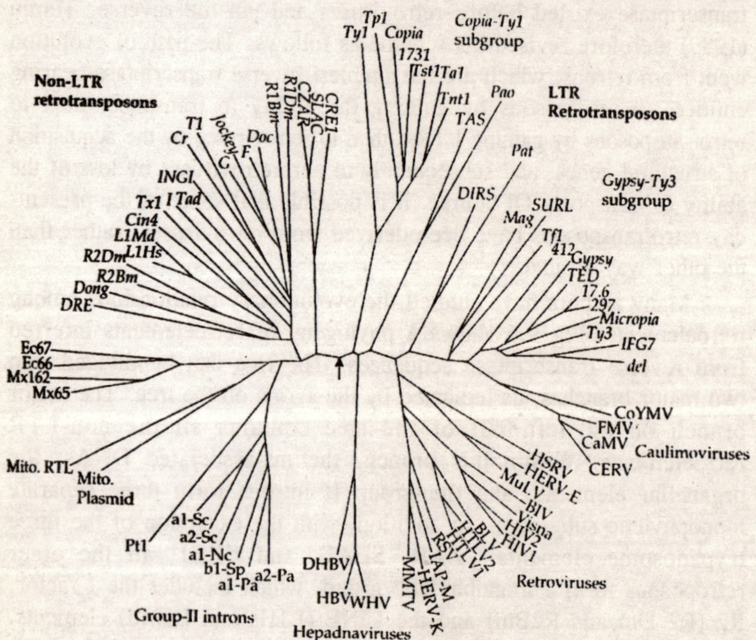

Fig. 9.6. Unrooted phylogenetic tree of retroelements.

hepadnaviruses (as represented by the hepatitis B viruses) were derived from retroviruses. (As mentioned above, gypsy and its relatives might be retroviruses.) Note, however, that Fig. 9.6 is based on very limited sequence data, so that the branching order of the subgroups and the branching order of the elements within each subgroup should be taken with caution. Further data are needed for obtaining a more reliable tree.

RETROGENES

The classification includes a class of *retrosequences*, or *retrotranscripts*, which are genomic sequences that have been derived through the reverse transcription of RNA and subsequent integration into the genome but lack the ability to produce reverse transcriptase. The template RNA is usually the transcript of a gene. Some authors call retrosequences "nonviral retroposons". If a gene is not transcribed within any germline cells, the creation of a retrosequence requires the RNA to cross cell barriers. This can happen when an RNA molecule becomes encapsulated within the virion particle of a retrovirus and is then transported to a germline cell where it is reverse-transcribed. This process is referred to as *retrofection*.

Since retrosequences originated from RNA sequences, most of them bear marks of RNA processing and are, hence also referred to as *processed retrosequences*. The diagnostic features of such sequences

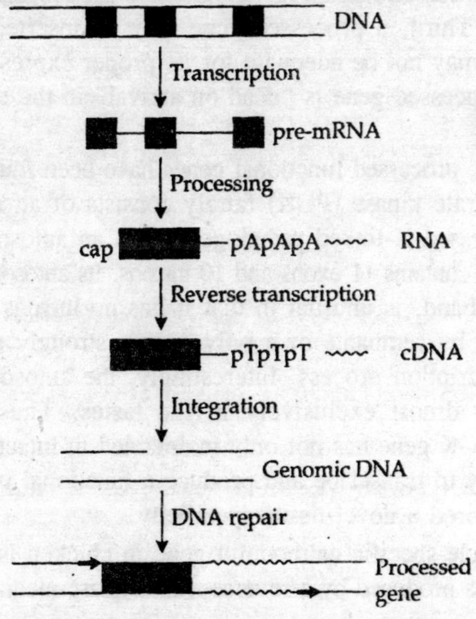

Fig. 9.7. Creation of a processed retrosequence.

include: (1) lack of introns, (2) precise boundaries coinciding with the transcribed regions of genes, (3) stretches of poly-A at the 3' end (there are exceptions, as in most of snRNA pseudogenes), (4) short direct repeats at both ends, indicating that transposition may have been involved, (5) various post transcriptional modifications, such as the addition or removal of short stretches of nucleotides, and (6) chromosomal position different from the locus of the original gene from which the RNA was transcribed.

A retrosequence is called a *retrogene* (a processed gene) if it is functional but a *retropseudogene* if it is nonfunctional. Examples of retrogenes are discussed below, while examples of retropseudogenes will be presented in the next section.

A retrogene is a functional retrosequence producing a protein that is identical or nearly identical to that produced by the gene from which the retrogene has been derived. There are several reasons why it is highly unlikely that a reverse-transcribed gene will retain its functionality. First, the process of reverse transcription is error prone and many differences (mutations) between the RNA template and the cDNA can occur. Second, unless a processed gene has been derived from a gene transcribed by RNA polymerase III, it usually does not contain regulatory sequences that reside in the untranscribed regions and is therefore likely to be inactive, even if its coding region is intact. The processed human metallothionein pseudogene appears to be such a case. Third, a processed gene may be inserted at a genomic location that may not be adequate for its proper expression. For these reasons, a processed gene is "dead on arrival" in the vast majority of cases.

However, processed functional genes have been found. The human phosphoglycerate kinase (PGK) family consists of an active X-linked gene, a processed X-linked pseudogene, and an autosomal gene. The X-linked gene contains 11 exons and 10 introns. Its autosomal homologue, on the other hand, is unusual in that it has no introns and is flanked on its 3' end by remnants of a poly-A tail, strongly suggestive of a reverse transcription process. Interestingly, the autosomal PGK gene is expressed almost exclusively in the tastes. Thus, the reverse-transcribed PGK gene has not only maintained in intact reading frame and the ability to transcribe and produce a functional polypeptide, but also has acquired a novel tissue-specificity.

The muscle-specific calmodulin gene in chicken is intronless and apparently was produced by a reverse-transcriptase-mediated event. This might also be the case for the intronless globin gene of *Chironomus*. All but one of the 17-20 intronless actin genes in *Dictyostelium* are functional, and these might also be retrogenes.

The rat and mouse preproinsulin I gene may represent an instance of a semi-processed retrogene. The gene contains a single small intron in the 5' untranslated region. In comparison, its homologue, preproinsulin II contains the same small intron, as well as an additional larger intron within the coding region for the C peptide. All preproinsulin genes from other mammals, including other rodents, contain two introns as well. Moreover, the preproinsulin I gene is flanked by short repeats, and the polyadenylation signal is followed by a short poly-A tract. These features suggest that preproinsulin I gene might be a semiprocessed retrogene derived from a partially processed preproinsulin II mRNA. Indeed, based on comparisons between the two preproinsulin genes, preproinsulin I appears to have been derived from an abberant mRNA transcript that was initiated at least 500 base pairs upstream of the normal cap site and from which only the first intron has been spliced out. It is likely that the aberrant transcript contained 5' regulatory sequences not normally transcribed, so that the retrogene has maintained its function following its integration into a new genomic location.

Retropseudogenes

A retropseudogene is a nonfunctional retrosequence. Most retropseudogenes are apparently derived from processed RNA transcripts (in this case they are called *processed pseudogenes*), though some seem to have been derived from unprocessed RNA transcripts. In principle, retropseudogenes can be derived from all RNA transcripts that are present in germ cells or can retroinfect germ cells. Indeed, retropseudogenes derived from RNA polymerases (POL) II and III transcripts (e.g., mRNA, snRNA, tRNA, and 7SL RNA) are known. However, so far there seems to be no example of a retropseudogene that is derived from a RNA polymerase I transcript (e.g., 18S, 28S, 5.8S, and 5S rRNAs). It is not clear why this class of RNA transcripts cannot be or are rarely retrotransposed.

Pseudogenes Derived from POL II Transcripts

Most POL II retropseudogenes are derived from mRNA sequences; the others are mainly from snRNAs (small nuclear RNAs). They bear all the hall-marks of a functional retrosequence but have molecular defects that prevent them from being expressed. Many processed pseudogenes get truncated during retrofection. 5' truncation of processed mRNA is particularly common, though 3' truncation is also known. Truncation of the 5' end can occur during (1) transcription (e.g., initiation downstream of the normal initiation site), (2) RNA processing (e.g., faulty splicing), and (3) reverse transcription (e.g., failure of the enzyme to complete the reverse transcription to the 3' end of the RNA molecule, which corresponds to the 5' end of the DNA).

Table 9.1. The number of retropseudogenes and the number of parental functional genes

Species	*Gene*	*Number of genes*	*Number of retropseudogenes*
Human	Argininosuccinate synthetase	1	14
	β-actin	1	~20
	β-tubulin	2	15-20
	Cu/Zn superoxide dismutase	1	≥4
	Cytochrome c	2	20-30
	Dihydrofolate reductase	1	~5
	Nonmuscle tropomyosin	1	≥3
	Glyceraldehyde-3-phosphate dehydrogenase	1	~25
	Phosphoglycerate kinase	2	1
	Ribosomal protein L32	1	~20
	Triosephosphate isomerase	1	5-6
Mouse	α-globin	2	1
	Cytokeratin endo A	1	1
	Glyceraldehyde-3-phosphate dehydrogenase	1	~200
	Myosin light-chain	1	1
	Propiomelanocortin	1	1
	Ribosomal protein L7	1-2	≥20
	Ribosomal protein L30	2	≥15
	Ribosomal protein L32	1	16-20
	Tumor antigen p53	1	1
Rat	α-tubulin	2	10-20
	Cytochrome c	1	20-30

Pseudogenes derived from mRNAs are abundant in mammals. Table 9.1 shows a list of such pseudogenes in humans and rodents from which both the number of functional genes and the number of processed pseudogenes are known or have been estimated. Note that there is a tendency to underestimate the number of pseudogenes because old pseudogenes may have diverged in sequence from their parental gene to such an extent that they are no longer detectable by molecular probes derived from the functional homologue. Interestingly, in some cases the number of processed pseudogenes is much greater than the number of functional genes. For example, in the genome of mouse about 200 processed pseudogenes have been produced from a single gene for glyceraldehyde-3-phosphate dehydrogenase.

Although processed pseudogenes are abundant in mammals, they are relatively rare in other organisms, including chicken, amphibia, and *Drosophila*. For example, mammals have 20-30 tubulin retropseudogenes while chicken and *Drosophila* have none, though, like mammals, chicken and *Drosophila* have many α- and β-tubulin genes. It is not clear why such a conspicuous difference occurred. One hypothesis is that it is due to differences in gametogenesis between mammals and other organisms. While spermatogenesis is very similar among mammals, chicken, amphibia, and *Drosophila*, mammalian oogenesis differs from that in the other organisms by prolonging the lampbrush stage from birth to ovulation. This state of relatively suspended animation may 40 years in humans, but for only a few months in amphibians, and for less than three weeks in birds, and is virtually absent in *Drosophila*. It retroposition in mammals in fact occurs predominantly in the female germline, retrosequences would be under-represented on the Y chromosome.

Pseudogenes Derived from POL III Transcripts

The genomes of mammals and other higher eukaryotes contain several types of highly repetitive interspersed sequences. These sequences, originally detected as rapidly reannealing components of genomic DNA, have been divided into two major classes referred to as short and long interspersed elements (SINEs and LINEs, respectively). The SINEs typically range from 75 to 500 bp in length, while the LINEs can extend to 700 bp. DNA-sequencing studies have indicated that the SINEs are derived from POL III transcripts with the LINEs from POL II transcripts. LINEs are classified as retroposons because they contain a region coding for reverse transcriptase; actually most LINEs are nonfunctional and such LINEs should be classified as retrosequences. In contrast, SINEs generally do not code for reverse transcriptase and are retrosequences, though they have often been classified as retroposons.

A typical (or generic) SINE sequence contains an internal POL III promoter, an A-rich 3' end, and flanking direct repeats. The promoter has a typical bipartite structure as represented by the A and B boxes. The 3' A-rich end regions vary from less than 8 to longer than 50 bp. The flanking direct repeats are produced by duplication of target sequences at the site of integration. These repeats are generally A-rich, suggesting a preference of insertion at A-rich regions.

Many different SINE families are known. The Alu family was the first to have been well characterized, and so in the past other SINE families were often referred to as Alu-like families. The Alu family was originally defined as the fraction of rapidly reannealing repetitive sequences that contain a distinctive Alu1 site. The human Alu family has all of the features of the genetic SINE but is a head-to-tail dimer of two similar sequences ~130 bp long; some monomers have also been found. The left half has a POL III promoter, while the right half contains a 31-bp insert inside the POL III promoter.

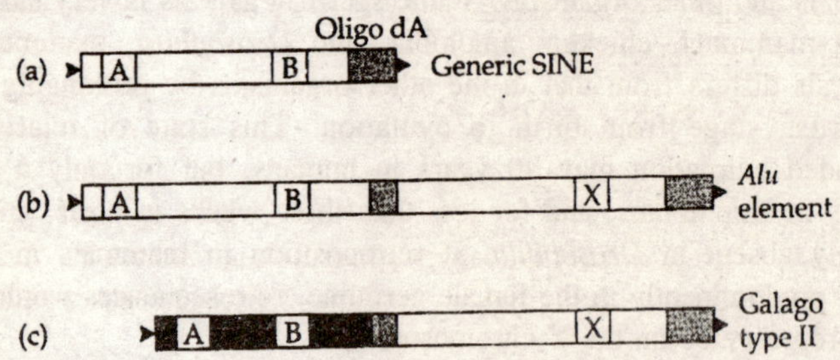

Fig. 9.8. (a) Features of the generic SINE. (b) The Alu family. (c) The galago type II family.

A related family, the type II family, is found in the genome of *Galago crassicaudatus*, a prosimian. The right half of this family is almost identical to the right half of the Alu family. The left half is entirely different from the Alu family but also contains a POL III promoter. The type II family probably arose from a fusion of two different SINE elements.

The discovery of SINEs raised the question of their origin. The first clue in this regard was the finding that the human Alu family has high sequence similarity to the 7SL RNA gene. 7SL RNA is an abundant cytoplasmic RNA that functions in protein secretion as a component of the signal recognition particle. Ullu et al. (1982) noted that approximately 100 nucleotides at the 5' end and 45 nucleotides at the 3' end of the RNA are homologous with the human Alu right monomer consensus sequence. The central portion of 155 nucleotides is unique to 7SL RNA and shows no similarity with the human Alu

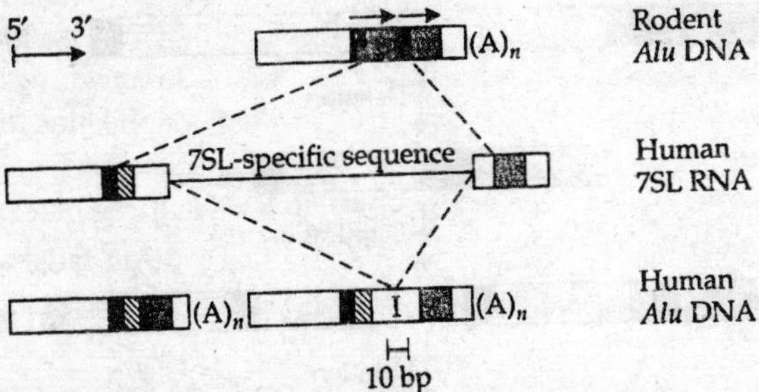

Fig. 9.9. The structural relationship of human 7SL RNA to the consensus sequences of human and rodent Alu DNA.

DNA or the rodent Alu equivalent, the B1 element, which is a monomer.

The next important finding concerning the origin of SINEs was the discovery that the Monomer family in the prosimian galago, which is closely related to the left-half sequence of the type II family, is related to a methionine-tRNA gene. The two sequences exhibit nearly 70% similarity. Since the Monomer family is not present in other species, not even in other primates, it must have been derived from the tRNA gene relatively recently. There are other SINE families that appear to be amplified tRNA pseudogenes.

However, SINEs are obviously not directly derived from 7SL RNA or tRNA transcripts because the sequence dissimilarity between any element and the parental gene is considerable. For example, the B1 element not only has a deletion relative to the 7SL RNA gene, but also contains a small internal duplication.

Actually, something must occur to make a sequence much more effective in multiplying itself for it to produce a SINE family. An important requirement for the formation of a new SINE family appears to be a POL III promoter that is functional and allows strong expression in the germline, for only retroposition in the germline can result in heritable sequence amplification. An important difference between a POL II and a POL III transcript is that the latter contains an internal promoter whereas the former does not, and this could be the major season why POL III retropseudogenes are generally much more abundant than POL II retropseudogenes. Another important requirement would be to have an A-rich 3' end. For a retroposition to occur, a SINE RNA molecule must be reverse-transcribed into DNA, and having an A-rich 3' end may enable the molecule to self-prime its own reverse transcription.

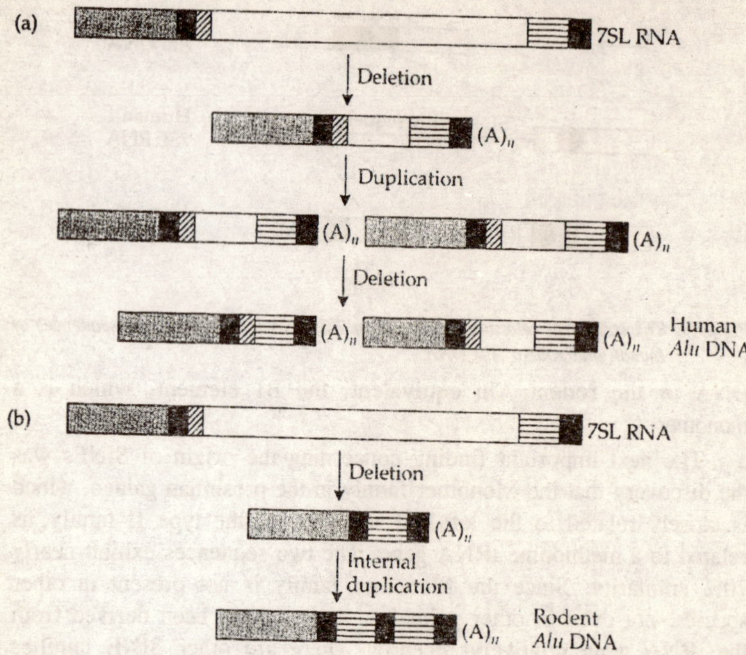

Fig. 9.10. Origin of Alu sequences in (a) humans and other primates and (b) rodents.

Evolutionary Fate of Retropseudogenes

Due to the ubiquity of reverse transcription, the genomes of mammals are literally bombarded with copies of reverse-transcribed sequences. The vast majority of these copies are nonfunctional from the moment they are integrated into the genome. Moreover, such sequences cannot be easily rescued by gene conversion, since they are mostly located at great chromosomal distances from the parental functional gene. The phenomenon of a functional locus pumping out defective copies of itself and dispersing them all over the genome has been likened to a volcano generating larva, and the process has been termed the *Vesuvian mode* of evolution.

As soon as a retropseudogene is established as a chromosomal sequence in the genome, it is affected by two evolutionary processes. The first involves the rapid accumulation of point mutations. This accumulation eventually obliterates the sequence similarity between the pseudogene and its functional homologue, which evolves much more slowly. The nucleotide composition of the pseudogene will come to resemble more and more its nonfunctional surroundings, and it will eventually "blend" into its surrounding regions. This process has been called *compositional assimilation*.

The second evolutionary process is characterized by a pseudogene becoming increasingly shorter compared to the functional gene. This *length abridgment* is caused by the excess of deletions over insertions. It has been estimated that a processed pseudogene loss about half of its DNA in about 400 million years. This process is so slow that the human genome, for instance, still contains major chunks of pseudogenic DNA that were found in very distant ancestors. Obviously, these ancient pseudogenes have by now lost almost all similarity to the functional genes.

In summary, it seems that processed genes are created at a faster rate than the rate by which they are obliterated by deletions. It can therefore be concluded that abridgment is too slow a process to offset the increase in genome size following the continuous Vesuvian bombardment. So, the limitation on the number of pseudogenes in a genome is probably due to other factors such as natural selection.

Evolution of Alu Sequences

The evolution of Alu sequences has received much attention, because the Alu repeats represent one of the most abundant families of retrosequences and because Alu sequence data are abundant. A major issue is how Alu repeats have become so abundant.

Noting that a typical Alu sequence contains an internal POL III promoter and an A-Rich 3' end, Jagadeeswaran et al. (1981) and Van Arsdell et al. (1981) proposed a model for the transposition of Alu sequences (and other repetitive sequences with a similar structure). The model assumes that the complementary strand of an Alu sequence is transcribed into an RNA sequence, which contains the entire Alu sequence, including its A-rich 3' end, and ends in a U-rich sequence characteristic of the site of termination of POL III transcription. The 3' oligo(U) sequence of the RNA transcript airs with the internal oligo(A) to prime cDNA synthesis, and the cDNA sequence is inserted back into the genome. In this model, retroposition is a cascade process, because new sequences may become additional sources for producing Alu sequences. This model was supported by a phylogenetic analysis of 59 human Alu sequences by Bains (1986), who found that all these sequences were about equally divergent from the center of a dendrogram, i.e., there was no subgroup of Alu elements with significantly greater similarity between members of the subgroup than between any of these members and other Alu elements. Based on this observation, Bains

(1986) suggested that the Alu family derives from a large pool of precursors.

However, Slagel et al. (1987) identified a subfamily in an analysis of 22 Alu sequences. Four of the sequences appeared to be closely related because they not only formed one cluster in a dendrogram constructed from distances based on nucleotide mismatches but also shared a two-base deletion that was not shared by any of the other sequences in the sample. In a subsequent statistical analysis of more sequences, Willard et al. (1987) proposed a subdivision of Alu sequences into at least three subfamilies, each of which was founded by a distinct master sequence. This suggestion was supported by a study using the correspondence analysis, which is a powerful technique for identifying clusters.

The existence of subfamilies has also been suggested by extensive analyses of "shared differences". In this approach, a consensus sequence is constructed from the sequences under study. Shared (diagnostic) differences are defined as the differences (substitution, insertions, or deletions) shared by a subset of sequences but differing from the consensus. For example, the shared differences of 30 Alu sequences. The diagram focuses on a few highly shared nucleotides by rearranging the sequences first to show off the diagnostic positions and second to isolate the CpG dinucleotides that are hot spots of mutation. The consensus nucleotide is shown at the left column and the Alu repeats are aligned approximately in order of divergence from the consensus. The diagram immediately identifies diagnostic positions that define three classes of Alu repeats, denoted as IV, III, and II. For example, at position 93 all class-IV sequences have T, while almost all class-III and -II sequences have C, and at position 76 all class-IV and -III sequences have A, while all class-II sequences have T. Class-I sequences are even more divergent than class-II sequences from the consensus sequence and are not shown. Jurka and Milosavljevic (1991) regard each of classes I, III, and IV as one subfamily and divide class II, the major class, into three subfamilies. Recently, Britten (1994) added a fifth class of represent newly and very recently inserted elements.

Britten et al. (1988) proposed that the patterns of sharing arose from a series of different source (master) genes producing each class of Alu repeats in turn and being sequentially replaced during evolution rather than from a set of coexistent sources producing all classes at all times. If several different sources operate concurrently, they would produce sets of Alu repeats of all classes, each initially identical to the respective source. Then each class would age and would include sequences with a range of divergence from each other and from their source. However, observation shows the opposite. Members of the classes that are more divergent from the consensus are more divergent from each other. For example, the average pairwise divergence between class-II members (the rightmost 9 Alu repeats) is 56.3 mutations (20% substitutions, insertions, and deletions, including CpG positions), whereas that between the leftmost 13 Alu repeats of class IV is only 31.6 mutations (11%). Apparently, the rightmost 9 Alu repeats were copied from earlier source sequences and inserted into the genome earlier in the past. Interestingly, the average sequence similarity with the 7SL RNA gene decreases in the order of classes I, II, III, IV and V sequences. The first source gene was the most similar to the 7SL RNA gene and subsequent source genes were derived from preceding ones by mutation, becoming less similar to the 7SL RNA gene. Each source gene served at one time or another as the predominant source

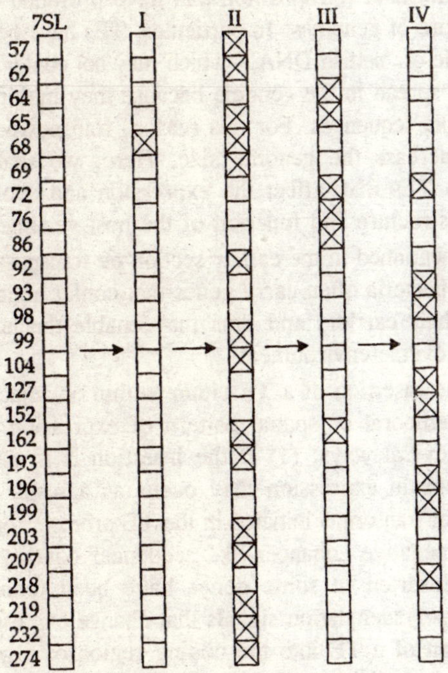

Fig. 9.11. Sequence of mutations at diagnostic positions between different subfamilies of Alu sequences.

of Alu sequences and was superseded by a descendant line. The successive waves of fixation did not occur in sudden bursts, but consecutive subfamilies continued to coexist within the genome for long periods of time. For example, from the degree of sequence divergence and the rate of evolution of noncoding sequences in primates, Britten (1994) estimated that most of the class-II members were produced between 50 to 30 million years ago and that most of the class-III members were produced between 46 and 26 million years ago.

The source (master) gene model is suggested to be applicable also to other SINE families and to LINE families. However, whether there are only a few or many source genes active at each time remains unclear. Since no source gene has yet been identified, the nature and function of source genes are unknown. Also, how source genes have evolved in succession is unclear.

Effects of Transposition on the Host Genome

Transposition and retroposition can have profound effects on the size and structure of genomes. In particular, TEs have been considered the best example of "selfish DNA," which may not confer any advantage to the host but spread in the genome because they multiply faster than the host genomic sequences. For this reason, transposition of TEs can significantly increase the genome size. Here, we shall discuss the ways in which TEs may affect the expression and evolution of host genes and the structure and function of the host genome.

First, as mentioned in the earlier section on transposable elements, transposons in bacteria often carry genes that confer antibiotic or other resistance to their carriers and thus may enable the host species to survive in an adverse environment.

Second, the insertion of a TE either within or adjacent to a gene may alter its temporal or spatial pattern of expression. This change can occur in several ways: (1) If the insertion is in the 5' flanking region, a change in expression may occur as a result of either the "readthrough" of transcripts initiated in the TE promoter or the presence of positive or negative enhancer-like sequences contained within the TE. (2) TEs inserted in some genes have been found to contain termination or polyadenylation signals that change expression patterns. (3) The insertion of a TE into the coding region of a gene can alter the pattern of gene splicing. Although the vast majority of such changes are likely to be deleterious, occasionally a change may be advantageous or only slightly deleterious so that it can survive. Several successful

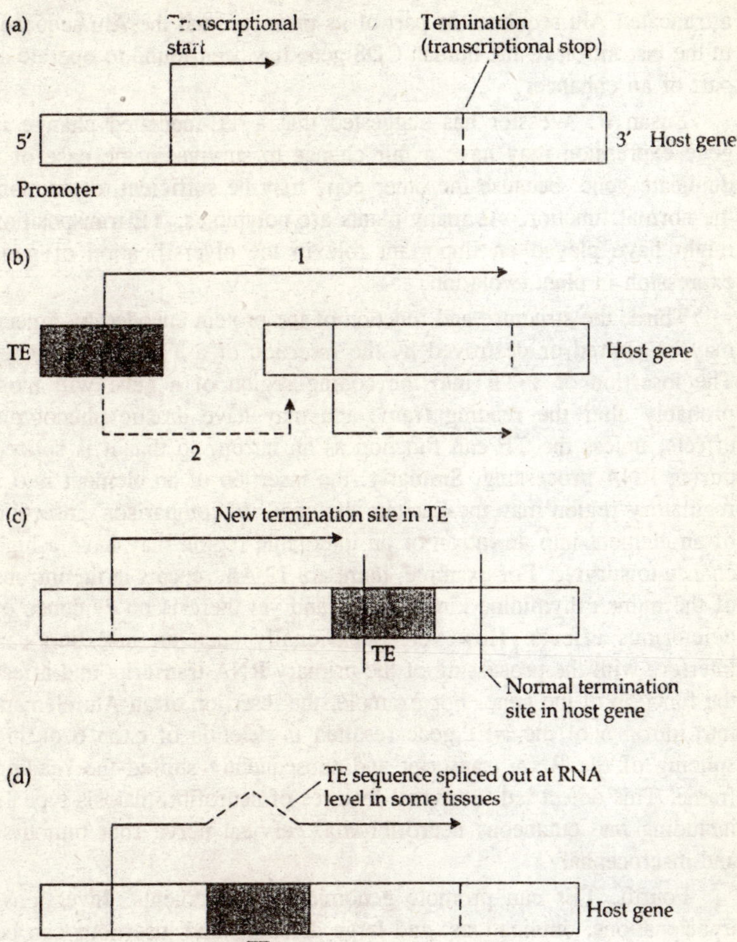

Fig. 9.12. Mechanisms by which TEs may alter the regulation of gene expression.

examples have been found. One such example is the mouse sex limited protein (Slp) gene, which is a member of the histocompatibility complex and arose by tandem duplication of the fourth component of the complement C4 gene. Although Slp and C4 are highly homologous, they are markedly different in their tissue-specific expression. These regulatory differences have been shown to be attributable to an enhancer sequence contained within a cryptic retrotransposon, which was inserted adjacent to the ancestral Slp gene several million years ago. As two additional examples, the human θ1 α globin has been shown to utilize a truncated Alu sequence as part of its promoter and the Alu sequence in the last intron of the human CD8 gene has been found to operate as part of an enhancer.

Susan R. Wessler has suggested that a TE-mediated change in gene expression may have a fair chance to survive in the case of a duplicate gene, because the other copy may be sufficient to maintain the normal function. As many plants are polyploids, TE transposition might have played an important role in the diversification of gene expression in plant evolution.

Third, the structure and function of the protein encoded by a gene may be altered or destroyed by the insertion of a TE into the gene. The insertion of a TE into the coding region of a gene will most probably alter the reading frame and may have drastic phenotypic effects, unless the TE can function as an intron, so that it is spliced during RNA processing. Similarly, the insertion of an element into a regulatory region may incapacitate the gene. In comparison, insertion of an element into an intron or an intergenic region may have a high chance to survive. For example, there are 13 Alu repeats in the introns of the human thymidine kinase gene, and yet there is no evidence of deleterious effects. However, occasionally such an insertion can interfere with the processing of the primary RNA transcript and affect the function of the gene. For example, the insertion of an Alu element into intron 5 of the NF1 gene resulted in deletion of exon 6 during splicing of the RNA transcript and subsequently shifted the reading frame. This defect led to several features of neurofibromatosis type 1, including one cutaneous neurofibroma, cervical nerve root tumours, and macrocephaly.

Fourth, TEs can promote genomic rearrangements. Inversions, translocations, duplications, and large deletions and insertions can be mediated by transposable elements. These rearrangements can take place, for example, as a direct result of transposition, i.e., by moving pieces of DNA from one genomic location to another, thus altering the milieu in which they are expressed. A more indirect effect will ensure if as a result of transposition, two sequences that had previously had little similarity with each other are now sharing a similar TE, so that crossing-over between them is possible. This type of "nonhomologous recombination" may occasionally produce a beneficial gene duplication or gene rearrangement. For example, a duplication of the entire growth hormone gene early in human evolution might have occurred via Alu-Alu recombination, and the duplication giving rise to

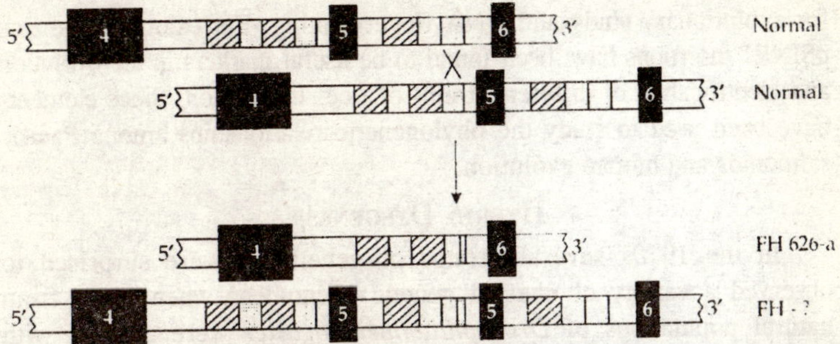

Fig. 9.13. Unequal crossing-over in the low-density-lipoprotein receptor gene.

the human $^{G}\gamma$ and $^{A}\gamma$ globin genes may have resulted from recombination between short L1 elements. In the majority of cases, however, it may result in a deleterious mutation. Fig. 9.13 illustrates how an unequal crossing-over event, facilitated by the presence of multiple Alu sequences within the introns flanking exon 5 of the low density lipoprotein receptor gene, gave rise to a mutant gene lacking the exon. The deletion of exon 14 by the same mechanism was also observed. Patient homozygous for these deletions have a high level of cholesterol in the blood (hypercholesterolemia). Recombination between two Alu elements has also been shown to be responsible for the deletion of the promoter and the first exon of the adenosine deaminase gene in patients with adenosine deaminase deficiency. In general, genomic instability has been demonstrated for all regions containing Alu repeat sequences.

Fifth, insertion of TEs into members of a multigene family can reduce the rate and limit the extent of gene conversion between members and therefore promote the rate of divergence between duplicate genes. The reason is that the process of gene conversion involves pairing of homologous DNA, and nonhomologous regions created by insertion of TEs might reduce the chance of such pairing and thus the chance of gene conversion between duplicate genes.

Sixth, there is evidence that some TEs may cause an increase in the rate of mutation. For example, strains of *E. coli* that contain the transposable element Tn10 were found to have elevated rates of mutation. Under most conditions this trait will be deleterious to the carrier. However, under severe environmental stress it is possible that an elevated rate of mutation might be advantageous in creating mutations, some of which may be better adapted to the new circumstances.

Since insertion of an element at a site is a rather event and thus can be used as a genetic marker, SINE and LINE elements are useful for evolutionary study and DNA fingerprinting. For example, the rice pSINE1 insertions have been found to be useful markers in the evolution and identification of different strains of rice. In addition, Since elements have been used to study the phylogenetic relationships among Pacific salmonids and human evolution.

Hybrid Dysgenesis

In the 1970s several groups of geneticists were surprised to observed a variety of unusual genetic phenomena when males from natural populations of *Drosophila melanogaster* were crossed with females from long-established laboratory strains. These phenomena include high frequencies of partial or complete sterility, male recombination (in *Drosophila* recombination is usually restricted to females), visible and lethal mutation, reversion of mutation, chromosomal rearrangement, transmission ratio distortion (including sex ratio distortion), and chromosomal nondisjunction. These correlated abnormal genetic traits have been collectively called *hybrid dysgenesis*. The dysgenic effects are usually asymmetrical, that is, they occur in one type of hybrid but usually not in the reciprocal hybrid. The syndrome has later been found to be caused by transposable elements.

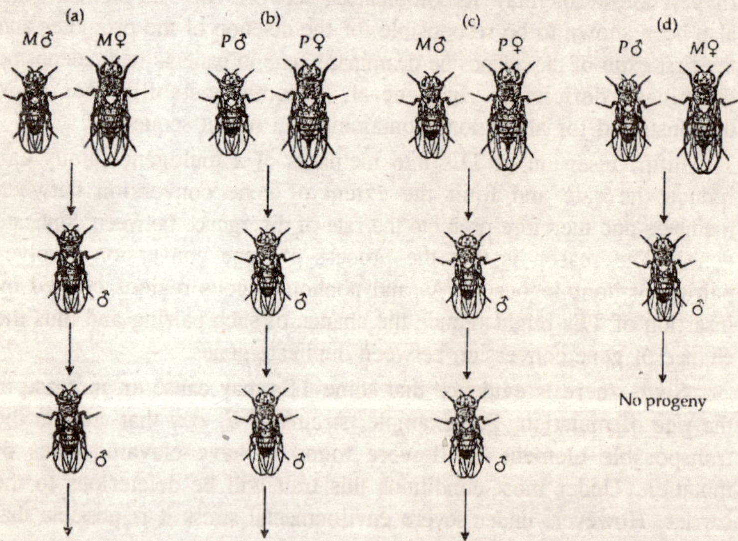

Fig. 9.14. Hybrid dysgenesis in Drosophila. (a) Matings within M strains and (b) matings within P strains produce normal progeny. (c) Normal progeny are also produced in crosses between M males and P females. (d) In the crosses between P males and M females, dysgenic offspring are produced, many of which are sterile.

There are two well-documented dysgenic systems in *Drosophila*, the P-M and I-R systems. We shall only discuss the P-M system; P and M stand for paternal and maternal, respectively. When a male from a P strain mates with a female from a M strain, the offspring are dysgenic, whereas in the reciprocal mating, the offspring are normal.

The cause of P-M dysgenesis is a family of transposable elements called P elements. In P strains, there are 30-50 P elements in the genome; however, many of them may be defective because of deletions. They are distributed throughout all chromosomes, although in some strains transposition shows a preference for the X chromosomes. M strains do not carry P elements. The asymmetry of the hybrid dysgenesis system is thought to result from maternal inheritance of a P element-encoded repressor in F_1 progeny of P female × M male matings and the absence of such a repressor in F_1 progeny of the reciprocal M female × P male matings. Dysgenic traits develop because of the transposition of P elements in germline cells that contain no repressors in cytoplasm, which is maternally derived. In the presence of repressors, P-element transposition may be wholly or partially inhibited. Transposition of P elements normally does not take place in somatic cells, because the third intron is not excised as it is in germline cells.

An interesting observation was made by Kidwell (1983) concerning the distribution of P-carrying strains. P characteristics were not found in any *D. melanogaster* strains collected before 1950, and collections made subsequently showed increasing frequencies of P with decreasing age. A similar observation was made with the I-R dysgenic system. Two hypotheses were suggested to explain the distribution of P elements. Engels (1981) proposed that most strains in nature are of the P type but that they tend to lose the transposon in laboratory populations (the stochastic loss hypothesis). The second hypothesis postulates a recent introduction of P transposons into *D. melanogaster* populations followed by a rapid spread of P elements in formerly M populations. There are several reasons why the second hypothesis, the recent invasion hypothesis, is more plausible. First, laboratory strains lacking active P elements may be only 30 years old or younger, whereas theoretical studies showed that the mean time for the stochastic loss of elements is at least the order of $1/\upsilon$, where υ is the excision rate and is of the order of 10^{-4} per generation, or at most 10^{-3} per year. The stochastic-loss hypothesis is also not supported by the observation that P-carrying laboratory strains that have been monitored for almost 20 years did not lose their P characteristic. Second, there is a temporal-

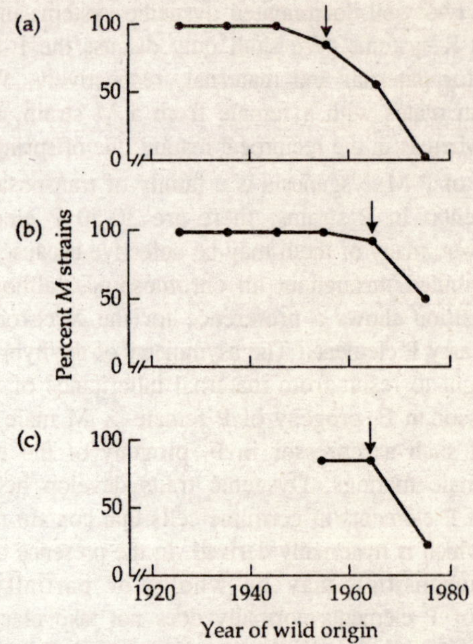

Fig. 9.15. Changes in the frequencies of M strains in natural population.

geographical cline in the distribution of P strains, suggesting that the invasion of the Americas occurred approximately 10 years earlier than that in Europe, Africa, and the rest of the world. Third, active P elements are highly invasive when they are introduced into susceptible populations either by crossing or by experimental germline transformation. Lastly, there is evidence that P elements have been recently acquired by *D. melanogaster* from *D. willistoni*, a distantly related species. This acquisition appears to have occurred by horizontal transfer.

Hybrid dysgenesis has been thought for a while to represent an early stage in the process of speciation, by acting as a postmating reproductive isolation mechanism between different populations belonging to the same species. Indeed, the sterility of hybrids produced by crosses between the sibling species *D. melanogaster* and *D. simulans* is very similar to dysgenesis (e.g., rudimentary gonads, segregation distortion). There are, however, problems with this view. First, although hybrids exhibit reduced fitness and are, therefore, partially isolated in terms of reproduction, the transposition of P elements in the germline virtually ensures that most of the chromosomes transmitted to the hybrids will bear P elements, and the cytotype will eventually change to the P type too. Thus, provided the reduction in fitness of the hybrids is not too great, the P element will spread through the entire population. Indeed, the hybrid must be almost completely sterile for an effective reproductive isolation to last. Second, P elements can be transposed horizontally as an infectious agent from individual to individual. Thus, an entire population may rapidly be taken over by P, so that hybrid dysgenesis is likely to last for very short periods of time in nature. Indeed, many species of *Drosophila* are known in which all individuals and all populations carry P elements or P-like elements, and yet hybrid dysgenesis does not occur in any of these species. Finally, hybrid dysgenesis seems to be restricted to *Drosophila* and may not represent a universal phenomenon in nature. Even in *Drosophila* no evidence has been found for the involvement of mobile elements as barriers to gene flow between sibling species.

Maintenance of Transposable Elements in Populations

There are two opposing views regarding the widespread existence of TEs in natural populations. One view is that TEs become widespread because they confer selective advantages to their hosts. As mentioned above, a TE may carry a gene for drug or metal resistance and may thus confer a selective advantage to its host bacterium under certain adverse conditions. Moreover, in chemostat competition experiments Chao et al. (1983) found that transposon Tn10 was able to increase the growth rate of *E. coli* cells by increasing the mutation rate of its host genome; the mutagenic effect was apparently due to transposition of Tn10 or its flanking insertion sequence IS10. The same advantage was also found for the insertion sequence IS50; since IS50 improved the growth rate of its host without transposition, the advantage was probably not mutational in origin but might have been derived from the effect of the transposase of IS50 on one or more cellular processes such as DNA replication or repair. In addition to the above observations, it has also often been argued that TEs can generate rearrangements that increase evolutionary versatility. However, such a long-term evolutionary advantage cannot explain the initial spread of TEs within a population, though it may be able to explain the persistence of the phenomenon in population, though it may be able to explain the persistence of the phenomenon in a population. Although the examples of drug and metal resistance and the chemostat experiments do provide support for the selectionist view, there is much reservation that this view can be a general explanation for the spread of TEs in natural populations, particularly in higher organisms.

The other view is that TEs are intragenomic parasites or "selfish DNA". The argument is that insertion of a TE into a chromosomal site is more likely to cause deleterious rather than beneficial effects because it may interfere with the expression or may disrupt the coding frame of a gene. Despite such disadvantages, a TE can spread through a population because it replicates faster than the host genome. Indeed, Hickey (1982) showed that a high rate of transposition can overcome even extremely severe deleterious effects of transposition. Furthermore, Ginzburg et al. (1984) and Uyenoyama (1985) showed that a hybrid dysgenesis factor can rapidly spread through a population, despite its selective disadvantages.

The preceding models are essentially single-locus models. Sawyer and Hartl (1986) studied models with elements dispersed over multiple genomic sites but with limited possibilities of sexual transmission between individuals (and recombination between elements at different sites); such models are suitable for bacterial transposable elements. Models of randomly mating, diploid populations with elements dispersed over multiple genomic sites and with relatively high frequencies of recombination between different sites have been studied by Charlesworth and Charlesworth (1983), Langley et al. (1983), Charlesworth (1985), Brookfield (1986), and others.

Let us consider a randomly mating population with enough recombination so that linkage disequilibrium can be ignored. Let u be the probability that a transposable element produces a new genomic copy, i.e., the probability of replicative transposition. Let υ be the probability that the element is excised. The values of u and υ have been determined experimentally for several TEs in *Drosophila melanogaster*. Transposition rates were found to vary among the TEs studied but on the average were in the order of 10^{-4} per element per generation. Excision rates were about one order of magnitude lower. Therefore, in the absence of selection against the TE, the number of copies in the genome is expected to increase indefinitely. A simple way to take selection into account is to assume that the fitness of an individual, w, decreases with the copy number, n. The justification for this assumption is that TE insertion frequently alters the expression of adjacent genes and that, with increasing numbers of TEs, the probability of a deleterious alternation of gene expression increases. Charlesworth showed that as long as w decreases with n, then regardless of the exact relationship between n and w, the mean fitness ($\overline{W}$) of a population at equilibrium relative to an individual lacking TE is

$$\overline{W} = e^{-n(u-\upsilon)} \qquad ...(1)$$

In the case of *Drosophila melanogaster* there are approximately 50 families of TEs, each appearing in the genome on the average 10 times. Thus, $n \approx 500$. Since υ is smaller than u by at least an order of magnitude, $u - u \approx u = 10^{-4}$. Solving equation (1), we obtain $\overline{W}$ = 0.95. Therefore, the reduction in fitness is s = 1 - 0.95 = 0.05. The stability of the equilibrium given by Equation 1 requires that the logarithm of fitness declines more steeply than linearly with increasing n. For simplicity, however, if we assume linearity, then the reduction in fitness with each additional copy is approximately $0.05/500 = 10^{-4}$. Such a small selective coefficient suggests that the number of copies of transposable elements within an organism is strongly affected by random genetic drift. Equation 1 is deterministic.

The above calculations shows that a very weak intensity of selection is sufficient to check the spread of elements. On the other hand, the studies of Mukai and colleagues on viability mutations indicate a mean selection coefficient of the order of 2% against homozygous detrimental mutations and 0.7% against heterozygotes. If all insertions result in a mutation with this magnitude of reduction in fitness, elements would never be able to spread in a population, unless transposition rates are higher than seem realistic for most elements. Of course, there probably exist many different classes of insertion sites. For example, sites that are sufficiently remote from coding regions may be selectively neutral or subject to only very weak selection, whereas sites in coding regions are likely to be subject to strong selection. Obviously, elements are more likely to be found in sites where insertions have little or no negative effects on fitness than in sites where insertion have deleterious effects. This uneven distribution of elements will be conspicuous in large populations because selection is effective when the population size is large. This distribution is in fact observed in restriction-map surveys of segments of the *Drosophila* genome.

Noting that the selection intensity required to check the spread of elements is very weak, Charlesworth (1985, 1988) and Charlesworth and Langley (1989) suggested that selection is not the only factor involved in stabilizing element frequencies in natural populations. Another possible factor is the regulation of transposition rates; if the rate of transposition is high when the copy number is low, then an element may be able to spread at least to some extent. For example, in the case of P element transposition, events proceed rapidly when a genome has only one or a small number of copies. However, the evidence for transposition regulation is equivocal for other elements in *Drosophila*. Another possible factor is crossing-over between homologous elements located at different chromosomal sites (ectopic exchange). This type of crossing-over can lead to deleterious chromosomal rearrangements and, as its rate increases with the number of elements in the genome, it can retard the spread of an element.

Reviewing theoretical models and data from *Drosophila*, bacteria, and yeast, Charlesworth (1988) and Charlesworth and Langley (1989) concluded that there was nothing strikingly inconsistent with the view that TEs are maintained in populations as a result of transpositional increase in copy number in spite of negative selection. In fact, this view can provide detailed explanations for a range of phenomena associated with TEs. In particular, the fact that elements in bacteria, yeast, and *Drosophila* are usually present at low frequencies at individual chromosomal sites into which they can insert can be convincingly explained by this view, but is almost impossible to be accommodated under the hypothesis that elements persist as a result of favourable mutations associated with their transpositional activities. Under the latter hypothesis, the frequencies would have been considerably higher than observed.

Horizontal Gene Transfer: Mechanisms

Horizontal or *lateral gene transfer* is defined as the transfer of genetic material from one genome to another, specifically between two species. It is distinct from the normal mode of transmission from parents to offspring, which is commonly known as vertical transfer. As is well known, horizontal transfer can occur between an organelle and a nuclear genome or between different organelles in the same species. However, we shall not discuss this latter situation but shall focus on between-species transfers. In this section we shall discuss possible mechanisms of horizontal transfer, and in the next two sections we shall discuss how to detect gene transfer, and some plausible examples of horizontal transfer.

In bacteria, a commonly known mechanism of horizontal transfer is *transformation*, the uptake of DNA by competent cells. A substantial number of bacterial species are naturally competent for transformation and recalcitrant species can be rendered competent by physical or chemical treatments. Transformation consists of the following steps:

1. Release or appearance of DNA in environment.
2. Induction of competent state in recipient host cells.

3. Interaction of cell and DNA.
4. Entry of DNA and processing in cell.
5. Functional integration and expression of entering DNA into cell operations.

Note that, except for the first step, these steps are common to all types of gene transfer.

The other two known mechanisms of gene transfer in bacteria are *conjugation* and *transduction*, which, unlike transformation, may require a vehicle (vector) to transport the genetic information between organisms. In bacteria, two types of vectors are known: plasmid and bacteriophages (i.e., bacterial viruses). Conjugation requires the help of a plasmid. For example, the F fertility factor (plasmid) in *E. coli* can initiate conjugation between a male (F^-) and a female (F^-) bacterium that leads to the replication of the F^+ chromosome and the transfer of a progeny F^+ chromosome to the F^- cell. In this process a piece of host DNA carried by a plasmid can be transferred from one bacterial strain to another. In contrast, transduction occurs via bacteriophages. A phage particle may encapsulate a section of the host DNA, and when the particle attaches to another host cell, the fragment of the bacterial chromosome is injected into the cell. It can then engage in crossing over with the host chromosome, being integrated into the host genome.

Horizontal gene transfers from eukaryotes to prokaryotes may occur by transformation, as is commonly practiced in molecular biology laboratories. Gene transfers from prokaryotes to eukaryotes may occur by conjugation. The best studied case is the transfer of the *T-DNA* (transferred DNA), from the tumor inducing (Ti) plasmid within the bacterium *Agrobacterium tumefaciens* to a plant cell genome. The T-DNA, which contains oncogenic functions for tumor initiation and maintenance, is the cause of crown gall tumors. Another path of transfer is via symbiosis, as is well known in the transfer of organelle genes to the nuclear genome. It is not clear how horizontal gene transfers between eukaryotes can occur except via retroviruses as vectors. Retroviruses are capable of both incorporating chromosomal DNA into their genomes and crossing species boundaries. In fact, there are several examples of endogeneous retroviral sequences being transferred between vertebrates species. In the case of the transfer of the P transposable element between *Drosophila* species, mites were suggested to have been the vector.

Horizontal Gene Transfer: Detection Methods

One may be able to detect a horizontal-transfer event through the discovery of an outstanding discontinuity in the phylogenetic distribution of the gene in question. For example, the bacterium *Salmonella typhimurium* contains a histone-like gene which, as far as we know, has no counterpart in other bacteria. Also, inferring horizontal transfers that have occurred in the recent past is not difficult, especially if the donor species and the selective advantage of the transfer, such as in the case of the transfer of an antibiotic-resistant gene are known. In other situations, however, horizontal gene transfers are difficult to prove, and a case can be made only if it is supported by a rigorous test. Three tests are described below:

First, we consider the situation where the phylogeny of he species from which the gene sequences were obtained is known. For example, suppose that the true phylogenetic relationships among species A, B, and C is as shown in Fig. 9.16a; D is a known outgroup. However, a phylogenetic analysis of the sequences of a gene gives the tree in Fig. 9.16b. This disparity does not necessarily mean that a horizontal transfer has occurred, because factors other than horizontal transfer can cause a discrepancy between the species tree and a gene tree. Therefore, the case is plausible only if the tree in Fig. 9.16b is statistically better than that in Fig. 9.16a. Such a test can be conducted using the methods. For example, using the bootstrap technique, one may examine whether there is statistical support for a horizontal transfer, i.e., whether the bootstrap proportion for the B-C clade is 95% or higher.

Second, the phylogeny of the taxa from which the sequence data were obtained is not known. In this situation, the problem of detecting a horizontal-transfer event becomes more complex. In the past, a claim

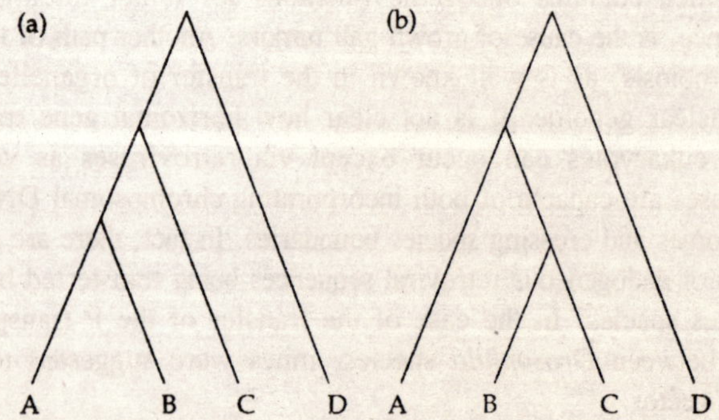

Fig. 9.16. (a) True tree. (b) Inferred tree.

of horizontal transfer was often made when the phylogenies inferred from different genes were different. To make such a claim statistical, Lawrence and Hartl (1992) proposed a method for testing whether the inconsistencies between the taxonomic relationships inferred from two data sets are statistically significant. Suppose that two sets of nucleotide sequences are available from two different genes; the two sets are arbitrarily called the reference data set and the test data set, respectively. First, a matrix of identity coefficient is generated for each data set, in which the i, j-th entry is the proportion of sits that are identical between taxa i and j. To make the matrices commensurate, the magnitudes of the relationships are ranked within each row to create corresponding matrices of ranks. The two matrices of ranks are compared row by row using the Spearman rank-correlation statistic. This statistic ranges from +10, indicating a perfect correlation between ranks, to -1.0, indicating a perfect negative correlation between the ranks. The Spearman statistics for all rows in the matrices of ranks are averaged to yield an overall similarity coefficient. A similarity coefficient less than 1.0 indicates some discrepancy in the taxonomic relationships inferred from the two genes. Second, to assess the statistical significance of a discrepancy, the bootstrap resampling technique is used. Positions from the aligned nucleotide sequences of the reference data set are chosen at random with replacement to create a simulated data set with the same sample size, that is, to create a new test data set. A matrix of ranks is computed. This process is repeated to compile a distribution of similarity coefficients, which is used to determine if the similarity coefficient of the two original data set is significantly smaller than the distribution of similarity coefficients generated from the bootstrapped data sets. A case of horizontal transfer is plausible only if the test is significant. This procedure may also be used to identify the taxon in which the inferred transfer event had occurred, because exclusion of this taxon should eliminate the discrepancy, whereas exclusion of any other taxon would not. In some cases it might make a difference which of the two original data sets is chosen as the reference data set. So, significance should be corroborated by retesting after interchange of the reference and test data sets.

The above two test are based on phylogenetic analysis and several cautions should be made. First, as unequal rates of evolution among lineages may positively mislead phylogenetic analysis, one should make sure that an "out-of-place" lineage in the inferred tree is not due to the unequal-rate effect. Second, one should try to use an extensive data set and also a suitable outgroup to root the tree. Third, one should make sure that no paralogous genes are mistaken as orthologous genes. The importance of such cautions can be seen from the following two examples. One is the case of ponyfish Cu-Zn superoxide dismutase (SOD). Initially, it was claimed that the symbiotic bacterium *Photobacterium leiognathi* acquired the gene from its ponyfish host, but the availability of additional sequence data suggested that the *P. leiognathi* sequence was actually of prokaryotic origin. The second example is that the glyceraldehyde-3-phosphate dehydrogenase gene in *E. coli* was originally suggested to be a eukaryote-to-prokaryote transfer because it was the sole prokaryote in a eukaryotic clade. However, the finding of an orthologous sequence in a distantly related bacterial species, *Anabaena*, suggested that the direction of transfer might have been from an ancestral bacterium to the eukaryotes, rather than the other way around.

Third, Stephens (1985), Sawyer (1989), and Maynard Smith (1992) have developed methods for testing whether a gene consisting of a mosaic of regions was derived from different ancestral genes by either recombination or gene conversion. We discuss the method by Maynard Smith because it allows an determination of the crossover points. The method considers three situations of the availability of sequence data: (1) two parental sequences and a derived sequence, (2) a set of sequences with no information about ancestry, and (3) just two sequences.

The method, called the maximum chi-squared (MCS) method, will be explained by reference to Fig. 9.17. Suppose that the DNA region under study contains n polymorphic sites, sites that show variation among sequences. Let us compare sequences 1 and 3, which differ at s sites. Suppose that for an arbitrary cut after the *k*th site, the two sequences differ at r sites. Then the proportion of site differences between the two sequences is r/k before the cut and (s - r)/(n - k) after the cut. The value, c, of the $2 \times 2\chi^2$ is calculated. Among all the possible cuts, find the site, k_{max}, for which the value of c i maximum, c_{max}. If the difference between sequence 1 and 3 has a mosaic structure with only two blocks, then optimal boundary between the blocks occurs after site k_{max}.

To test whether this division is significant, construct T sequence pairs of length n, differing at s randomly distributed sites, and for each trial pair find c_{max}. If the value of c_{max} for the real data is greater than the value for every trial pair, then the observed mosaic structure is significant at the level $P < 1/T$.

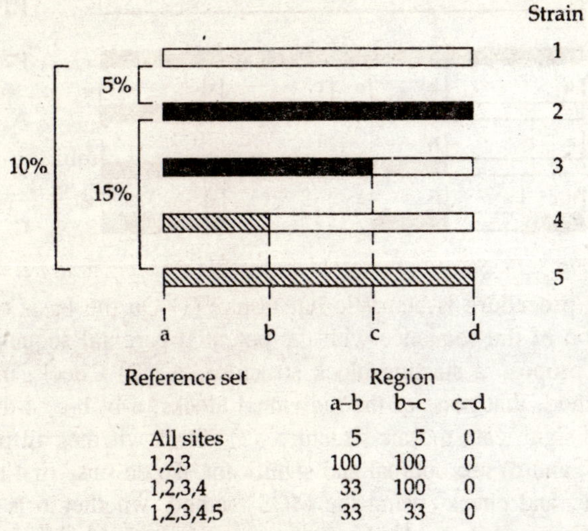

Reference set	Region a–b	Region b–c	Region c–d
All sites	5	5	0
1,2,3	100	100	0
1,2,3,4	33	100	0
1,2,3,4,5	33	33	0

Fig. 9.17. Schematic presentation of sequence comparison.

One difficulty in using this method is in making an appropriate choice of polymorphic sites. Suppose that we are comparing sequences 1 and 3, but we do not have sequence 5. If we treat as polymorphic any site that varies in the set of sequences 1, 2, 3, and 4, we will be in danger of identifying a false division at site b, because in this case the divergence at polymorphic sites between sequences 1 and 3 is only 33% in the a-b block but 100% in the b-c block. This danger may be avoided if we consider all sites, polymorphic or not. There are, however, two drawbacks to this: the computations take much longer and the test becomes less powerful. The most satisfactory procedure is to consider only the two parental and the derived sequences (1, 2, and 3) and to treat as polymorphic only those sites that vary within this set of three. However, if we do not know the ancestry, we should consider all sequences, and if only two sequences are available, we should consider all sites. Note also that in determining the block structure of sequence 3, it can be compared either with parent 1 or parent 2. These two comparisons need not give the same answer. If only one of them is statistically significant, one may accept that division as being optimal. If both comparisons are significant but the cuts identified are different, then an ambiguity occurs, but one may assume that the true cut is somewhere in between.

In the above discussion we assumed that the region consists of only two blocks. In principle, the method can be extended to more than two blocks, but the computation time may be prohibitive. The

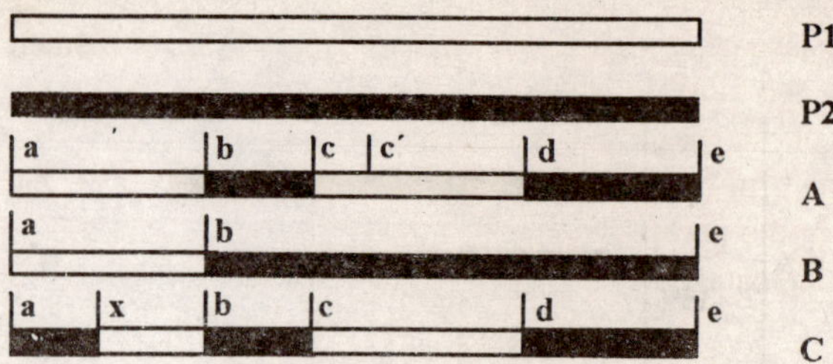

Fig. 9.18. Successive stages in the analysis of mosaic structure.

following procedure is heuristic but faster. (1) On the basis of visual comparison of the sequence with the potential parental sequences (P1 and P2), propose a starting block structure, A. (2) Check, using the MCS method, that none of the individual blocks, a-b, b-c, c-d, and d-e, contain significant mosaic structure. (3) Check whether all proposed cuts (b, c, and d) are optimal and significant. To do this, first consider region a-c, and check, using the MCS method, whether b is optimal and significant. If it is, check whether cut c is optimal and significant for region b-d. It is, check whether cut d is optimal and significant for region c-e. Suppose that one of the cuts proves not to be optimal. For example, suppose that cut c' instead of c is optimal for region b-d. If so, replace c by c' and start the procedure again at the beginning. This process of successive adjustment can be continued until a starting structure is found for which all the proposed cuts are both optimal and significant. (4) Suppose that one of the proposed cuts proves not to be statistically significant. For example, suppose that b is optimal and significant, but that no significant cut exists in region b-d. Then the proposed starting structure A must be abandoned, and replaced by B.

Note that step 2, checking that the proposed individual blocks do not themselves contain mosaic structure, is important. Thus, suppose that C is the true structure, and that cuts x, b, c and d are optimal and significant. If the existence of cut x was overlooked, and A was taken as the starting structure, then calculations on region a-c might fail to find a significant cut. This failure in turn can lead to the structure of the rest of the sequence being wrongly identified or missed altogether.

Horizontal Gene Transfer: Examples

Many cases of horizontal transfer have been proposed. We present two examples below.

Evolution of Penicillin Resistance in Neisseria Species

Penicillin-resistant strains of *Neisseria gonorrhoeae* have become increasingly prevalent during the last two decades. In many of these isolates, resistance is due to the acquisition of a β-lactamase, which inactivates the antibiotic In other isolates, resistance is due to alterations of the chromosomal genes encoding two high-molecular-weight penicillin-binding proteins (PBPs 1 and 2), combined with reductions in the permeability of the outer membrane. These PBPs are enzymes that catalyze the final stages of peptidoglycan (cell wall) biosynthesis: in penicillin-susceptible strains, penicillin binds to these PBPs and inactivates them leading to impairment of cell wall synthesis and cell death. In resistant strains, the affinities of these PBPs for penicillin are reduced, so that higher concentrations of the antibiotic are required for their inactivation.

Isolates of the other pathogenic member of the genus *Neisseria*, *N. meningitidis*, that have increased levels of resistance to penicillin have been reported in the last few years. These low-level penicillin-resistant isolates have reductions in the affinity of PBP 2, but not of PBP 1. Penicillin resistance due to alterations of PBP 2 has also emerged recently in some of the commensal *Neisseria* species.

Using restriction-enzyme analysis, Spratt et al. (1989) found that the PBP 2 genes (penA) of penicillin-susceptible strains of *N. meningitidis* wee rather uniform, whereas the penA genes of 42 penicillin-resistant isolates were heterogenous and all had very different patterns of restriction fragments from those of the susceptible strains. These results suggested that the pen A genes of penicillin-resistant isolates of *N. meningitidis* are very different in nucleotide sequence from those of susceptible isolates. The authors obtained the sequence of penA from a resistant strain of *N. meningitidis* and compared it with those from susceptible isolates. They found that the penA gene of the resistant isolate has acquired blocks of DNA from the penA genes of a closely related commensal *Nesseria* species, *N. flavescens*. *Nesseria* is naturally competent for transformation, so it is plausible that these blocks of DNA have been acquired by this recombinational mechanism. Similar mosaic pen A genes have also been found in penicillin-resistant isolates of *N. gonorrhoeae* and *N. lactamica*.

N. flavescens was identified as the donor of the blocks of DNA in the penA genes of the above resistant isolates. *N. flavescens* isolates, including those from the preantibiotic era, fortuitously produce a PBP 2 that has a lower affinity for penicillin than PBP 2 of *N. meningitidis*, *N. gonorrhoeae*, and *N. lactamica*. Apparently, genetic transformation has provided a mechanism by which the latter *Neisseria* species can obtain increased levels of resistance to penicillin by acquiring the PBP 2 genes from *N. flavescens*.

In a more detailed study, Spratt et al. (1992) obtained the sequences of the penA genes from 23 *Neisseria* isolates and found complex mosaic structures. The penA genes of susceptible strains of *N. meningitidis* were uniform: the two most dissimilar strains differed by only 0.5%. These DNA sequences are not distinguished and are denoted as NmS1; the *N. flavescens* sequence (strain U) is denoted as Nf1.

Using the maximum chi-squared method described in the preceding section with a very stringent criterion, i.e., P=0.001, Spratt et al. (1992) found that the penA genes of the resistant *N. meningitidis* strains B, C, D, E and F have acquired blocks of DNA from isolates that are very similar to the type strain of the commensal species *N. flavescens*. For example, strain B has acquired two blocks from *N. flavescens*. Interestingly, strains C and D, although isolated from different countries, have mosaic penA genes that are identical in sequence.

The penA gene of the susceptible *N. lactamica* strain (Q) has a mosaic structure. The region between sites 870 and 1254 differs from susceptible *N. meningitidis* by 58/385 (15.1%) nucleotides, whereas the two species differ by only 26/992 (2.6%) in the regions before and after this central block. The presence of this block in the penA gene of all three of the *N. lactamica* isolates that were sequenced (Q, R, and S), and in the closely related species *N. polysaccharea* (strain T), suggests that it represents an ancient recombinational event that occurred in the common ancestor of three strains.

The two penicillin-resistant *N. lactamica* strains (R and S) have acquired a block of DNA from *N. flavescens*. The penA genes have identical block structures, but they differ in sequence at 26 sites. The penA gene of the type strain of *N. polysaccharea* (T) closely resembles that of the susceptible *N. lactamica* (Q), but has acquired blocks of DNA from *N. flavescens*. In recent years, isolates of *N. polysaccharea* with increased levels of resistance to penicillin have become common in some countries.

Spratt et al. (1992) also gave examples of mosaic penA gene structures in *N. gonorrhoeae* and *N. cinerea*. In conclusion, it is clear that horizontal gene transfer involving the penA gene has occurred many times in *Neisseria* species.

Horizontal Transfer of P Transposable Elements

Another example of horizontal gene transfer involves the P elements in *Drosophila melanogaster*. As mentioned previously, P elements have rapidly spread through natural populations of *D. melanogaster* within

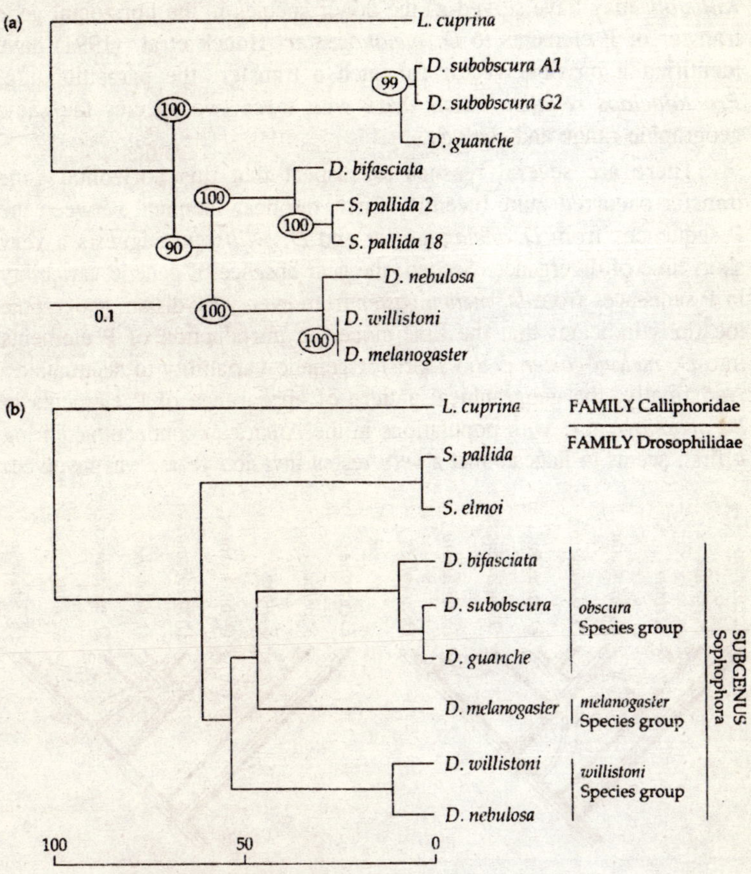

Fig. 9.19. (a) Phylogeny of P-element nucleotide sequences inferred by the neighbour-joining method. (b) Phylogeny of the nine Diptera species established from molecular, morphological and fossil data.

the last four decades. P elements do not exist in closely related species of the *melanogaster* group, such as *D. mauritiana*, *D. sechellia*, *D. simulans*, and *D. yakuba*. Where then did these elements come from? An extensive survey of hundreds of *Drosophila* species by Daniels et al. (1990) has shown that, with the exception of *D. melanogaster*, P sequences are not found in any other species of the *melanogaster* subgroup. In contrast, all species of the distantly related *willistoni* and *saltans* groups contain P elements. In particular, the P element from *D. willistoni* was found to be identical to the one in *D. melanogaster* with the exception of a single base substitution, indicating that *D.*

willistoni may have served as the donor species in the horizontal gene transfer of P elements to *D. melanogaster*. Houck et al. (1991) have identified a possible vector for such a transfer: the parasitic mite, *Proctolaelaps regalis*, which share with these two species the same geographic range and ecology.

There are several reasons to suspect that this horizontal gene transfer occurred quite recently. First, the near identical between the P sequences from *D. melanogaster* and *D. willistoni* suggests a very short time of divergence. Second, the near absence of genetic variability in P sequences from *D. melanogaster* from even very distant geographic locations indicates that the time since the introduction of P elements into *D. melanogaster* is too short for genetic variability to accumulate. And finally, the geographical pattern of appearance of P elements in *D. melanogaster*, with populations in the American continent acquiring it first, seems to indicate that a very recent invasion years, was involved.

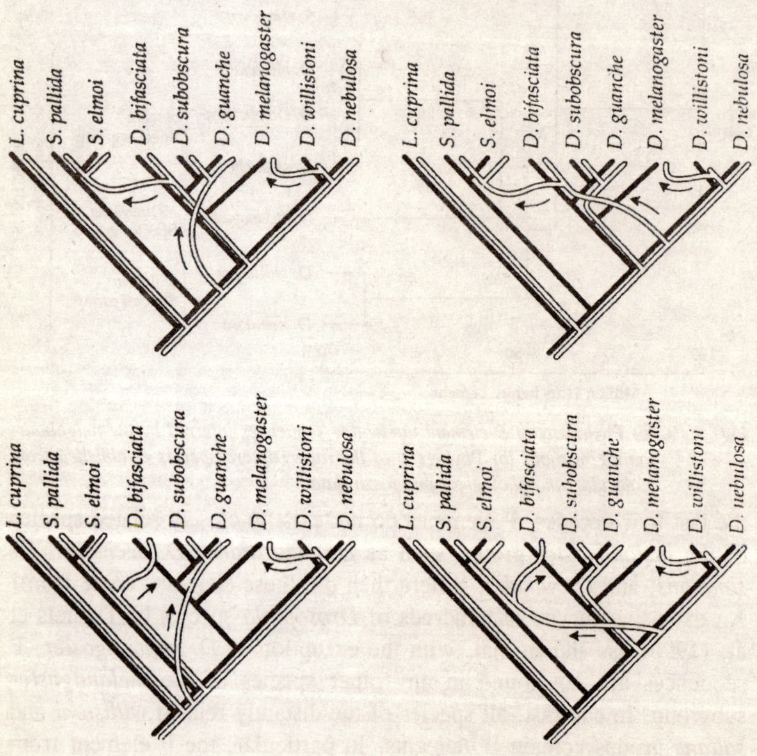

Fig. 9.20. Four possible scenarios, involving horizontal transfer, to explain the

Clark et al. (1994) have recently conducted a phylogenetic analysis of the P-element nucleotide sequences from nine species of *Drosophila* and other Diptera. The neighbour-joining tree obtained is well supported by bootstrap resampling and the tree is also supported by both parsimony and maximum-likelihood analyses. However, it is incongruent with the phylogeny of the nine species reliably inferred from molecular, morphological, and fossil data. A minimum of three horizontal transfer events would be required to explain the discrepancy between the two trees. In each, there is both a transfer between *D. willistoni* and *D. melanogaster* and an independent transfer between the ancestors of *Scaptomyza* and *D. bifasciata*. A third transfer, which occurs deeper in the phylogeny, is of ambiguous position. With the exception of the *D. willistoni—D. melanogaster* transfer, which has apparently occurred relatively recently, it is difficult to unequivocally identify when, and in what direction, any of the other horizontal transfers occurred. For example, solely on the basis of the comparisons presented here, it is not possible to determine whether there was a transfer from *D. bifasciata* to *Scaptomyza* or, as proposed by Hagemann et al. (1992), from *Scaptomyza* to *D. bifasciata*. Note that other explanations of the discrepancy between a and b are possible but less plausible.

10

Selection in Molecular Evolution

In this final chapter we come back to the long-standing controversy on the relative roles of mutation and selection in molecular evolution. Mutation here is taken in the broad sense, including not just point mutation but also genetic changes caused by other mechanisms such as insertion, deletion, duplication, and transposition. Also, it should be emphasized that we are concerned with molecular evolution, not the evolution of form, function, or fitness, where presumably selection reigns. The presentation will not doubt be biased by my own view of the issue, but I shall provide data to support my argument.

As described already, the essence of the neutralist-selectionist controversy is whether mutation or selection is the major driving force of molecular evolution. One approach to this question is to see whether the upper limit of the rate of evolution is determined by mutation or selection. Considerable insight might also be gained by considering the relative roles of mutation and selection in the emergence of new functions, in the evolution of genetic code, and in the switch in the most preferred synonymous codons between species. It will also be interesting to see whether directional mutation pressure changes the amino acid composition of proteins. At the end of the chapter we shall also examine whether neo-Darwinism is adequate for explaining adaptive evolution at the molecular level. That is, we shall examine whether the rate of molecular evolution is determined by the rate of environmental changes as postulated by neo-Darwinism, even if we consider only the class of advantageous evolutionary changes.

Is Sequence Evolution Mutation or Selection Driven?

The selectionist view of evolution, which is essentially the same as the neo-Darwinism, contains the following two key elements: First, natural populations contain a sufficient amount of genetic variability to respond to almost any kind of selection. Second, evolution is determined mainly by environmental changes and natural selection. Since there is enough genetic variability, no new mutations are required for a population to respond to an environmental change. Therefore, there is no relationship between the rate of mutation and the rate of evolutionary change. In this view, natural selection is the driving factor of evolution. In contrast, neutralists believe that mutation plays a more important role in molecular evolution than does selection, so that the rate of molecular evolution is mainly determined by the rate of mutation. In other words, mutation is the driving factor in the rate of molecular evolution.

As the two view give very different predictions for the relationship between the rate of mutation and the rate of evolution, we can test the predictions using molecular data. Under the neutralist view, the rate of nucleotide substitution should be higher for an organism with a higher mutation rate than for an organism with a lower mutation rate. In contrast, no such relationship exists under the selectionist view. Note that the controversy is concerned mainly with nucleotide changes that cause amino acid changes (i.e., nonsynonymous substitutions), not with synonymous or silent nucleotide changes. Also, the issue here is the average rate of nonsynonymous substitution in the genome of an organism, not the rate in a single gene or certain codons.

The effect of neutral selection on the nonsynonymous rate can be inferred by the variation in nonsynonymous rate among genes in the same genome. This variation is likely to be largely due to the effect of natural selection because the differences in mutation rate among regions in a genome should be relatively small. The highest nonsynonymous rate is less than 1000 times the lowest one in the same genome. As explain already, the variation in nonsynonymous rate among genes in a genome is mainly due to differences in functional constraints among genes, that is, mainly due to differences in the intensity of negative selection among genes. Therefore, the variation in nonsynonymous rate that can be attributed to differences in the frequency or intensity of positive selection among genes is likely to be much smaller than 1000 times. On the other hand, the effect of mutation on the nonsynonymous rate can be inferred from comparisons between different organisms. Here we use the synonymous rate as a rough estimate of the mutation rate in an organism, though the synonymous rate maybe significantly reduced by codon usage constraints. It is seen that the synonymous rate and the nonsynonymous rate increase approximately in parallel, by about one million times, from animals to retroviruses. One might argue that the nonsynonymous rate in retroviruses is very high because natural selection is very effective in an organism with rapid generations. However, the increase in the nonsynonymous rate from animals to retroviruses is only of the same order of magnitude as the increase in the synonymous rate, and so both increases can be largely attributed to the increase in mutation rate. At any rate, the data clearly show a strong positive correlation between the rate of mutation and the rate of nonsynonymous substitution. This correlation is predicted by the neutralist view but not by the selectionist view.

Since the nonsynonymous rate in an organism is low when the mutation rate is low and is high when the mutation rate is high, the upper limit of the nonsynonymous rate is determined by the rate of mutation. Because negative selection is much more prevalent than positive selection, the presence of selection actually reduces rather than elevates the upper limit of the rate of evolution. In fact, the highest nonsynonymous rate in a genome is still lower than that of the synonymous rate, which is likely to be lower than the mutation rate. Therefore, it appears that mutation is the major driving force in molecular evolution, including protein sequence evolution.

Mutation Pressure versus Amino Acid Composition

As mentioned above, selectionist believe that evolution occurs by environmental changes and natural selection. Under this view, the amino acid composition of a protein is determined by natural selection and is not or is little affected by the pattern of mutation. In contrast, under the neutralist view the amino acid composition of a protein may be affected by the pattern of mutation because amino acid changes in evolution are mainly caused by mutation. It is well known that many bacterial species have extreme nucleotide compositions (GC contents i.e., G+C%), which are thought to have arisen from biased mutation pressures. These observations can be used to test the predictions of the two views. The neutralist view predicts that directional mutation pressure will affect the amino acid composition of proteins, whereas the selectionist view predicts no such effect.

Support for the neutralist prediction actually already existed in Sueoka's 1961 finding that the levels of alanine and glycine tend to increase but those of isoleucine and tyrosine tend to decrease with the genomic GC content of the bacterial species studied. Because alanine (codons GCU, GCC, GCA, GCG) and glycine (GGU, GGC, GGA, GGG) are encoded by GC-rich codons, whereas isoleucine (AUU, AUC, AUA) and tyrosine (UAU, UAC) are encoded by GC-poor codons, Sueoka's finding suggested that the amino acid composition of proteins in a bacterium is affected by its genomic GC content. This suggestion was supported by Jukes and Bhushan's (1986) study of the trpA and trpB proteins and Osawa et al.'s (1992) study of eight ribosomal proteins in bacteria.

X. Gu, D. Hewett-Emmett, and W.-H. Li have used extensive DNA sequence data from bacteria to test the neutralist prediction. The 20 amino acids are classified into three groups according to the GC content of their codons. Group A consists of proline, alanine, glycine, and tryptophan, which are encoded by codons with a high GC content; e.g., alanine is encoded by GCU, GCC, GCA, or GCG. Group B consists of valine, threonine, histidine, glutamic acid, aspartic acid, glutamine cysteine, and serine, which are encoded by codons with an intermediate GC content; e.g., aspartic acid is encoded by either GAU or GAC. Group C consists of phenylalanine, tyrosine, asparagine, lysine, isoleucine, and methionine, which are encoded by codons with a high AT (AU) content; e.g., phenylalanine is encoded by either UUU or UUC. Note that arginine and leucine are not included in these groups because each of them is encoded by two families of codons that belong to two different groups. Denote the frequencies of groups A, B, and C in a protein by f_A, f_B, and f_C, respectively. Under the neutralist hypothesis, f_A should increase whereas f_C should decrease with increasing GC content of the genome, while f_B would not be much affected by changes in GC content.

Fig. 10.1 shows the result for the dnaA gene. In this figure GC_4 denotes the GC content at the fourfold degenerate sites of the gene and is used as a measure of the intensity of the GC mutational pressure. Clearly, f_A increases whereas f_C decreases with increasing GC_4 though f_B is little affected by change in GC_4. The results are: (1) f_A is positively correlated with GC_4; the correlation coefficient r ranges from 0.66 to 0.93, (2) f_C is negatively correlated with GC_4; r ranges from –0.76 to –0.97, and (3) the correlation between f_B and GC_4 is weak—11 of the 15 genes show no significance correlation. All these results are as predicted by the neutralist hypothesis.

The observed effects of mutation pressure on the amino acid composition of proteins are not predicted by the selectionist view.

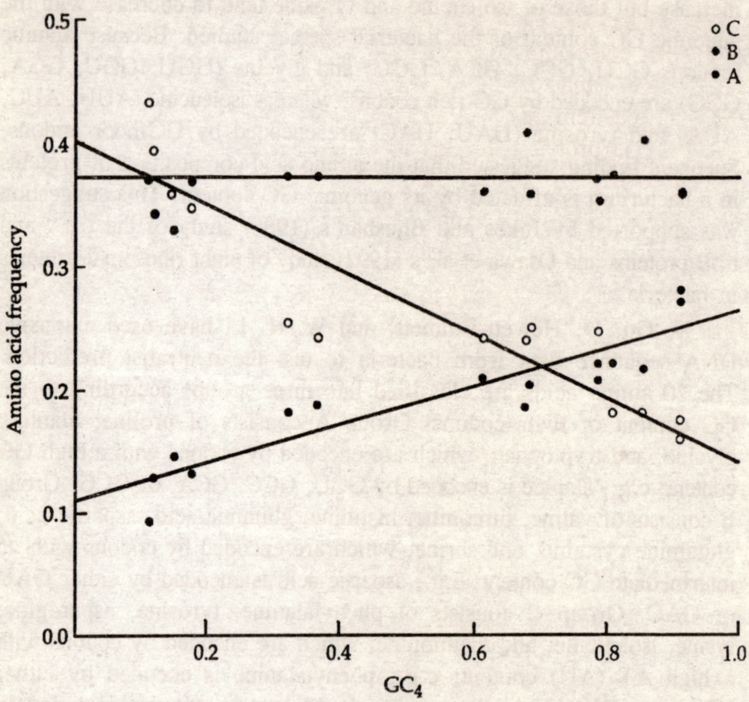

Fig. 10.1. Correlation between amino acid frequencies and the GC content (GC_4) at the fourfold degenerate sites of the gene (dnaA) in bacteria.

Under this view one will have to assume that amino acids encoded by GC-rich codons are advantageous in species with a high GC genome, whereas amino acids encoded by GC-poor codons are favoured in species with a GC-poor genome. It is doubtful that such a nice correlation occurs in so many genes in so many species. In contrast, these observations can be easily explained under the neutralist view by assuming that either most of these amino acid changes are nearly neutral or their selective disadvantages are so small that substitutions can occur under strong mutation pressure. The effect of mutation pressure on the amino acid composition is expected to be stronger on proteins that are not subject to strong functional constraints (and thus can evolve relatively fast) than on proteins that are subject to strong functional constraints. Indeed, the correlation coefficient between f_A (or f_C) and GC_4 is, on average, higher for the faster evolving proteins than for the more slowly evolving proteins. For example, dnaN has evolved at a relatively high rate and shows a strong correlation between f_A and GC_4 ($r=0.90$), whereas hsp70 is highly conservative and shows a much weaker correlation between f_A and GC_4 (r=0.66). The reason for a weak correlation in highly conservative proteins is that in such proteins few amino acid substitutions would be allowed because of functional requirements. Viewed from this perspective, it is remarkable that all the 15 proteins studied show a significant correlation between f_A (f_C) and GC_4.

Switches in Species-specific Codon Preferences

We noted that the usage of synonymous codons is highly nonrandom in most of the species studied to date. We further noted that the preferred (i.e., the most frequently used) codon for an amino acid may differ among species. For example, the preferred codon for leucine (Leu) is CUG in *Escherichia coli* but UUG in the yeast *Saccharomyces cerevisiae*. Therefore, a switch in codon preference must have occurred either in the *E. coli* lineage or in the *S. cerevisiae* lineage since their separation. Many such switches appear to have occurred in evolution. The question is then, What causes such switches?

One possible explanation is as follows. We note that in an organism the preferred codon for an amino acid is usually recognized by the tRNA species that is the most abundant among the cognate tRNAs for that amino acid. For example, the preferred codon and the most abundant cognate tRNA species for Leu are CUG and $tRNA_1^{Leu}$ in *E. coli*, but are UUG and $tRNA_3^{Leu}$ in *S. cerevisiae*. Thus, it seems that species-specific codon preferences are determined by tRNA abundances, as a result of selection for translational efficiency and accuracy. So, intuitively a simple explanation for a witch in codon preference is that it is caused by a switch in the abundances of tRNAs. For example, one may assume that the switch in the preferred Leu codon from CUG in *E. coli* to UUG in *S. cerevisiae* was caused by the switch in the most abundant cognate tRNA species from $tRNA_1^{Leu}$ in *E. coli* to $tRNA_3^{Leu}$ in *S. cerevisiae*. Such a switch in tRNA abundance, however, has the following problems: Since the codon UUG was originally rare, a large increase in the expression level of $tRNA_3^{Leu}$ would be a waste of energy and disadvantageous, and since the codon CUG was originally abundant, a large reduction in the expression level of $tRNA_1^{Leu}$ would be highly disadvantageous. Of course, a gradual switch may be possible. For example, a small elevation in the expression level of $tRNA_3^{Leu}$ may be nearly neutral. This evolution would allow an increase in the frequency of UUG, a reduction in the frequency of CUG and a reduction in the expression level of $tRNA_1^{Leu}$. Repetition of such a gradual shift could eventually lead to a complete switch. Note, however, until the codon UUG became abundant, there would be no selective advantage for the proposed shifts to occur. So, it is doubtful that a switch in tRNA abundance is the major cause of switches in codon preference.

A more likely cause for a codon preference switch is a change in mutation pressure that exerts a strong pressure for change in codon frequencies. We note that changes in codon frequencies may exert selection pressure on tRNA abundance. So, suppose that there was greater mutation pressure toward a higher AT content in the *S. cerevisiae* lineage than in the *E. coli* lineage. Then the codon CUG would have a tendency to change to the codon UUG and there would be a selective advantage for $tRNA_3^{Leu}$ to become abundant and a selective advantage for $tRNA_3^{Leu}$ to become less abundant. (There would have been a pressure for UUG to change to UUU, but UUU codes for phenylalanine, not leucine.) This explanation provides a simple scenario for the switch in the preferred leucine codon from CUG in *E. coli* to UUG in *S. cerevisiae*. In this respect, it is interesting to note that the AT content of the *S. cerevisiae* genome (62%) is indeed higher than that of the *E. coli* genome (50%). The following observations also support the importance of mutation pressure as a cause of switches in codon preference. *Micrococcus luteus*, which has a genomic GC content of 74%, uses almost exclusively CUC and CUG for leucine, whereas *Mycoplasma capricolm*, which has a genomic GC content of only 25%, prefers UUA. For arginine, *Micrococcus luteus* prefers CGC and CGG, whereas *Mycoplasma capricolm* prefers AGA.

In both of the above schemes the switch in codon preference is initiated by mutation, not by selection, though selection may indeed play a role at a later stage. Note that in the process of a switch numerous synonymous substitutions would have occurred because there are many genes in a genome. Indeed, a change in mutation pressure may cause switches in many codon families and, as a consequence, a very large number of synonymous substitutions.

Emergence of Nonuniversal Genetic Codes

Almost all living organisms, from bacteria to man, use the universal genetic code. A plausible explanation for this striking phenomenon is the frozen accident theory, which states that the code is universal because any change would alter the amino acid sequences of many highly evolved proteins and would therefore be highly detrimental to the organism. The persistence of the universal code in millions of lines of life is a good example of the view that natural selection often inhibits further evolution of a system or the emergence of novelty. However, the code is not really frozen and a number of exceptions to the universal code have been found. The question here is how a change in the code may evolve. The best answer to this question seems to be the codon capture hypothesis, which postulates that a codon may disappear during evolution from the coding sequences of the genome of an organism or organelle, and then may later reappear with a different assignment. We consider two examples below to show how this hypothesis may explain the emergence of a new code.

The first example is that *Mycoplasma capricolum* uses UGA (a stop codon) as well as UGG for tryptophan. *Mycoplasma* DNA is very high in AT (about 75%), indicating the existence of directional mutation pressure from GC to AT. Under this pressure, all UGA stop codons mutated to UAA and thus disappeared from the genome; a rare codon may also become lost by random drift. UGA then became an unassigned codon. (This situation caused no harm because UGA was absent from the genome.) The gene for tryptophan tRNA with anticodon CCA duplicated, and one of the duplicates became UCA, which pairs with regular tryptophan codon UGG by wobble pairing between G and U. Then some UGG tryptophan codons mutated to UGA, which pairs with anticodon UCA. The result is that UGA became "captured" by tryptophan, no longer being used as a stop codon in *M. capricolum*. Various experimental findings substantiate this evolutionary scenario. First, the genes in *M. capricolum* for two tryptophan tRNAs, one with anticodon CCA and one with UCA, are arranged in tandem, close together on the chromosome. This suggests gene duplication followed by a mutation of anticodon CCA to UCA. Next, tRNA with anticodon CCA in *Mycoplasma*, which is no longer needed, appears to be disappearing. Yamao et al. (1988) reported failure to find tRNA (CCA) appreciably changed in vivo with tryptophan, and Inamine et al. (1990) found that the gene for tRNA Trp (CCA) has disappeared in some of the mycoplasma species such as *M. pneumoniae* and *M. genitalium*. It seems probable that the change from tryptophan coding from UGG to UGA in *Mycoplasma* is an example of "evolution in action."

The second example is that AAA codes for asparagine instead of lysine in mitochondrial DNA of some animals such as platyhelminth (*Fasciola hepatica*) and sea urchin (*Strongylocentrotus purpuratus*). Normally, lysine is coded by AAA with anticodon UUU and by AAG with anticodon CUU. GC pressure would change all AAA codons to AAG. Anticodon UUU, which pairs with AAA, is thiolated, so that it does not pair strongly with AAG, and it would tend to disappear under GC pressure, which was assumed to be exerted. Anticodon that have been converted by evolution to INN in eukaryotes. I in anticodon INN pairs with U, C, and A in third positions of codons. AAA codons, reappeared by mutations of AAU and AAC (asparagine), would pair with IUU and code for Asn. Note that although some codons have changed, the amino acid sequence remains unchanged.

In the above two examples we assumed that the disappearance of a codon is due to mutation pressure. Note, however, that a codon may also disappear because of selection against the codon (with or without the assistance of random drift). So, loss of codon can be initiated either by mutation or selection. However, almost all the other steps in the evolution of a new code such as the loss of a tRNA gene from the genome, the duplication of a tRNA gene, mutation of an anticodon, and the reappearance of a codon are all due to mutation. Moreover, since the amino acid sequence remained unchanged during the process of the reassignment of a codon, no natural selection was involved in the evolution of the DNA sequence. Therefore, one may conclude that mutation plays a more important role in the emergence of nonuniversal genetic code than does selection.

Emergence of New Functions

The emergence of a new function in a DNA or protein sequence is supposedly advantageous and is commonly believed to have occurred by advantageous mutations. However, acquiring a new function may require many mutational steps, and a point that needs emphasis is that the early steps might have been selectively neutral because the new function might not be manifested until a certain number of steps had already occurred.

Let us consider the emergence of the regulation of hemoglobin function by 2,3-diphosphoglycerate (DPG or 2,3-DPG). This is a relatively simple case, so that it would be easier to infer the likely steps and mechanisms involved. DPG is the most abundant organic phosphate in the red cells of most mammals. It can assist hemoglobin in unloading its oxygen to the tissues, because when it binds to hemoglobin, oxygen is released from hemoglobin. DPG binds between the ends of the two β chains of hemoglobin: The negative charges of DPG interact with the positive charges on the side chains of the histidine at position 2 (His2), the lysine at position 82 (Lys82), and the histidine at position 143 (His143), as well as the amino termini of the β chains. Therefore, the critical residues for binding DPG are

His2, Lys82, and His143 of the β chain; the protonated N-terminus is also important but less critical. For example, cat hemoglobin B does not respond to DPG because the second residue of the β chain is phenylalanine rather than histidine and because its N-terminal serine is blocked by an acetyl group. In contrast, glycine of the β chain compensates to some extent for the absence of histidine at position 2. The failure of ruminant hemoglobins to respond to DPG is due to the deletion of the second residue of the a chain; because of this deletion, the N termini of the β chains in the native deoxygenated hemoglobin apparently become too far apart to bind DPG. Human fetal globin ($\alpha_2\gamma_2$) has only a weak response to DPG because His 143 is replaced by serine in the γ chains.

DPG binding of hemoglobin apparently evolved after the divergence between the α and β globins because neither the α globin chain nor myoglobin binds to DPG. The second residue of the β globin is evidently an insertion that occurred after the divergence between the α- and β-globin genes because both the α chain and myoglobin lack this residue. A comparison of the β chain with the α chain and myoglobin suggests that Lys82 and His143 of the β chain also evolved after the α-β globin divergence. One possible scenario for the evolution of these three critical residues is that Lys82 and His143 evolved before His2. In this case the first two changes would have no selective advantage because before the insertion of His2 the β chain would be equivalent to the situation in ruminant β globin, which has no DPG binding affinity. (Of course, one cannot rule out the possibility that either Lys82 or His 143 or both have evolved because of a selective advantage other than DPG binding.) Another possible scenario is that the insertion at the second position occurred before Lys82 and His 143 evolved. This insertion alone would have no DPG binding affinity and thus would have no selective advantage, even if the amino acid inserted was histidine. In short, regardless of the path via which the evolution has occurred, some early steps might have been selectively neutral.

Conclusions

It is now well recognized that the evolution of noncoding regions, especially non functional sequences, is mainly determined by mutation pressure. For example, mutation pressure provides a simple explanation for why the genome of vertebrates is AT-rich—it is due to the facts that in vertebrates G and C nucleotides are, on average, more mutable than A and T nucleotides and that the vertebrate genome is made up mostly of noncoding sequences. Although the origin of GC-rich isochores in the genome of warm-blooded vertebrates remains controversial, a simple explanation is that it arose from biased local mutation pressure.

Evidence was given above that mutation pressure is the major cause for switches in species-specific codon preference. The conclusion was that when directional mutation pressure is strong, it can force not only a switch in codon preference but also a switch in tRNA abundance, so that a new correspondence between tRNA abundance and codon usage is achieved.

The genetic code was thought to be frozen and invariable, but new codes have been found in some organisms and organelles. Yet, what was also surprising is that these seemingly impossible changes were apparently mainly caused by directional mutation pressure. The significance of mutation is difficult to appreciate because the rate of mutation is low, so that mutation cannot produce any significant change in a short time period. However, because it is ever-present, mutation pressure can produce unexpected evolutionary changes when the time sale is millions or billions of years.

The neutralist-selectionist controversy, however, pertains mainly to protein evolution, and a highly contentious point is whether mutation or selection is the driving factor of the rate of protein evolution. As discussed above, a contrast between the substitution rates in animals and viruses strongly suggested that mutation is the major determinant. It was concluded that the higher the mutation rate is, the higher the nonsynonymous rate will be. This conclusion and the fact that directional mutation pressure causes biased amino acid composition in proteins suggest that mutation rather than selection is the major driving force in protein evolution. This conclusion does not deny the existence of a advantageous substitutions but implies that such substitutions are less frequent than neutral or nearly neutral substitutions in protein evolution.

Given the above conclusion, one might wonder, What role does selection play in evolution? There are two types of selection, negative (purifying) and positive selection. Negative selection acts like a policeman. It eliminates disadvantageous mutants or genotypes from the population. As the vast majority of nonneutral mutations are deleterious, negative selection is the prevailing type of selection. For this reason, the presence of selection tends to slow down the rate of evolution, and this explains why in almost all genes the nonsynonymous rate is lower than the synonymous rate and the rates in nonfunctional sequences such as pseudogenes. The power of purifying selection is best seen in highly conserved proteins such as histones and ubiquitins, which have undergone virtually no amino acid substitution in the past billion years of evolution. In such proteins, negative selection has obviously greatly reduced the rate of evolution.

Positive selection is powerful in comparison with random drift. It can greatly increase the probability for an advantageous mutation to increase its frequency and become fixed in the population. For example, the probability of fixation of a mutations is $1/(2N)=0.00005$ for a neutral mutation in a population of $N=10,000$, whereas it is approximately $2s=0.002$ for an advantageous mutation with $s=0.001$. In this case, positive selection increases the probability of fixation by fortyfold! Positive selection also occurs when disadvantageous because of environmental change. Such switches in selective advantage as well as advantageous mutations can help a population cope with environmental changes.

Positive selection is very important for the evolution of a new function. Although it was stressed in the previous discussion that some early mutational steps in the evolution of a new function may be selectively neutral, sooner or later an advantageous mutation must occur, otherwise some already-established steps may become lost by mutation and random drift, and the process would be disrupted. When an advantageous mutation occurs, positive selection helps the new mutation to become established in the population. The emerging system is much less likely to be disrupted once an advantageous mutation is incorporated into it. The system will become even more advantageous, and positive selection becomes even stronger if another favourable mutational step occurs. Thus, positive selection may play a key role at a certain stage of the evolution of a new function.

Positive selection may have played an especially important role in evolution of duplicate genes. As discussed already, there are two evolutionary paths for a redundant duplicate—either it fixes a null mutation and becomes a pseudogene, or it fixes an advantageous mutation, so that it has a better chance of evolving a new function. The ratio (ρ) of the rate of advantageous mutation to that of null mutation is expected to be very low, so that the second path is unlikely to occur without the help of positive selection. Indeed, Walsh (1995) showed that the probability for the second path is close to ρ if the selective advantage (s) for the mutation is small, so that $S=4N_e s$ is considerably smaller than 1. In contrast, this probability is increased to ρS if $S>>1$ (but $\rho S<<1$). For example, if ρ is 0.001, then the probability of the fixation of an advantageous mutation before the fixation

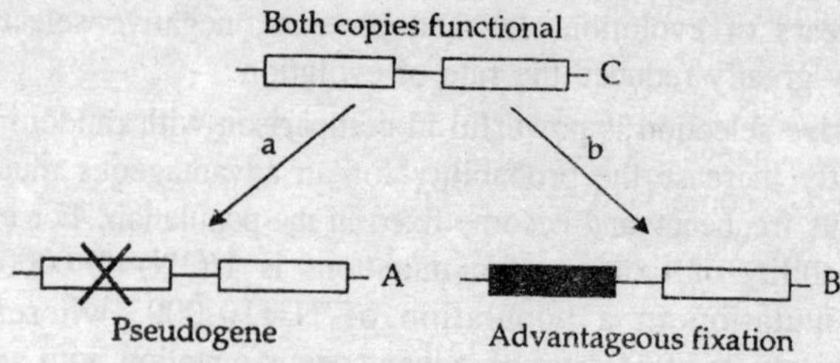

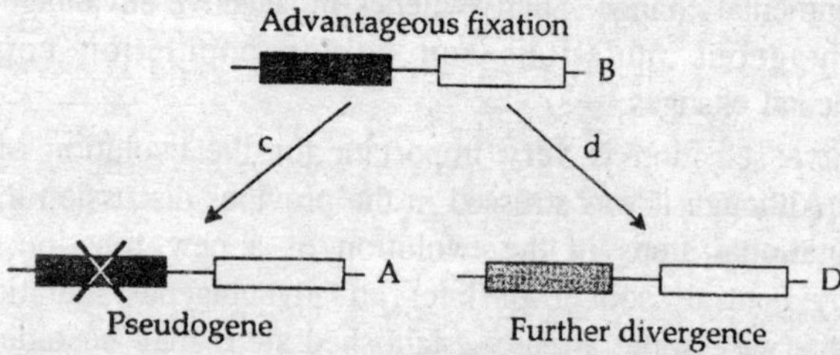

Fig. 10.2. Phase I: A duplicate gene (state C) may first fix a null mutation and become a pseudogene or may first fix an advantageous mutation. Phase II: Following the fixation of the first advantageous mutation, the duplicate locus may still fix a null mutation or it may fix another advantageous mutation.

of a null mutation is approximately only 0.001 if $S<1$ but is approximately 0.04 if $S=40$; the increase is 40 times. Note further that if $S<1$, the fixation of an advantageous mutation is of little help in retaining the duplicate gene, because the duplicate gene can still readily fix a null mutation and become a pseudogene. Only if $S>>1$, will there be sufficient selective advantage for retaining the duplicate gene, allowing time for it to accumulate further advantageous mutations. Clearly, positive selection is required in the early stage of the evolution of a new function from a duplicate gene.

Having described the relative roles of mutation and selection in molecular evolution in general, we now consider *adaptive* molecular evolution in particular. That is, we now consider only the class of *advantageous evolutionary changes*. It appears that even for this class of evolutionary changes, the rate of evolution is still mainly determined by the rate of mutation, not by the rate of environmental change as proposed by selectionists. As noted above, neo-Darwinian assumes that whenever the environment changes, the population has sufficient genetic variation, i.e., it has a "right" mutant variant, to respond to the change. Although this variant was originally disadvantageous, it has been maintained in the population by recurrent mutation; it becomes advantageous, and its frequency starts to increase when the environment changes. This view is unlikely to hold at the molecular level. Most molecular changes are unique rather than recurrent, and most of them will soon become lost if there is no selective advantage. Thus, it is likely that in the majority of cases when the environment changes, no "right" variant is there in the population and it will take sometime for a right mutation to occur. A more plausible view is that occasionally a mutation happens to be advantageous in the existing environment. As the mutation has an immediate selective advantage, it has a definite, though small, probability of becoming fixed in the population. In this view the rate of evolution is mainly determined by how often advantageous mutations occur. Positive selection also affects the rate of evolution in the sense that the probability of fixation of an advantageous mutation and the rate of substitution increase with the magnitude of the selective advantage. This view is similar to the view proposed by Morgan (1925, 1932) and elaborated in the framework of modern molecular genetics by Nei (1987); Nei has pointed out that Morgan's view is equivalent to the mutation-selection theory or classical theory of Dobzhansky's (1955) sense. To compare this view and neo-Darwinism, let us consider a few examples.

First, let us consider the emergence of trichromatic vision in higher primates. Trichromatic vision is helpful for detecting colour fruits and has evolved independently in New World monkeys (NWMs) and in the common ancestor of the Old World monkeys (OWMs), apes and humans. The two systems differ in that OWMs, apes and humans possess two X-linked opsin genes, which code for the red green pigments, respectively, whereas most NWMs possess only one X-linked opsin locus, but the locus has three alleles, two of which has spectral sensitivities similar to those of the human red and green pigments, respectively. In the evolution of either system there would have been at most only one relevant environmental change in monkeys, that is, a change from a diet without fruits to one with fruits. However, the emergence of each system required multiple mutational steps; for example, at least 5 of the 15 amino acid differences between the human red and green pigments are responsible for the differences in spectral sensitivity between the two pigments. Obviously, most, if not all, of these steps required new mutations, and the rate at which the system evolved would have depended on how fast these mutations has occurred, not on how fast the environment had changed.

Second, the emergence of regulation of hemoglobin function by organic phosphates such as DPG requires at least three mutational steps. However, the evolution of this regulation would require at most only one environmental change, that is, the physiology of vertebrates became ready to accommodate this regulation, i.e., to synthesize DPG (in mammals) or inositol pentaphasphate (in birds). Therefore, how fast the regulation had evolved would have depended on the rate of occurrence of the three critical residues, not on the rate of environmental change. Interestingly, this regulatory system is preadaptive to life at high altitudes. For example, it has enabled Bolivians to adapt to life at 15,000 ft in the Andes; these Bolivians have 20% higher levels of DPG in their red blood cells than do people who live at sea level. This example suggests that a new function can be preadaptive, rather than being a response to an environmental change, and may enable an organism to explore a new niche.

Third, as explained, a domain duplication or a domain shuffling may have an immediate selective advantage in the existing environment because it may immediately improve the function of the protein. Here, no environmental change is necessary. As emphasized, domain duplication and domain shuffling have played an important role in the evolution of proteins.

Fourth, whether a duplicate gene can be retained in the population depends on how often advantageous mutations occur and how large the selective advantages are, not on how often the environment changes. Subsequently, how fast the duplicate gene evolves into a new gene depends on how often advantageous novel mutations occur, not on how often the environment changes. For example, as mentioned above, during the evolution of a pair of duplicate genes into the red and green pigment genes in OWMs, apes and humans there would have been at most only one environmental change, but there were multiple amino acid substitutions. So, the rate of evolution of the two genes depended on the rate of occurrence of these mutations, not on the rate of environmental change.

The above examples revealed the inadequacy of neo-Darwinism even for explaining *adaptive* molecular evolution. This is not surprising because neo-Darwinism was developed long before the development of modern genetics and molecular biology. A good understanding of an evolutionary process or the emergence of a new system requires not only the theory of evolution and population genetics but also knowledge of the molecular steps involved.

Since the time of Darwin it has been apparent that the great shaping force in phenotypic evolution is natural selection. Selection, acting on the genetic variability in natural populations, is responsible for the wondrous diversity of animal and plant life that is so apparent as we look around us. Yet, molecular biology has revealed a deeper and more extensive level of genetic change among individuals or organisms. Most of these changes have no visible effect and the neo-Darwinian paradigm is inadequate. Such mutation-driven molecular changes for exceed selection-driven phenotypic changes. Thus, inside the "forest" of selectively determined phenotypic changes are an enormously greater number of "trees" mostly marching to different drummer—the tempo of mutation. Moreover, a new paradigm is needed even for adaptive evolution at the molecular level. In conclusion, molecular studies have greatly deepened our understanding of how evolution works, and the rapid progress in the past two decades is only the beginning.

INDEX